原版影印说明

1.《凝聚态物质与材料数据手册》（6册）是*Springer Handbook of Condensed Matter and Materials Data* 的影印版。为使用方便，由原版1卷改为6册：

第1册 通用表和元素

第2册 材料类：金属材料

第3册 材料类：非金属材料

第4册 功能材料：半导体和超导体

第5册 功能材料：磁性材料、电介质、铁电体和反铁电体

第6册 特种结构

2. 全书目录、作者信息、缩略语表、索引在各册均完整呈现。

本手册数据全面准确，1 025个图和914个表使查阅更加方便，是非常实用的案头参考书，适于材料及相关专业本科生、研究生、专业研究人员使用。

材料科学与工程图书工作室

联系电话 0451-86412421
0451-86414559

邮　　箱 yh_bj@aliyun.com
xuyaying81823@gmail.com
zhxh6414559@aliyun.com

Springer 手册精选原版系列

凝聚态物质与材料数据手册

功能材料：磁性材料、电介质、铁电体和反铁电体

【第5册】

Springer Handbook of

Condensed Matter and Materials Data

W.Martienssen

H.Warlimont

Editors

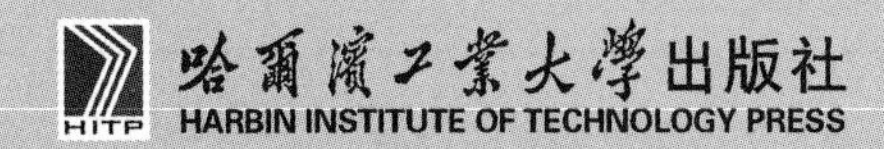

黑版贸审字08-2014-009号

Reprint from English language edition:
Springer Handbook of Condensed Matter and Materials Data
by Werner Martienssen and Hans Warlimont

图书在版编目（CIP）数据

凝聚态物质与材料数据手册. 第5册，功能材料：磁性材料、电介质、铁电体和反铁电体：英文 /（德）马蒂安森（Martienssen，W.），（德）沃利蒙特（Warlimont，H.）主编. —哈尔滨：哈尔滨工业大学出版社，2014.3
（Springer手册精选原版系列）
ISBN 978-7-5603-4459-1

Ⅰ. ①凝… Ⅱ. ①马… ②沃… Ⅲ. ①凝聚态-材料-技术手册-英文 ②功能材料-技术手册-英文 Ⅳ. ①O469-62 ②TB3-62

中国版本图书馆CIP数据核字（2013）第291540号

材料科学与工程
图书工作室

责任编辑 许雅莹 张秀华 杨 桦
出版发行 哈尔滨工业大学出版社
社 址 哈尔滨市南岗区复华四道街10号 邮编 150006
传 真 0451-86414749
网 址 http://hitpress.hit.edu.cn
印 刷 哈尔滨市石桥印务有限公司
开 本 787mm × 960mm 1/16 印张 15
版 次 2014 年 3 月第 1 版 2014年 3 月第 1 次印刷
书 号 ISBN 978-7-5603-4459-1
定 价 68.00元

（如因印刷质量问题影响阅读，我社负责调换）

Springer Handbook of Condensed Matter and Materials Data

W. Martienssen and H. Warlimont (Eds.)

With 1025 Figures and 914 Tables

Springer Handbook provides a concise compilation of approved key information on methods of research, general principles, and functional relationships in physics and engineering. The world's leading experts in the fields of physics and engineering will be assigned by one or several renowned editors to write the chapters comprising each volume. The content is selected by these experts from Springer sources (books, journals, online content) and other systematic and approved recent publications of physical and technical information.

The volumes will be designed to be useful as readable desk reference book to give a fast and comprehensive overview and easy retrieval of essential reliable key information, including tables, graphs, and bibliographies. References to extensive sources are provided.

Preface

The Springer Handbook of Condensed Matter and Materials Data is the realization of a new concept in reference literature, which combines introductory and explanatory texts with a compilation of selected data and functional relationships from the fields of solid-state physics and materials in a single volume. The data have been extracted from various specialized and more comprehensive data sources, in particular the Landolt–Börnstein data collection, as well as more recent publications. This Handbook is designed to be used as a desktop reference book for fast and easy finding of essential information and reliable key data. References to more extensive data sources are provided in each section. The main users of this new Handbook are envisaged to be students, scientists, engineers, and other knowledge-seeking persons interested and engaged in the fields of solid-state sciences and materials technologies.

The editors have striven to find authors for the individual sections who were experienced in the full breadth of their subject field and ready to provide succinct accounts in the form of both descriptive text and representative data. It goes without saying that the sections represent the individual approaches of the authors to their subject and their understanding of this task. Accordingly, the sections vary somewhat in character. While some editorial influence was exercised, the flexibility that we have shown is deliberate. The editors are grateful to all of the authors for their readiness to provide a contribution, and to cooperate in delivering their manuscripts and by accepting essentially all alterations which the editors requested to achieve a reasonably coherent presentation.

An onerous task such as this could not have been completed without encouragement and support from the publisher. Springer has entrusted us with this novel project, and Dr. Hubertus von Riedesel has been a persistent but patient reminder and promoter of our work throughout. Dr. Rainer Poerschke has accompanied and helped the editors constantly with his professional attitude and very personable style during the process of developing the concept, soliciting authors, and dealing with technical matters. In the later stages, Dr. Werner Skolaut became a relentless and hard-working member of our team with his painstaking contribution to technically editing the authors' manuscripts and linking the editors' work with the copy editing and production of the book.

Prof. Werner Martienssen

Prof. Hans Warlimont

We should also like to thank our families for having graciously tolerated the many hours we have spent in working on this publication.

We hope that the users of this Handbook, whose needs we have tried to anticipate, will find it helpful and informative. In view of the novelty of the approach and any possible inadvertent deficiencies which this first edition may contain, we shall be grateful for any criticisms and suggestions which could help to improve subsequent editions so that they will serve the expectations of the users even better and more completely.

September 2004
Frankfurt am Main, Dresden

Werner Martienssen,
Hans Warlimont

About the Authors

Wolf Assmus

Johann Wolfgang Goethe-University
Physics Department
Frankfurt am Main, Germany
assmus@physik.uni-frankfurt.de
http://www.rz.uni-frankfurt.de/piweb/kmlab/Leiter.html

Chapter 1.3

Dr. Wolf Assmus (Kucera Professor) is Professor of Physics at the University of Frankfurt and Dean of the Physics-Faculty. He is a solid state physicist, especially interested in materials research and crystal growth. His main research fields are: materials with high electronic correlation, quasicrystals, materials with extremely high melting temperatures, magnetism, and superconductivity.

Stefan Brühne

Johann Wolfgang Goethe-University
Physics Department
Frankfurt am Main, Germany
bruehne@physik.uni-frankfurt.de

Chapter 1.3

Dr. Stefan Brühne, née Mahne, a chemist by education in Germany and England, received his PhD in 1994 from Dortmund University, Germany, on giant cell crystal structures in the Al–Ta system. Following a post doc position at the Materials Department (Crystallography) at ETH Zurich he spent seven years in the ceramics industry. His main activity was R&D of glasses, frits and pigments for high-temperature applications, thereby establishing design of experiment (DoE) techniques. Since 2002, at the Institute of Physics at Frankfurt University he has been investigated X-ray structure determination of quasicrystalline, highly complex and disordered intermetallic materials.

Fabrice Charra

Commissariat à l'Énergie Atomique, Saclay
Département de Recherche sur l'État Condensé, les Atomes et les Molécules
Gif-sur-Yvette, France
fabrice.charra@cea.fr
http://www-drecam.cea.fr/spcsi/

Chapter 5.3

Fabrice Charra conducts researche in the emerging field of nanophotonics, in the surface physics laboratory of CEA/Saclay. The emphasis of his work is on light emission and absorption form single nanoscale molecular systems. His area of expertise also extends to nonlinear optics, a domain to which he contributed several advances in the applications of organic materials.

Gianfranco Chiarotti

University of Rome "Tor Vergata"
Department of Physics
Roma, Italy
chiarotti@roma2.infn.it

Chapter 5.2

Gianfranco Chiarotti is Professor Emeritus, formerly Professor of General Physics, Fellow of the American Physical Society, fellow of the Italian National Academy (Accademia Nazionale dei Lincei). He was Chairman of the Physics Committee of the National Research Council (1988–1994), Chair Franqui at the University of Liège (1975), Assistant Professor at the University of Illinois (1955–1957), Editor of the journal Physics of Solid Surfaces, and Landolt–Börnstein Editor of Springer-Verlag from 1993 through 1996. He has worked in several fields of solid state physics, namely electronic properties of defects, modulation spectroscopy, optical properties of semiconductors, surface physics, and scanning tunnelling microscopy (STM) in organic materials.

Claus Fischer

Formerly Institute of Solid State and Materials Research (IFW)
Dresden, Germany
A_C.FischerDD@t-online.de

Chapter 4.2

Claus Fischer recieved his PhD from the Technical University Dresden (Since his retirement in 2000 he continues to work as a foreign scientist of IFW in the field of high-T_c superconductors.) His last position at IFW was head of the Department of Superconducting Materials. The main areas of research were growth of metallic single crystals in particular of magnetic materials, developments of hard magnetic materials, of materials for thick film components of microelectronics and of low-T_c and high-T_c superconducting wires and tapes. Many activities were performed in cooperation with industrial manufacturers.

Günter Fuchs

Leibniz Institute for Solid State and Materials Research (IFW) Dresden
Magnetism and Superconductivity in the Institute of Metallic Materials
Dresden, Germany
fuchs@ifw-dresden.de
http://www.ifw-dresden.de/imw/21/

Chapter 4.2

Dr. Günter Fuchs studied physics at the Technical University of Dresden, Germany, and received his PhD in 1980 on the pinning mechanism in superconducting NbTi alloys. Since 1969 he has been at the Institute of Solid State and Materials Research (IFW) in Dresden. His activities are in superconductivity (HTSC, MgB_2, intermetallic borocarbides) and the applications of superconductors. He received the PASREG Award for outstanding scientific achievements in the field of bulk cuprate superconductors in high magnetic fields in 2003.

Frank Goodwin

International Lead Zinc Research Organization, Inc.
Research Triangle Parc, NC, USA
fgoodwin@ilzro.org
http://www.ilzro.org/Contactus.htm

Chapter 3.1

Frank Goodwin received his Sc.D. from the Massachusetts Institute of Technology in 1979 and is responsible for all materials science research at International Lead Zinc Research Organization, Inc. where he has conceived and managed numerous projects on lead and zinc-containing products. These have included lead in acoustics, cable sheathing, nuclear waste management and specialty applications, together with zinc in coatings, castings and wrought forms.

Susana Gota-Goldmann

Commissariat à l'Energie Atomique (CEA)
Direction de la Recherche Technologique (DRT)
Fontenay aux Roses, France
susana.gota-goldmann@cea.fr

Chapter 5.3

Dr. Susana Gota-Goldmann received her PhD in Materials Science form the Université Pierre et Marie Curie (Paris V) in 1993. After her PhD, she was engaged as a researcher in the Materials Science Division of the CEA (Commissariat à l'Energie Atomique, France). She has focused her scientific activity on the growth and characterisation of nanometric oxide layers with applications in spin electronics and photovoltaics. In parallel she has developed the use of synchrotron radiation techniques (X-ray absorption magnetic dicroism, photoemission, resonant reflectivity) for the study of oxide thin layers. Recently she has moved from fundamental to technological research. Dr. Gota-Goldmann is now working as a project manager at the scientific affairs direction of the Technology Research Division (CEA/DRT).

Sivaraman Guruswamy

University of Utah
Metallurgical Engineering
Salt Lake City, UT, USA
sguruswa@mines.utah.edu
http://www.mines.utah.edu/metallurgy/MML

Chapter 3.1

Dr. Guruswamy is a Professor of Metallurgical Engineering at the University of Utah. He obtained his Ph.D. degree in Metallurgical Engineering from the Ohio State University in 1984. He has made significant contributions in several areas including magnetic materials development, deformation of compound semiconductors, and lead alloys. His current work focuses on magnetostrictive materials and hybrid thermionic/thermoelectric thermal diodes.

Gagik G. Gurzadyan

Technical University of Munich
Institute for Physical and Theoretical Chemistry
Garching, Germany
gurzadyan@ch.tum.de
http://zentrum.phys.chemie.tu-muenchen.de/gagik

Chapter 4.4

Gagik G. Gurzadyan, Ph.D., Dr. Sci., has extensive experience in nonlinear optics and crystals, laser photophysics and spectroscopy. He has authored several books including the Handbook of Nonlinear Optical Crystals published by Springer-Verlag. He worked in the Institute of Spectroscopy (USSR), CEA/Saclay (France), Max-Planck-Institute of Radiation Chemistry (Germany). At present he works at the Technical University of Munich with ultrafast lasers in the fields of nonlinear photochemistry of biomolecules and femtosecond spectroscopy.

Hideki Harada

High Tech Association Ltd.
Higashikaya, Fukaya,Saitama, Japan
khb16457@nifty.com
http://homepage1.nifty.com/JABM

Chapter 4.3

Dr. Hideki Harada is chief advisor of magnetic materials and their application and President of High Tech Association Ltd., Saitama, Japan. He is Chairman of the Japan Association of Bonded Magnet Industries (JABM) and received his Ph.D. in 1987 with a work on electrostatic ferrite materials. He worked in research and development of magnetic materials and cemented carbide tools at Hitachi Metals where he also was on the Board of Directors. He received the Japanese National Award for Industries Development Contribution.

Bernhard Holzapfel

Leibniz Institute for Solid State and Materials Research Dresden – Institute of Metallic Materials
Superconducting Materials
Dresden, Germany
B.Holzapfel@ifw-dresden.de
http://www.ifw-dresden.de/imw/26/

Chapter 4.2

Dr. Bernhard Holzapfel is head of the superconducting materials group at the Leibniz Institute for Solid State and Materials Research (IFW) Dresden, Germany. His main area of research is pulsed laser deposition of functional thin films and superconductivity. Currently he works on the development of HTSC high J_c coated conductors using ion beam assisted deposition or highly textured metal substrates. His work is supported by a number of national and European founded research projects.

Karl U. Kainer

GKSS Research Center Geesthacht
Institute for Materials Research
Geesthacht, Germany
karl.kainer@gkss.de
http://www.gkss.de

Chapter 3.1

Professor Kainer is director of Institute for Materials Research at GKSS-Research Center, Geesthacht and Professor of Materials Technology at the Technical University of Hamburg-Harburg. He obtained his Ph.D. in Materials Science at the Technical University of Clausthal in 1985 and his Habilitation in 1996. In 1988 he received the Japanese Government Research Award for Foreign Specialists. His current research activities are the development of new alloys and processes for magnesium materials.

Catrin Kammer

METALL – Intl. Journal for Metallurgy
Goslar, Germany
Kammer@metall-news.com
http://www.giesel-verlag.de

Chapter 3.1

Catrin Kammer received her Ph.D. in materials sciences from the Technical University Bergakademie Freiberg, Germany, in 1989. She has been working in the field of light metals and is author of several handbooks about aluminium and magnesium. She is working as author for the journal ALUMINIUM and is teaching in material sciences. Since 2001 she is editor-in-chief of the journal METALL, which deals with all non-ferrous metals.

Wolfram Knabl

Plansee AG
Technology Center
Reutte, Austria
wolfram.knabl@plansee.com
http://www.plansee.com

Chapter 3.1

Dr. Wolfram Knabl studied materials science at the Mining University of Leoben, Austria and received his Ph.D. at the Plansee AG focusing on the development of oxidation protective coatings for refractory metals. Between 1996 and 2002 he was responsible for the test laboratories at Plansee AG and since October 2002 he is working in the field of refractory metals, especially material and process development in the technology center of Plansee AG.

Alfred Koethe

Leibniz-Institut für Festkörper- und Werkstoffforschung
Institut für Metallische Werkstoffe (retired)
Dresden, Germany
alfred.koethe@web.de

Chapter 3.1

Dr. Alfred Koethe is physicist and professor of Materials Science. He retired in 2000 from his position as head of department in the Institute of Metallic Materials at the Leibniz Institute of Solid State and Materials Research in Dresden, Germany. His main research activities were in the fields of preparation and properties of ultrahigh-purity refractory metals and, especially, of steels (stainless steels, high strenght steels, thermomechanical treatment, microalloying, relations chemical composition/microstructure/properties).

Dieter Krause

Schott AG
Research and Technology-Development
Mainz, Germany
dieter.krause@schott.com

Chapter 3.4

Dieter Krause studied physics at the universities of Erlangen and Munich, Germany, where he received his Ph.D. for work on magnetism and metal physics. He was professor in Tehran, Iran, lecturer in Munich and Mainz, Germany. As scientist and director of Schott's corporate research and development centre he was involved in research on optical and mechanical properties of amorphous materials, thin films, and optical fibres. Now he is consultant, chief scientist, and the editor of the "Schott Series on Glass and Glass Ceramics – Science, Technology, and Applications" published by Springer.

Manfred D. Lechner

Universität Osnabrück
Institut für Chemie – Physikalische Chemie
Osnabrück, Germany
lechner@uni-osnabrueck.de
http://www.chemie.uni-osnabrueck.de/pc/index.html

Chapter 3.3

Professor Lechner has a PhD in chemistry from the University of Mainz, Germany. Since 1975 he is Professor of Physical Chemistry at the Institute of Chemistry of the University of Osnabrück, Germany. His scientific work concentrates on the physics and chemistry of polymers. In this area he is mainly working on the influence of high pressure on polymer systems, polymers for optical storage and waveguides as well as synthesis and properties of superabsorbers from renewable resources.

Gerhard Leichtfried

Plansee AG
Technology Center
Reutte, Austria
gerhard.leichtfried@plansee.com
http://www.plansee.com

Chapter 3.1

Dr. Gerhard Leichtfried received his Ph.D from the Montanuniversität Leoben and is qualified for lecturing in powder metallurgy. For 20 years he has been working in various senior positions for the Plansee Aktiengesellschaft, a company engaged in refractory metals, composite materials, cemented carbides and sintered iron and steels.

Werner Martienssen

Universität Frankfurt/Main
Physikalisches Institut
Frankfurt/Main, Germany
Martienssen@Physik.uni-frankfurt.de

Chapters 1.1, 1.2, 2.1, 4.1

Werner Martienssen studied physics and chemistry at the Universities of Würzburg and Göttingen, Germany. He obtained his Ph.D. in Physics with R.W. Pohl, Göttingen, and holds an honorary doctorate at the University of Dortmund. After a visiting-professorship at the Cornell University, Ithaca, USA in 1959 to 1960 he taught physics at the University of Stuttgart and since 1961 at the University of Frankfurt/Main. His main research fields are condensed matter physics, quantum optics and chaotic dynamics. Two of his former students and coworkers became Nobel-laureates in Physics, Gerd K. Binnig for the design of the scanning tunneling microscope in 1986 and Horst L. Störmer for the discovery of a new form of quantum-fluid with fractionally charged excitations in 1998. Werner Martienssen is a member of the Deutsche Akademie der Naturforscher Leopoldina, Halle and of the Akademie der Wissenschaften zu Göttingen. Since 1994 he is Editor-in-Chief of the data collection Landolt–Börnstein published by Springer, Heidelberg.

Toshio Mitsui

Osaka University
Takarazuka, Japan
t-mitsui@jttk.zaq.ne.jp

Chapter 4.5

Toshio Mitsui is an emeritus professor of Osaka University. He studied solid state physics and biophysics at Hokkaido University, Pennsylvania State University, Brookhaven National Laboratory, the Massachusetts Institute of Technology, Osaka University and Meiji University. He was the first to observe the ferroelectric domain structure in Rochelle salt with a polarization microscope. He proposed various theories on ferroelectric effects and biological molecular machines.

Manfred Müller

Dresden University of Technology
Institute of Materials Science
Dresden, Germany
m.mueller33@t-online.de

Chapter 4.3

Dr.-Ing. habil. Manfred Müller is a Professor emeritus of Special Materials at the Institute of Materials Science of the Dresden University of Technology. Before his retirement he was for many years head of department for special materials at the Central Institute for Solid State Physics and Materials Research of the Academy of Sciences in Dresden, Germany. His main field was the research and development of metallic materials with emphasis on special physical properties, such as soft and hard magnetic, electrical and thermoelastic properties. His last field of research was amorphous and nanocrystalline soft magnetic alloys. He is a member of the German Society of Materials Science (DGM) and was a member of the Advisary Board of DGM.

Sergei Pestov

Moscow State Academy of Fine Chemical Technology
Department of Inorganic Chemistry
Moscow, Russia
pestovsm@yandex.ru

Chapter 5.1

Dr. Pestov is a docent of the Inorganic Chemistry Department and a head of group on liquid crystals (LC) at the Moscow State Academy of Fine Chemical Technology. He earned his Ph.D. in physical chemistry in 1992. His research is focused on thermal analysis and thermodynamics of systems containing LC and physical properties of LC. He is an author of a Landolt–Börnstein volume and two books devoted to liquid crystals.

Günther Schlamp

Metallgesellschaft Ffm and Degussa Demetron (retired)
Steinbach/Ts, Germany

Chapter 3.1

Günther Schlamp received his Ph.D. from the Johann-Wolfgang-Goethe University of Frankfurt/Main, Germany, in Physical Chemistry. His industrial activities in research include the development and production of refractory material coatings, high purity materials and parts for electronics, and sputter targets for the reflection-enhancing coating of glas. He has contributed to several Handbooks with repoprts on properties and applications of noble metals and their alloys.

Barbara Schüpp-Niewa

Leibniz-Institute for Solid State and Materials Research Dresden
Institute for Metallic Materials
Dresden, Germany
b.schuepp@ifw-dresden.de
http://www.ifw-dresden.de

Chapter 4.2

Barbara Schüpp-Niewa studied chemistry in Gießen and Dortmund where she received her Ph.D. in 1999. Since 2000 she has been a scientist at the Leibniz-Institute for Solid State and Materials Research Dresden with a focus on crystal structure investigations of oxometalates with superconducting or exciting magnetic ground states. Her current research activities include coated conductors.

Roland Stickler

University of Vienna
Department of Chemistry
Vienna, Austria
roland.stickler@univie.ac.at

Chapter 3.1

Professor Stickler received his master and Dr. degree from the Technical University in Vienna. From 1958 to 1972 he was manager of physical metallurgy with the Westinghouse Research Laboratory in Pittsburgh, Pa. In 1972 he accepted a full professorship at the University of Vienna heading a materials science group in the Institute of Physical Chemistry, and from 1988 he was head of this institute until his retirement as professor emeritus in 1998. He was involved in research and engineering work on superalloys, semiconductor materials and high melting point materials, investigating the relationship between microstructure and mechanical behavior, in particular fatigue and fracture mechanics properties. He was leader of a successful project on brazing under microgravity conditions in the Spacelab-Mission. Further activities included the participation in European COST projects, in particular as chairman of actions on powder metallurgy and light metals. He has authored and coauthored more than 250 publications in scientific journals and proceedings.

Pancho Tzankov

Max Born Institute for Nonlinear Optics and Short Pulse Spectroscopy
Berlin, Germany
tzankov@mbi-berlin.de
http://staff.mbi-berlin.de/tzankov/

Chapter 4.4

Pancho Tzankov studied laser physics at Sofia University, Bulgaria, and received his Ph.D. in physical chemistry from the Technical University of Munich, Germany. He is now a postdoctoral fellow at the Max Born Institute in Berlin, Germany. His research activities involve development of new nonlinear optical parametric sources of ultrashort pulses and their application for time-resolved spectroscopy.

Volkmar Vill

University of Hamburg
Department of Chemistry, Institute of Organic Chemistry
Hamburg, Germany
vill@chemie.uni-hamburg.de
http://liqcryst.chemie.uni-hamburg.de/

Chapter 5.1

Professor Volkmar Vill received his Diploma in Chemistry in 1986, his Diploma in Physics in 1988 and his Ph.D. in Chemistry in 1990 from the University of Münster, Germany. In 1997 he earned his Habilitation in Organic Chemistry from the University of Hamburg where he is Professor of Organic Chemistry since 2002. He is the author of the LiqCryst – Database of Liquid Crystals and the Editor of the Handbook of Liquid Crystals, of Landolt–Börnstein, Organic Index, and Vol. VIII/5a, Physical Properties of Liquid Crystals.

Hans Warlimont

DSL Dresden Material-Innovation GmbH
Dresden, Germany
warlimont@ifw-dresden.de

Chapters 3.1, 3.2, 4.2, 4.3

Hans Warlimont is a physical metallurgist and has worked on numerous topics in several research institutions and industrial companies. Among them were the Max-Planck-Institute of Metals Research, Stuttgart, and Vacuumschmelze, Hanau. He was Scientific Director of the Leibniz-Institute of Solid State and Materials Research Dresden and Professor of Materials Science at Dresden University of Technology. Recently he has established DSL Dresden Material-Innovation GmbH to industrialise his invention of electroformed battery grids.

Acknowledgements

2.1 The Elements

by Werner Martienssen

We thank Dr. G. Leichtfried, Plansee AG, A-6600 Reutte/Tirol for recently determined new data on the refractory metals Nb, Ta, and Mo, W.

4.1 Semiconductors

by Werner Martienssen

In selecting the "most important information" from the huge data collection in Landolt–Börnstein, the author found great help in the new *Semiconductors: Data Handbook* [1]. Again, the data in this Springer Handbook of Condensed Matter and Materials Data represent only a small fraction of the information given in *Semiconductors: Data Handbook*, which is about 700 pages long. I am much indebted to my colleague O. Madelung for kindly presenting me the manuscript of that Handbook prior to publication.

[1] O. Madelung (Ed.): *Semiconductors: Data Handbook*, 3rd Edn. (Springer, Berlin, Heidelberg 2004)

4.5 Ferroelectrics and Antiferroelectrics

by Toshio Mitsui

The author of this subchapter thanks the coauthors of LB III/36 for their helpful discussions and suggestions. Especially, he is much indebted to Prof. K. Deguchi for his kind support throughout the preparation of the manuscript.

Contents

List of Abbreviations ... 19

第1册 通用表和元素

Part 1 General Tables

1 **The Fundamental Constants**
Werner Martienssen ... 3
1.1 What are the Fundamental Constants and Who Takes Care of Them? ... 3
1.2 The CODATA Recommended Values of the Fundamental Constants ... 4
References ... 9

2 **The International System of Units (SI), Physical Quantities, and Their Dimensions**
Werner Martienssen ... 11
2.1 The International System of Units (SI) ... 11
2.2 Physical Quantities ... 12
2.3 The SI Base Units ... 13
2.4 The SI Derived Units ... 16
2.5 Decimal Multiples and Submultiples of SI Units ... 19
2.6 Units Outside the SI ... 20
2.7 Some Energy Equivalents ... 24
References ... 25

3 **Rudiments of Crystallography**
Wolf Assmus, Stefan Brühne ... 27
3.1 Crystalline Materials ... 28
3.2 Disorder ... 38
3.3 Amorphous Materials ... 39
3.4 Methods for Investigating Crystallographic Structure ... 39
References ... 41

Part 2 The Elements

1 **The Elements**
Werner Martienssen ... 45
1.1 Introduction ... 45
1.2 Description of Properties Tabulated ... 46
1.3 Sources ... 49
1.4 Tables of the Elements in Different Orders ... 49
1.5 Data ... 54
References ... 158

第2册 材料类：金属材料

Part 3 Classes of Materials

1 **Metals**
Frank Goodwin, Sivaraman Guruswamy, Karl U. Kainer, Catrin Kammer, Wolfram Knabl, Alfred Koethe, Gerhard Leichtfried, Günther Schlamp, Roland Stickler, Hans Warlimont 161
1.1 Magnesium and Magnesium Alloys 162
1.2 Aluminium and Aluminium Alloys 171
1.3 Titanium and Titanium Alloys 206
1.4 Zirconium and Zirconium Alloys 217
1.5 Iron and Steels 221
1.6 Cobalt and Cobalt Alloys 272
1.7 Nickel and Nickel Alloys 279
1.8 Copper and Copper Alloys 296
1.9 Refractory Metals and Alloys 303
1.10 Noble Metals and Noble Metal Alloys 329
1.11 Lead and Lead Alloys 407
References 422

第3册 材料类：非金属材料

2 **Ceramics**
Hans Warlimont 431
2.1 Traditional Ceramics and Cements 432
2.2 Silicate Ceramics 433
2.3 Refractory Ceramics 437
2.4 Oxide Ceramics 437
2.5 Non-Oxide Ceramics 451
References 476

3 **Polymers**
Manfred D. Lechner 477
3.1 Structural Units of Polymers 480
3.2 Abbreviations 482
3.3 Tables and Figures 483
References 522

4 **Glasses**
Dieter Krause 523
4.1 Properties of Glasses – General Comments 526
4.2 Composition and Properties of Glasses 527
4.3 Flat Glass and Hollowware 528
4.4 Technical Specialty Glasses 530
4.5 Optical Glasses 543
4.6 Vitreous Silica 556
4.7 Glass-Ceramics 558

4.8 Glasses for Miscellaneous Applications 559
References 572

第4册 功能材料：半导体和超导体

Part 4 Functional Materials

1 **Semiconductors**
Werner Martienssen 575
1.1 Group IV Semiconductors and IV–IV Compounds 578
1.2 III–V Compounds 604
1.3 II–VI Compounds 652
References 691

2 **Superconductors**
Claus Fischer, Günter Fuchs, Bernhard Holzapfel, Barbara Schüpp-Niewa, Hans Warlimont 695
2.1 Metallic Superconductors 696
2.2 Non-Metallic Superconductors 711
References 749

第5册 功能材料：磁性材料、电介质、铁电体和反铁电体（本册）

3 **Magnetic Materials**
Hideki Harada, Manfred Müller, Hans Warlimont 755
3.1 Basic Magnetic Properties 755
3.2 Soft Magnetic Alloys 758
3.3 Hard Magnetic Alloys 794
3.4 Magnetic Oxides 811
References 814

4 **Dielectrics and Electrooptics**
Gagik G. Gurzadyan, Pancho Tzankov 817
4.1 Dielectric Materials: Low-Frequency Properties 822
4.2 Optical Materials: High-Frequency Properties 824
4.3 Guidelines for Use of Tables 826
4.4 Tables of Numerical Data for Dielectrics and Electrooptics 828
References 890

5 **Ferroelectrics and Antiferroelectrics**
Toshio Mitsui 903
5.1 Definition of Ferroelectrics and Antiferroelectrics 903
5.2 Survey of Research on Ferroelectrics 904
5.3 Classification of Ferroelectrics 906
5.4 Physical Properties of 43 Representative Ferroelectrics 912
References 936

第6册 特种结构

Part 5 Special Structures

1 **Liquid Crystals**
Sergei Pestov, Volkmar Vill 941
1.1 Liquid Crystalline State 941
1.2 Physical Properties of the Most Common Liquid Crystalline Substances 946
1.3 Physical Properties of Some Liquid Crystalline Mixtures 975
References 977

2 **The Physics of Solid Surfaces**
Gianfranco Chiarotti 979
2.1 The Structure of Ideal Surfaces 979
2.2 Surface Reconstruction and Relaxation 986
2.3 Electronic Structure of Surfaces 996
2.4 Surface Phonons 1012
2.5 The Space Charge Layer at the Surface of a Semiconductor 1020
2.6 Most Frequently Used Acronyms 1026
References 1029

3 **Mesoscopic and Nanostructured Materials**
Fabrice Charra, Susana Gota-Goldmann 1031
3.1 Introduction and Survey 1031
3.2 Electronic Structure and Spectroscopy 1035
3.3 Electromagnetic Confinement 1044
3.4 Magnetic Nanostructures 1048
3.5 Preparation Techniques 1063
References 1066

Subject Index
Periodic Table of the Elements
Most Frequently Used Fundamental Constants

List of Abbreviations

2D-BZ 2-dimensional Brillouin zone
2P-PES 2-photon photoemission spectroscopy

A

AES Auger electron spectroscopy
AFM atomic force microscope
AISI American Iron and Steel Institute
APS appearance potential spectroscopy
ARUPS angle-resolved ultraviolet photoemission spectroscopy
ARXPS angle-resolved X-ray photoemission spectroscopy
ASTM American Society for Testing and Materials
ATR attenuated total reflection

B

BBZ bulk Brillouin zone
BIPM Bureau International des Poids et Mesures
BZ Brillouin zone

C

CB conduction band
CBM conduction band minimum
CISS collision ion scattering spectroscopy
CITS current imaging tunneling spectroscopy
CMOS complementary metal–oxide–semiconductor
CODATA Committee on Data for Science and Technology
CVD chemical vapour deposition

D

DFB distributed-feedback
DFG difference frequency generation
DOS density of states
DSC differential scanning calorimetry
DTA differential thermal analysis

E

EB electron-beam melting
ECS electron capture spectroscopy
EELS electron-energy loss spectroscopy
ELEED elastic low-energy electron diffraction
ESD electron-stimulated desorption
EXAFS extended X-ray absorption fine structure

F

FEM field emission microscope/microscopy
FIM field ion microscope/microscopy

G

GMR giant magnetoresistance

H

HAS helium atom scattering
HATOF helium atom time-of-flight spectroscopy
HB Brinell hardness number
HEED high-energy electron diffraction
HEIS high-energy ion scattering/high-energy ion scattering spectroscopy
HK Knoop hardness
HOPG highly oriented pyrolytic graphite
HPDC high-pressure die casting
HR-EELS high-resolution electron energy loss spectroscopy
HR-LEED high-resolution LEED
HR-RHEED high-resolution RHEED
HREELS high-resolution electron energy loss spectroscopy
HRTEM high-resolution transition electron microscopy
HT high temperature
HTSC high-temperature superconductor
HV Vicker's Hardness

I

IACS International Annealed Copper Standard
IB ion bombardment
IBAD ion-beam-assisted deposition
ICISS impact ion scattering spectroscopy
ICSU International Council of the Scientific Unions
IPE inverse photoemission
IPES inverse photoemission spectroscopy
ISO International Organization for Standardization
ISS ion scattering spectroscopy
IUPAC International Union of Pure and Applied Chemistry

J

JDOS joint density of states

K

KRIPES K-resolved inverse photoelectron spectroscopy

L

LAPW linearized augmented-plane-wave method
LB Langmuir–Blodgett
LCM liquid crystal material
LCP liquid crystal polymer
LCs liquid crystals
LDA local-density approximation
LDOS local density of states
LEED low-energy electron diffraction
LEIS low-energy ion scattering/low-energy ion scattering spectroscopy
LPE liquid phase epitaxy

M

MBE molecular-beam epitaxy
MD molecular dynamics
MEED medium-energy electron diffraction
MEIS medium-energy ion scattering/medium-energy ion scattering spectroscopy
MFM magnetic force microscopy
ML monolayer
MOCVD metal-organic chemical vapor deposition
MOKE magneto-optical Kerr effect
MOSFET MOS field-effect transistor
MQW multiple quantum well

N

NICISS neutral impact collision ion scattering spectroscopy
NIMs National Institutes for Metrology

O

OPO optical parametric oscillation

P

PDS photothermal displacement spectroscopy
PED photoelectron diffraction
PES photoemission spectroscopy
PLAP pulsed laser atom probe
PLD pulsed laser deposition
PSZ stabilized zirconia
PZT piezoelectric material

R

RAS reflectance anisotropy spectroscopy
RE rare earth
REM reflection electron microscope/microscopy
RHEED reflection high-energy electron diffraction
RIE reactive ion etching
RPA random-phase approximation
RT room temperature
RTP room temperaure and standard pressure

S

SAM self-assembled monolayer
SAM scanning Auger microscope/microscopy
SARS scattering and recoiling ion spectroscopy
SAW surface acoustic wave
SBZ surface Brillouin zone
SCLS surface core level shift
SDR surface differential reflectivity
SEM scanning electron microscope
SEXAFS surface-sensitive EXAFS
SFG sum frequency generation
SH second harmonic
SHG second-harmonic generation
SI Système International d'Unités
SIMS secondary-ion mass spectroscopy
SNR signal-to-noise ratio
SPARPES spin polarized angle-resolved photoemission spectroscopy
SPIPES spin-polarized inverse photoemission spectroscopy
SPLEED spin-polarized
SPV surface photovoltage spectroscopy
SQUIDS superconducting quantum interference devices
SS surface state
STM scanning tunneling microscope/microscopy
STS scanning tunneling spectroscopy
SXRD surface X-ray diffraction

T

TAFF thermally activated flux flow
TEM transmission electron microscope/microscopy
TFT thin-film transistor
TMR tunnel magnetoresistance
TMT thermomechanical treatment
TOF time of flight
TOM torsion oscillation magnetometry

TRS truncation rod scattering
TTT time-temperature-transformation

U

UHV ultra-high vacuum
UPS ultraviolet photoemission spectroscopy
UV ultraviolet

V

VBM valence band maximum
VLEED very low-energy electron diffraction

X

XPS X-ray photoemission spectroscopy

4.3. Magnetic Materials

Magnetic materials consist of a wide variety of metals and oxides. Their effective properties are given by a combination of two property categories: *intrinsic* properties which are the atomic moment per atom p_{at}, Curie temperature T_c, magnetocrystalline anisotropy coefficients K_i, and magnetostriction coefficients λ_i; and *extrinsic* properties which are essentially their coercivity H_c and their magnetisation M or magnetic Induction J as a function of the applied magnetic field H. Moreover, the effective properties are depending decisively on the microstructural features, texture and, in most cases, on the external geometric dimensions such as thickness or shape of the magnetic part. In some cases non-magnetic inorganic and organic compounds serve as binders or magnetic insulators in multiphase or composite magnetic materials.

4.3.1 **Basic Magnetic Properties** 755
4.3.1.1 Atomic Moment 755
4.3.1.2 Magnetocrystalline Anisotropy .. 756
4.3.1.3 Magnetostriction 757

4.3.2 **Soft Magnetic Alloys** 758
4.3.2.1 Low Carbon Steels 758
4.3.2.2 Fe-based Sintered and Composite Soft Magnetic Materials 759
4.3.2.3 Iron–Silicon Alloys 763
4.3.2.4 Nickel–Iron-Based Alloys 769
4.3.2.5 Iron–Cobalt Alloys 772
4.3.2.6 Amorphous Metallic Alloys 772
4.3.2.7 Nanocrystalline Soft Magnetic Alloys 776
4.3.2.8 Invar and Elinvar Alloys 780

4.3.3 **Hard Magnetic Alloys** 794
4.3.3.1 Fe–Co–Cr 795
4.3.3.2 Fe–Co–V 797
4.3.3.3 Fe–Ni–Al–Co, Alnico 798
4.3.3.4 Fe–Nd–B 800
4.3.3.5 Co–Sm 803
4.3.3.6 Mn–Al–C 810

4.3.4 **Magnetic Oxides** 811
4.3.4.1 Soft Magnetic Ferrites 811
4.3.4.2 Hard Magnetic Ferrites 813

References 814

4.3.1 Basic Magnetic Properties

Basic magnetic properties of metallic systems and materials are treated by *Gignoux* in [3.1]. Extensive data on magnetic properties of metals can be found in [3.2]. Magnetic properties of ferrites are treated by *Guillot* in [3.3]. Extensive data on magnetic and other properties of oxides and related compounds can be found in [3.4] and [3.5].

4.3.1.1 Atomic Moment

The suitability of a metal or oxide to be used as a magnetic material is determined by its mean atomic moment (p_{at}). For metals the Bethe–Slater–Pauling curves, Fig. 4.3-1, indicate how p_{at} depends on the average number (n) of 3d and 4s electrons per atom, and on the crystal structure, i. e., face-centered cubic (fcc) or body-centered cubic (bcc) structure. Alloys based on Fe, Co, and Ni are most suitable from this point of view, corresponding to their actual use. The characteristic temperature dependence of the spontaneous magnetization $I_s(T)$, shown for Fe in Fig. 4.3-2, the Curie temperature T_c and the spontaneous magnetization at room temperature I_s (see Table 4.3-1), are the ensuing properties.

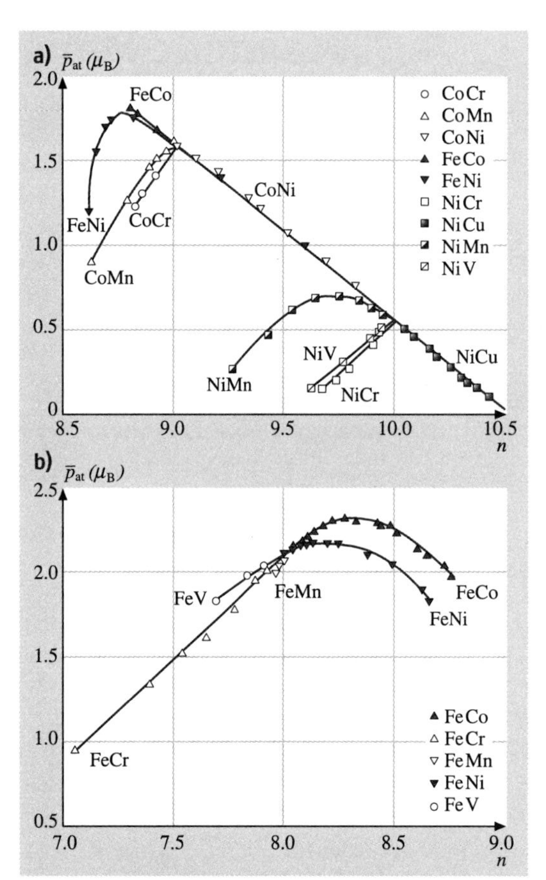

Fig. 4.3-1a,b Bethe–Slater–Pauling relation indicating the dependence of the mean magnetic moment per atom p_{at} on the average number n of 3d and 4s electrons per atom for binary alloys with (**a**) fcc structure and (**b**) bcc structure [3.6]

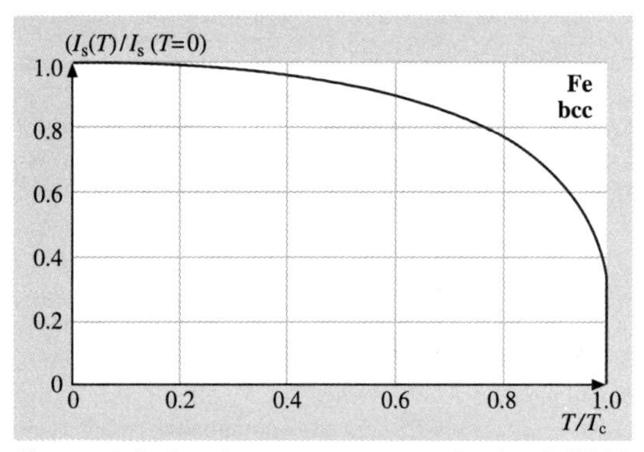

Fig. 4.3-2 Reduced spontaneous magnetization $I_S(T)/I_s$ ($T = 0$) vs. reduced temperature T/T_c for Fe [3.6]

4.3.1.2 Magnetocrystalline Anisotropy

Since the magnetic moment arises from the exchange coupling of neighboring ions, it is also, coupled to their positions in the crystal structures. Basically this is the origin of the magnetocrystalline anisotropy which plays a major role as an intrinsic property for the optimization of both soft and hard magnetic materials because it determines the crystallographic direction and relative magnitude of easy magnetization. As an example, Fig. 4.3-3a shows the magnetization curves for a single crystal of Fe in the three major crystallo-

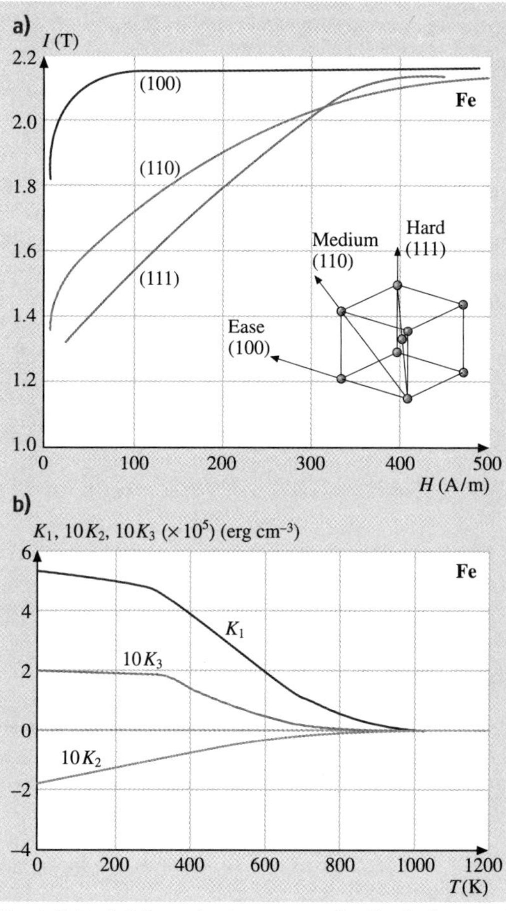

Fig. 4.3-3a,b Magnetization curves of an Fe single crystal indicting the typical characteristics of (**a**) magneto-crystalline anisotropy and (**b**) the temperature dependence of the magnetocrystalline anisotropy constants K_i of Fe [3.6]

Table 4.3-1 Curie temperatures and intrinsic magnetic properties at room temperature

Element	T_c (K)	I_s (T)	K_1 (10^4 J m^{-3})	K_2 (10^4 J m^{-3})	K_3 (10^4 J m^{-3})	λ_{100} (10^{-6})	λ_{111} (10^{-6})	λ_{0001} (10^{-6})	λ_{1010} (10^{-6})	λ_s (10^{-6})
Fe, bcc	1044	2.15	4.81	0.012	−0.012	22.5	−18.8	–	–	−4
Co, hcp	1388	1.62	41	120	–	–	–	−4	−22	−71
Ni, fcc	624	0.55	−0.57	−0.23	0	−46	−24.3	–	–	−34

graphic directions of the body-centered cubic crystal structure.

The anisotropy constants $K_1, K_2, \ldots$ are defined for cubic lattices, such as Fe and Ni, by expressing the free energy of the crystal anisotropy per unit volume as

$$E_a = K_0 + K_1 S + K_2 P + K_3 S^2 + K_4 SP + \ldots ,$$

with

$$S = \alpha_1^2\alpha_2^2 + \alpha_2^2\alpha_3^2 + \alpha_3^2\alpha_1^2 \quad \text{and} \quad P = \alpha_1^2\alpha_2^2\alpha_3^2 ,$$

where α_i, α_j, and α_k are the direction cosines of the angle between the magnetization vector and the crystallographic axes. Corresponding definitions pertain to the anisotropy constants for other crystal lattices.

In practice it can be important to know not only the room temperature value of the magnetocrystalline anisotropy constants K_i but also their temperature dependence, which is shown for Fe in Fig. 4.3-3b. For many technical considerations it suffices to take the dominating anisotropy constant K_1 into account.

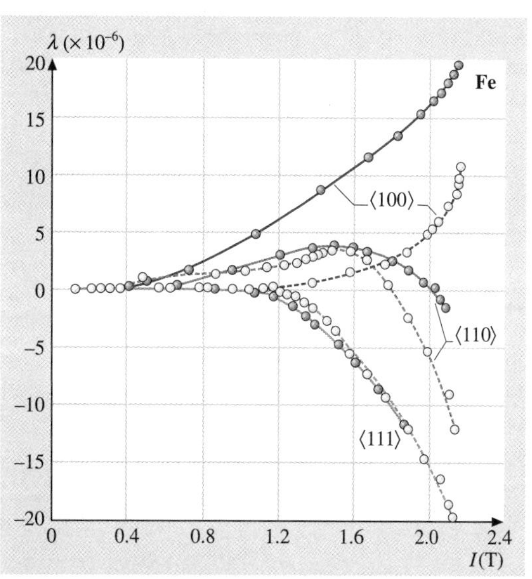

Fig. 4.3-4 Dependence of the magnetostrictive elongation on the magnetization intensity in three different crystallographic directions of Fe single crystals according to two different sources (*open* and *closed circles*) [3.7]

4.3.1.3 Magnetostriction

Magnetostriction is the intrinsic magnetic property which relates spontaneous lattice strains to magnetization. It is treated extensively by *Cullen* et al. in [3.3]. As a crystal property magnetostriction is purely intrinsic. The phenomenon of macroscopic magnetostrictive strains of a single or polycrystalline sample is due to the presence of magnetic domains whose reorientation occurs under the influence of magnetic fields or applied stresses and, thus, is a secondary property which has to be distinguished.

Magnetostriction plays a role in determining a number of different effects:

- As a major factor of influence on the coercivity of soft magnetic materials because it determines the magnitude of interaction of internal stresses of materials with the movement of magnetic domain walls. These boundaries are associated with strains themselves unless the material is free of magnetostriction altogether. This is practically impossible because the components of this property vary with the applied magnetic field, as shown for the example of Fe in Fig. 4.3-4. Moreover, they vary with temperature and alloy composition.
- As a decisive variable determining the properties of invar and elinvar alloys (Sect. 4.3.2.8).
- As a property of anomalously high magnitude in cubic Laves phase compounds of rare earth metals such as $Fe_2(Tb_{0.3}Dy_{0.7})$, which are the basis of magnetostrictive materials serving as magnetostrictive transducers and sensors [3.3].

4.3.2 Soft Magnetic Alloys

The basic suitability of a metal or alloy as a soft magnetic material is provided if its Curie temperature T_c and saturation polarization I_s at room temperature are sufficiently high. The main goal in developing soft magnetic materials is to reduce the magnetocrystalline anisotropy constants K_i, in particular K_1, and the magnetostriction constants λ_i, to a minimum. This leads to a minimum energy requirement for magnetization reversal. The appropriate combination of the intrinsic magnetic properties can be achieved by alloying additions and, in some cases, by annealing treatments which induce atomic ordering and, thus, an additional variation of K_1 and/or λ_i. Surveys and data of soft magnetic alloys are given in [3.8].

4.3.2.1 Low Carbon Steels

Low carbon steels are the most widespread magnetic metallic materials for use in electric motors and inductive components such as transformers, chokes, and a variety of other AC applications which require high magnetic induction at moderate to low losses. This is mainly due to the high intrinsic saturation polarization of the base element Fe, $J_s = 2.15$ T, and to the low cost of mass produced steels.

Main processing effects on achieving materials with a high relative permeability is the minimization of the formation of particles (mainly carbides and sulfides) which may impede domain boundary motion, and induc-

Table 4.3-2 Standard IEC specification for non-alloyed magnetic steel sheet. The conventional designation of the different grades (first column) comprises the following order: (1) 100 times the maximum specified loss (W kg^{-1}) at 1.5 T; (2) 100 times the nominal sheet thickness in (mm); (3) the characteristic letter "D"; (4) one tenth of the frequency in Hz at which the magnetic properties are specified. The materials are delivered in the semi-processed state. The magnetic properties apply to test specimens heat-treated according to manufacturer's specifications

Grade	Nominal thickness (mm)	Maximum specific total loss for peak induction (W kg^{-1})			Minimum induction in a direct or alternating field at field strength given (T)			Stacking factor	Conventional density (kg dm^{-3})
		1.5 T at at 50 Hz	1.0 T at at 50 Hz	1.5 T at at 60 Hz	at 2500 A/m	at 5000 A/m	at 10 000 A/m		
660-50-D5	0.50	6.60	2.80	8.38	1.60	1.70	1.80	0.97	7.85
890-50-D5	0.50	8.90	3.70	11.30	1.58	1.68	1.79	0.97	7.85
1050-50-D5	0.50	10.50	4.30	13.34	1.55	1.65	1.78	0.97	7.85
800-65-D5	0.65	8.00	3.30	10.16	1.60	1.70	1.80	0.97	7.85
1000-65-D5	0.65	10.00	4.20	12.70	1.58	1.68	1.79	0.97	7.85
1200-65-D5	0.65	12.00	5.00	15.24	1.55	1.65	1.78	0.97	7.85

Table 4.3-3 Standard IEC specifictaion for cold rolled magnetic alloyed steel strip delivered in the semi-processed state. The conventional designation of the different grades comprises the following order (first column): (1) 100 times the maximum specified loss at 1.5 T peak induction in (W kg^{-1}); (2) 100 times the nominal strip thickness, in (mm); (3) the characteristic letter "E"; (4) one tenth of the frequency in Hz, at which the magnetic properties are specified. the magnetic properties apply to test specimens subjected to a reference heat treatment

Grade	Nominal thickness (mm)	Reference treatment temperature (°C)	Maximum specific total loss for peak induction (W kg^{-1})		Minimum induction in a direct or alternating field at field strength given (T)			Conventional density (10^3 kg m^{-3})
			1.5 T	1.0 T	2500 Am^{-1}	5000 Am^{-1}	10 000 Am^{-1}	
340-50-E5	0.50	840	3.40	1.40	1.52	1.62	1.73	7.65
390-50-E5	0.50	840	3.90	1.60	1.54	1.64	1.75	7.70
450-50-E5	0.50	790	4.50	1.90	1.55	1.65	1.76	7.75
560-50-E5	0.50	790	5.60	2.40	1.56	1.66	1.77	7.80

ing a texture – by suitable combinations of deformation and recrystallization – with a preponderance of ⟨100⟩ and ⟨110⟩ components, while suppressing the ⟨111⟩ and ⟨211⟩ components, such that the operating induction can be aligned most closely to the easiest direction of magnetization ⟨100⟩. More details are given by *Rastogi* in [3.9]. Tables 4.3-2 and 4.3-3 show characteristic data of non-alloyed magnetic steel sheet and cold rolled magnetic alloyed steel strip, respectively.

4.3.2.2 Fe-based Sintered and Composite Soft Magnetic Materials

Iron-based sintered and composite soft magnetic materials are treated and listed extensively in [3.10]. Iron and Fe based alloys are used as sintered soft magnetic materials because of the particular advantages of powder metallurgical processing in providing net-shaped parts economically. The sintering process serves to achieve diffusion-bonding of the powder particles with a uniform distribution of the alloying elements. The metallic bond and degree of particle contact determine the magnetic properties in each alloy system.

The operating frequency in an application is limited by the resistivity of the material, which is increased beyond that of pure and dense Fe, primarily by varying the type and concentration of alloying elements. Density and crystal structure have a major impact on the magnetic properties, but only a minor effect on the resistivity. The sintered materials have a resistivity ranging from 10 to 80 μΩ cm and are applicable in DC and very low frequency fields. The sintered materials comprise Fe and Fe–P, Fe–Si, Fe–Si–P, Fe–Sn–P, Fe–Ni, Fe–Cr, and Fe–Co alloys. Characteristic data are listed in Tables 4.3-4 to 4.3-11 [3.10].

It is useful to note that the evaluation of a large number of test data of sintered soft magnetic materials yields consistent empirical relations. For sintered Fe products these are:

$$B_{15}[\mathrm{T}] = 4.47\varrho - 10.38\,,$$

$$B_{\mathrm{r}}[\mathrm{T}] = 3.87\varrho - 17.23\,,$$

$$H_{\mathrm{c}}[\mathrm{A/m}] = 11.47L^{-0.59}\,,$$

$$\rho[\mu\Omega\,\mathrm{cm}] = -4.34\varrho + 44.77\,,$$

and

$$\mu_{\max} = 0.21L^{0.68} \times 10^3\,,$$

where $\varrho =$ density ($\mathrm{g\,cm^{-3}}$) and $L =$ grain diameter (average intercept length) (μm).

Dust core materials consist of Fe or Fe alloy particles which are insolated by an inorganic high resistivity barrier. They are applied in the 1 kHz to 1 MHz range. These are to be distinguished from soft magnetic Fe composite materials, which consist of pure Fe particles separated by an insolating organic barrier, providing a medium to high bulk resistance. Compositions and magnetic properties of Fe based composite materials are listed in Tables 4.3-12 and 4.3-13. These materials are applicable in the frequency range from 50 Hz to 1 kHz.

Table 4.3-4 Magnetic properties of sintered Fe products in relation to density [3.10]

Density ϱ ($\mathrm{g\,cm^{-3}}$)	Sintering conditions temp. (°C), atm.	Induction at 1200 $\mathrm{A\,m^{-1}}$ (T)	Coercitivity H_c ($\mathrm{A\,m^{-1}}$)	Remanence B_r (T)	Max. rel. permeability μ_{max}
6.6	1120, DA	0.90	170[a]	0.78[a]	1700
6.6	1120, H_2/Vac.	0.95	140[a]	0.82[a]	1800
6.6	1260, H_2/Vac.	1.05	120[a]	0.85[a]	2800
6.9	1120, DA	1.05	170[a]	0.90[a]	2100
6.9	1120, H_2/Vac.	1.05	140[a]	0.97[a]	2300
6.9	1260, H_2/Vac.	1.20	120[a]	1.0[a]	3300
7.2	1120, DA	1.20	170[a]	1.05[a]	2700
7.2	1120, H_2/Vac.	1.20	140[a]	1.10[a]	2900
7.2	1260, H_2/Vac.	1.30	120[a]	1.15[a]	3800
7.4		1.25	136	1.20	3500
7.4		1.30	112	1.30	5500
7.6		1.50	80	1.40	6000

[a] Measured from a maximum applied magnetic field strength of 1200 A/m.

Part 4 | 3.2

Table 4.3-5 Magnetic properties of sintered Fe–0.8 wt% P products in relation to density [3.10]

Density ϱ ($g\,cm^{-3}$)	Sintering conditions temp. (°C), atm.	Induction at 1200 $A\,m^{-1}$ (T)	Coercitivity H_c ($A\,m^{-1}$)	Remanence B_r (T)	Max. rel. permeability μ_{max}
6.8	1120, DA	1.05	120[a]	1.00[a]	3500
7.0	1120, H_2/Vac.	1.20	100[a]	1.05[a]	4000
7.2	1260, H_2/Vac.	1.25	95[a]	1.15[a]	4000
7.0	1120, DA	1.20	120[a]	1.10[a]	4000
7.2	1120, H_2/Vac.	1.25	100[a]	1.15[a]	4500
7.4	1260, H_2/Vac.	1.30	95[a]	1.20[a]	4500
7.0	1120, DA	1.25	120[a]	1.20[a]	4500
7.2	1120, H_2/Vac.	1.30	100[a]	1.30[a]	5000
7.4	1260, H_2/Vac.	1.35	95[a]	1.25[a]	5000

[a] Measured from a maximum applied magnetic field strength of 1200 A/m

Table 4.3-6 Magnetic properties of sintered Fe–3 wt% Si products in relation to density[a] [3.10]

Density ϱ ($g\,cm^{-3}$)	Sintering conditions temp. (°C), time, atm.	Coercitivity H_c ($A\,m^{-1}$)	Remanence B_r (T)	Sat. induction B_s (T)	Max. rel. permeability μ_{max}
7.3		64	1.15	1.90	8000
7.5		48	1.25	2.00	9500
7.2	1250, 30 min, H_2 [b]	80	1.0		4300
7.01	1120, 60 min, H_2 [b]	117[d]			2800
7.19	1200, 60 min, H_2 [b]	88[d]			4200
7.40	1200, 60 min, H_2 [c]	79[d]	1.25		5600
7.43	1300, 60 min, H_2	56[d]			8400
7.55	1371, DA[e]	51	1.21	1.50	
7.55	1371, DA[e]	57	1.07	1.45	

[a] Measured according to ASTM A 596
[b] Conventional compacting at 600 MPa
[c] Warm compacted at 600 MPa
[d] Defined as coercive force i. e. magnetized to a field strength well below saturation
[e] Formed by MIM

Table 4.3-7 Magnetic properties of sintered Fe–S–P in relation to density[a] [3.10]

Density ϱ ($g\,cm^{-3}$)	Sintering conditions temp. (°C), atm.	Coercitivity H_c ($A\,m^{-1}$)	Remanence B_r (T)	Sat. induction B_s (T)	Max. rel. permeability μ_{max}
7.3[b]		45[b]	1.30[b]	1.90[b]	10 800[b]
7.55[c]		33[c]		2.00[c]	12 500[c]
7.3[c]	1250 30 min, H_2	60[c]	1.1[c]		6100[c]
6.8[d]	1120 30 min, H_2	100[d]	0.6[d]		2200[d]

[a] Measured according to ASTM A 596
[b] Fe/3 wt% Si/0.45 wt% P
[c] Fe/2 wt% Si/0.45 wt% P
[d] Fe/4 wt% Si/0.45 wt% P

Table 4.3-8 Magnetic properties of sintered Fe–Sn–P in relation to density [3.10]

Density ϱ ($g\,cm^{-3}$)	Sintering temp. (°C), time, atm.	Coercitivity H_c ($A\,m^{-1}$)	Remanence B_r (T)	Max. rel. permeability μ_{max}
7.2[a]	1120 30 min, H_2 [b]	80[a]	1.1[a]	4800[a]
7.4[b]	1250 30 min, H_2 [b]	37[b]	1.0[b]	9700[b]

[a] Fe/5 wt% Sn/0.45 wt% P
[b] Fe/5 wt% Sn/0.5 wt% P

Table 4.3-9 Magnetic properties of sintered Fe–Ni in relation to density[a] [3.10]

Density ϱ ($g\,cm^{-3}$)	Sintering temp. (°C), time atm.	Coercitivity H_c ($A\,m^{-1}$)	Remanence B_r (T)	Sat. induction B_s (T)	Max. rel. permeability μ_{max}
8.0[b]		24[b]	0.25[b]	1.55[b]	6000[b]
8.0[c]		13[c]	0.85[c]	1.60[c]	30 000[c]
8.5[d]		2[d]	0.40[d]	0.80[d]	74 900[d]
6.99[e]	1260, H_2/Vac.	20[e]	0.75[e]		
6.99[f]		24[f]	0.76[f]		7800[f]
7.30[e]	1260, H_2/Vac.	19[e]	0.90[e]		
7.30[f]		32[f]	0.85[f]		8700[f]
7.50[e]	1260, H_2/Vac.	16[e]	0.94[e]		21 000[e]
7.50[f]		32[f]	0.90[f]		9100[f]
7.4[c]	1250, 30 min, H_2	25[c]	0.8[c]		13 000[c]
7.66	1371, DA[f]	16	0.42	1.27	

[a] Measured according to ASTM A 596
[b] Fe/35–40 wt% Ni
[c] Fe/45–50 wt% Ni
[d] Fe/72–82 wt% Ni/3–5 wt% Mo
[e] Fe/47–50 wt% Ni
[f] Formed by MIM

Table 4.3-10 Magnetic properties of sintered Fe–Cr in relation to density[a] [3.10]

Density ϱ ($g\,cm^{-3}$)	Coercitivity H_c ($A\,m^{-1}$)	Remanence B_r (T)	Sat. induction B_s (T)	Max. rel. permeability μ_{max}
7.1[b]	200[b]	0.50[b]		1200[b]
7.45[c]	100[c]		1.70[c]	2500[c]
7.35[d]	100[d]		1.55[d]	1900[d]

[a] Measured according to ASTM A 596
[b] Fe/16–18 wt% Cr/0.5–1.5 wt% Mo
[c] Fe/12 wt% Cr/0.2 wt% Ni/0.7 wt% Si
[d] Fe/17 wt% Cr/0.9 wt% Mo/0.2 wt% Ni/0.8 wt% Si

Table 4.3-11 Magnetic properties of sintered Fe–Co products in relation to density[a] [3.10]

Density ϱ ($g\,cm^{-3}$)	Coercitivity H_c ($A\,m^{-1}$)	Remanence B_r (T)	Sat. induction B_s (T)	Max. rel. permeability μ_{max}
7.9[b]	136[b]	1.10[b]	2.25[b]	3900[b]

[a] Measured according to ASTM A 596
[b] Fe/47–50 wt% Co/1–3 wt% V

Table 4.3-12 Magnetic properties of composite Fe products in relation to density and insulation [3.10]

Density ϱ ($g\,cm^{-3}$)	Insulation	Curing (°C) (min)	Coercitivity H_c ($A\,m^{-1}$)	Induction at 3183 $A\,m^{-1}$ (T)	Max. rel. permeability μ_{max}
5.7–7.26	Polymer		263–374	0.33–0.83	97–245
7.2	Oxide + 0.75 Polymer		374	0.77	210
7.4–7.45	0.75–0.6 % Polymer		381–374	1.09–1.12	400–425
7.54	Oxide				630
7.04	Inorganic (LCM™)				305
7.0	0.5 % Phenolic resin[a]	150, 60	400		270
7.13–6.84[b]	0.4–1.8 % Phenolic[a]	160.30			325–175[c]
7.2[b]	0.8 % Phenolic[a]	150–500.30	445–310		270–390
6.6	3 % resin				200
6.7–7.0	Inorganic & 2 % resin	?–500			200–400
7.4	Inorganic (Somaloy™500)	500			600

[a] Dry Mixing or wet mixing with phenolic resin
[b] Compacted at 620 MPa and 65 °C
[c] AC permeability at 60 Hz and 1.0 T

Table 4.3-13 Core loss of composite Fe products in relation to lubricant and heat treatment [3.10]

Lubricant	Heat treatment			Density ϱ	Total loss at	Total loss at
	(°C)	(min.)	(Atm.)	($g\,cm^{-3}$)	100 Hz, 1.5 T ($W\,kg^{-1}$)	1000 Hz, 1.5 T ($W\,kg^{-1}$)
0.1 % Kenolube™ [d]	500	30	air	7.40[b]	29	
0.5 % Kenolube™	500	30	air	7.20[a]	32	450
0.5–0.8 % Kenolube™	500	30	air	7.36–7.26[b]	29–31	330–450
0.5 % Kenolube™	500	30	nitrogen	7.35[b]	31	350
0.5 % Kenolube™	250–580	30	steam	7.35[b]	30–70	
0.6 % LBI™	275	60	air	7.27[b]		700
0.5 % Kenolube™	500	30	air	7.20[a]	35	480
0.5 % Kenolube™	500	30	air	7.34[a]	34	420

[a] Conventional compacting at 600 MPa
[b] Conventional compacting at 800 MPa
Water atomized iron powder, particle size > 150 µm < 400 µm, low inorganic insulation thickness, (Somatoy™ 550)

4.3.2.3 Iron–Silicon Alloys

The physical basis of the use of Fe–Si alloys, commonly called silicon steels, as soft magnetic materials is the fact that both the magnetocrystalline anisotropy K_1 and the magnetostriction parameters λ_{100} and λ_{111} of Fe approach zero with increasing Si content (see Fig. 4.3-5a). The lower the magnitude of these two intrinsic magnetic properties is, the lower are the coercivity H_c and the AC magnetic losses p_{Fe}. The total losses p_{Fe} consist of the static hysteresis losses p_h and the dynamic eddy current losses p_w which may be subdivided into a classical p_{wc} and an anomalous p_{wa} eddy current loss term,

$$\begin{aligned} p_{Fe} &= p_h + p_{wc} + p_{wa} \\ &= c_h(f) H_c B f \\ &\quad + c_{wc}(\pi d B f)^2 / 6\rho\gamma + c_{wa}(\alpha_R/\rho)(Bf)^{3/2} , \end{aligned}$$

where $c_h(f)$ is a form factor of the hysteresis which depends on the frequency f, H_c is the coercivity of the material, B is the peak operating induction, c_{wc} and c_{wa} are terms taking the wave form of the applied field into account, d is the sheet thickness, ρ is the resistivity, γ is the density of the material, and α_R is the Raleigh constant. These are the factors to be controlled to obtain minimal losses. The increase in electrical resistivity with Si content (Fig. 4.3-5b) adds to lowering the eddy current losses as shown by the relation above.

The Fe–Si equilibrium diagram shows a very small stability range of the γ phase, indicating that the ferromagnetic α phase can be heat-treated in a wide temperature range without interference of a phase transformation which would decrease the magnetic softness of the material by the lattice defects induced.

Next to low-carbon steels, Fe–Si steels are the most significant group of soft magnetic materials (30% of the world market). A differentiation is made between non-oriented, isotropic (NO), and grain-oriented (GO) silicon steels. Non-oriented steels are mainly applied in rotating machines where the material is exposed to varying directions of magnetic flux. Grain-oriented steels with GOSS-texture (110) ⟨001⟩ are used predominantly as core material for power transformers.

Since Fe–Si steels are brittle above about 4.0 wt% Si, conventional cold rolling is impossible at higher Si contents.

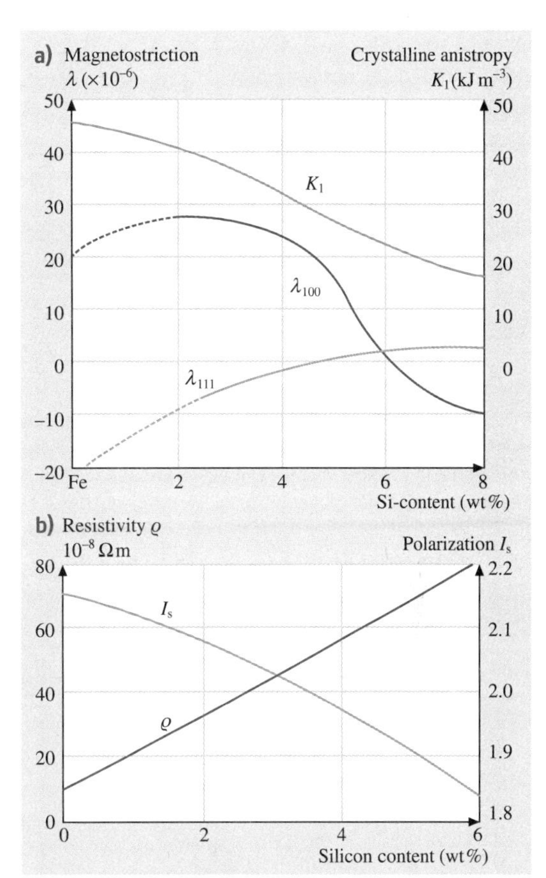

Fig. 4.3-5 (a) Magnetostriction λ_{100} and λ_{111} and magneto-crystalline anisotropy energy K_1. **(b)** Electrical resistivity ρ and saturation polarization I_s, as a function of the Si content in Fe–Si alloys

Non-oriented Silicon Steels (NO)

NO laminations are usually produced with thicknesses between 0.65 mm and 0.35 mm, and Si concentrations up to 3.5 wt%. According to their grade, NO silicon steels are classified in low grade (low Si content) alloys employed in small devices and high-grade (high Si content) alloys for large machines (motors and generators). Suitable microstructural features (optimum grain size) and a low level of impurities are necessary for optimum magnetic properties. Critical factors in processing are the mechanical behavior upon punching of laminations, the application of insulating coatings, and the build-up of stresses in magnetic cores. Table 4.3-14 lists the ranges of typical processing parameters.

In the case of low Si steels (< 1 wt% Si), the last two annealing steps are applied by the user after lamination punching (semi-finished sheet). Table 4.3-15 lists the specifications, including all relevant properties for non-oriented magnetic steel sheet.

Composition (wt%) Si: 0.9–3.4, Al: 0.2–0.6, Mn: 0.1–0.3
Melting, degassing, continous casting
Hot rolling to 1.8–2.3 mm (1000–1250 °C)
Cold rolling to intermediate gauge
Annealing (750–900 °C)
Cold rolling to final gauge (0.65–0.35 mm)
Decarburizing anneal (830–900 °C, wet H_2)
Recrystallization
Grain growth anneal (830–1100 °C)
Coating

Table 4.3-14 Schematic of NO silicon steel processing. Addition of small quantities (50–800 wt ppm) of Sb, Sn, or rare earth metals can be made to improve texture and/or control the morphology of the precipitates. Cold reduction in a single stage repesents a basic variant of the above scheme. The final grain growth annealing aims at an optimum grain size, leading to minimum losses. Coating provides the necessary interlaminar electrial insulation. Phosphate- or chromate-based coatings are applied, which ensure good lamination punchability [3.7]

Table 4.3-15 Standard IEC specification for nonoriented magnetic steel sheet delivered in the final sate. The conventional designation of the different grades comprises the following order (first column): (1) 100 times the maximum specified loss at 1.5 T peak induction in ($W\,kg^{-1}$); (2) 100 times the nominal sheet thickness; (3) the characteristic letter "A"; (4) one tenth of the frequency in Hz, at which the magnetic properties are specified. The anisotropy of loss, T, is specified at 1.5 T peak induction according to the formula $T = (P_1 - P_2)/(P_1 + P_2)100$, with P_1 and P_2 the power losses of samples cut perpendicular and parallel to the rolling direction, respectively [3.7]

Quality	Nominal thickness (mm)	Maximum specific total loss ($W\,kg^{-1}$) at peak induction		Minimum magnetic flux density (T) in direct or alternating field at field strength			Maximum anisotropy of loss (%)	Minimum stacking factor	Minimum number of bends	Conventional density ($10^3\,kg\,m^{-3}$)
		1.5 T	1.0 T	2500 Am^{-1}	5000 Am^{-1}	10 000 Am^{-1}				
250-35-A5	0.35	2.50	1.00	1.49	1.60	1.71			2	7.60
270-35-A5	0.35	2.70	1.10	1.49	1.60	1.71	±18	0.95	2	7.65
300-35-A5	0.35	3.00	1.20	1.49	1.60	1.71			3	7.65
330-35-A5	0.35	3.30	1.30	1.49	1.60	1.71			3	7.65
270-50-A5	0.50	2.70	1.10	1.49	1.60	1.71			2	7.60
290-50-A5	0.50	2.90	1.15	1.49	1.60	1.71	±18		2	7.60
310-50-A5	0.50	3.10	1.25	1.49	1.60	1.71			3	7.65
330-50-A5	0.50	3.30	1.35	1.49	1.60	1.71	±14		3	7.65
350-50-A5	0.50	3.50	1.50	1.50	1.60	1.71			5	7.65
400-50-A5	0.50	4.00	1.70	1.51	1.61	1.72		0.97	5	7.65
470-50-A5	0.50	4.70	2.00	1.52	1.62	1.73				7.70
530-50-A5	0.50	5.30	2.30	1.54	1.64	1.75				7.70
600-50-A5	0.50	6.00	2.60	1.55	1.65	1.76	±12		10	7.75
700-50-A5	0.50	7.00	3.00	1.58	1.68	1.76				7.80
800-50-A5	0.50	8.00	3.60	1.58	1.68	1.78				7.80
350-65-A5	0.65	3.50	1.50	1.49	1.60	1.71			2	7.65
400-65-A5	0.65	4.00	1.70	1.50	1.60	1.71			2	7.65
470-65-A5	0.65	4.70	2.00	1.51	1.61	1.72	±14		5	7.65
530-65-A5	0.65	5.30	2.30	1.52	1.62	1.73		0.97	5	7.70
600-65-A5	0.65	6.00	2.60	1.54	1.64	1.75				7.75
700-65-A5	0.65	7.00	3.00	1.55	1.65	1.76	±12		10	7.75
800-65-A5	0.65	8.00	3.60	1.58	1.68	1.76				7.80
1000-65-A5	0.65	10.00	4.40	1.58	1.68	1.78				7.80

Grain-Oriented Silicon Steels (GO)

Grain-oriented (GO) silicon steel is used mainly as core material for power transformers. Worldwide production is about 1.5 million tons per year. The increasing demand for energy-efficient transformers requiring still lower loss materials has led to continuous improvements of the magnetic properties over the years, where a decrease in the deviation from the ideal Goss texture (110)[001] has played a decisive role. Moreover, the eddy current losses have been reduced by decreasing the lamination thickness from 0.35 mm through 0.30 mm and 0.27 mm to 0.23 mm. Surface treatments of the laminations by mechanical scratching or laser scribing have been introduced. They increase the number of mobile Bloch walls and, thus, decrease the spacing between them, i. e., the domain size. Accordingly, the anomalous eddy current losses were reduced. The reduction of the total losses of grain oriented electrical steel due to these improvements is illustrated in Fig. 4.3-6.

The GO steels are classified in two categories: conventional grain-oriented (CGO) and high permeability (HGO) steels. The latter are characterized by a sharp crystallographic texture, with average misorientation of the [001] axes of the individual crystallites around the rolling direction (RD) on the order of 3°. For CGO the misorientation is about 7°. The relation between the angular deviation of grain orientation and the total loss reduction for HGO material is shown in Fig. 4.3-7.

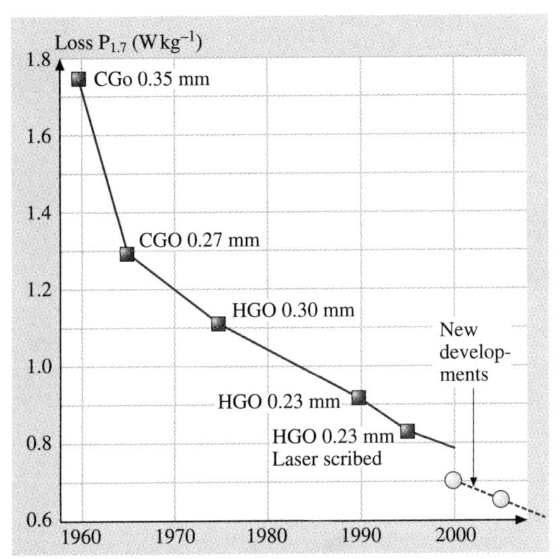

Fig. 4.3-6 Qualitative improvment of GO electrical steel [3.11]

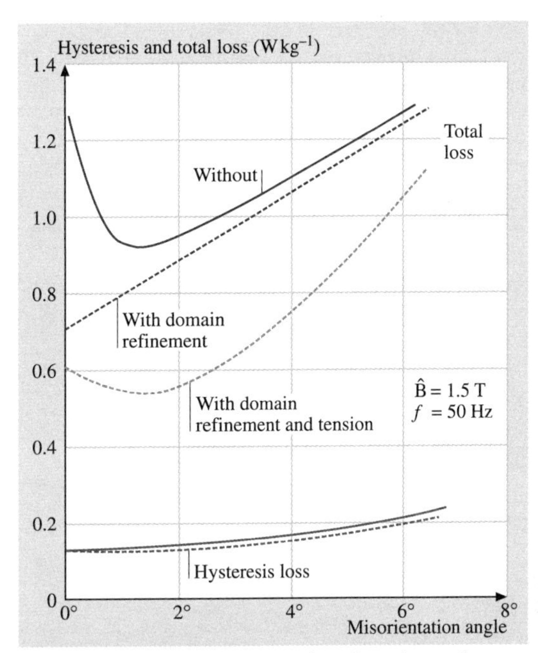

Fig. 4.3-7 Relation between grain orientation and power loss reduction for highly grain-oriented material (single crystal) (*Bölling* and *Hastenrath*) [3.12, 13]

The (GO) manufacturing route is an extraordinarily long sequence of hot and cold processing steps. The final magnetic properties are highly sensitive to even small parameter variations throughout this route. Some of these processes, their microstructure, and the inhibitor element influence are given in Table 4.3-16.

A key factor is the controlled development of the (110)[001] texture during secondary recrystallization. It requires the presence of large Goss textured grains in the surface layers of the annealed hot band, the presence of inhibitors as finely dispersed second-phase particles which strongly impede normal grain growth during primary recrystallization and a primary recrystallized texture having a suitable orientation relationship with respect to the Goss texture. This can be obtained through carefully-controlled chemistry and a precisely-defined sequence of thermomechanical treatments. Abnormal grain growth during secondary recrystallization may increase the size of magnetic domains and consequently greater energy dissipation under dynamic conditions. Refining the domain structure by laser or mechanical scribing core as mentioned above reduces the losses. Standard IEC specifications for grain-oriented magnetic

Table 4.3-16 Summary of the processing of grain-oriented silicon steel. The first column relates to the conventional grain-oriented (CGO) laminations. Process for three different types of high permeability (HGO) steels are outlined in columns 2–4. They basically differ for the type of grain growth inhibitors, the cold-rolling sequence, and the annealing temperatures. The processes CGO and HGO-1 adopt a two-stage cold reduction, with intermediate annealing, while HGO-2 and HGO-3 steels are reduced to the final thickness in a single step. Growth inhibition of the primary recrystallized grains is obtained by MnS precipitates in the CGO process. MnSe particles + solute Sb operate in process HGO-1, MnS + AlN particles in HGO-2, and solute B + N + S in HGO-3. Abnormal growth of (110)[001] grains occurs by final box annealing, which also promotes the dissolution of the precipitates [3.7]

Type of Steel							
CGO		**HGO-1**		**HGO-2**		**HGO-3**	
		Composition		(wt%)			
3–3.2	Si	2.9–3.3	Si	2.9–3.3	Si	3.1–3.3	Si
0.04–0.1	Mn	0.05	Mn	0.03	Al	0.02	Mn
0.02	S	0.02	Se	0.015	N	0.02	S
0.03	C	0.04	Sb	0.07	Mn	0.001	B
balance	Fe	0.03–0.07	C	0.03	S	0.005	N
		balance	Fe	0.05–0.07	C	0.03–0.05	C
				balance	Fe	balance	Fe
		Inhibitors					
MnS		MnSe + Sb		MnS + AlN		B + N + S	
		Melting, degassing and continuous casting					
		Reheating - hot rolling					
1320 °C		1320 °C		1360 °C		1250 °C	
		Annealing					
800–1000 °C		900 °C		1100 °C		870–1020 °C	
		Cold rolling					
70 %		60–70 %		87 %		80 %	
Annealing 800–1000 °C							
Cold rolling 55 %		65 %					
		Decarburizing anneal 800–850 °C (wet H_2)					
		MgO coating and coiling					
		Box-annealing					
1200 °C		820–900 °C + 1200 °C		1200 °C		1200 °C	
		Phospate coating and thermal flattening					

steel sheets are listed in Table 4.3-17. Basic properties of grain-oriented Fe–3.2 wt% Si alloys are given in Table 4.3-18.

A recent technology development in the production of GO electrical steel is the combination of thin slab casting, direct hot rolling, and acquired inhibitor formation. This practice combines the advantages of low temperatures, process-shortening, microstructural homogeneity, improved strip geometry, and better surface condition of the products. The slab thickness is on the order of 50–70 mm. Another future technology with remarkable process shortening is to produce (GO) hot strip in the thickness range of about 2–3 mm by direct casting from the steel melt using a twin-roll casting method. In pilot line tests, good workability and good magnetic properties have been achieved.

Further potential for cost and time saving is expected from replacing box annealing at the end of the cold process by short-time continuous annealing.

Table 4.3-17 Standard IEC specification for grain-oriented magnetic steel sheet: **(a)** normal material; **(b)** material with reduced loss; **(c)** high-permeability material. The conventional designation of the various grades (first column) includes, from left to right, (1) 100 times the maximum power loss, in ($W\,kg^{-1}$), at 1.5 T **(a)** or 1.7 T **(b,c)** peak induction; (2) 100 times the nominal sheet thickness, in (mm); (3) the letter "N" for the nominal material **(a)**, or "S" for material with reduced loss **(b)**, or "P" for high-permeability material **(c)**; (4) one tenth of the frequency in Hz, at which the magnetic properties are specified [3.7]

a) Grade	Thickness (mm)	Maximum specific total loss ($W\,kg^{-1}$) at peak induction		Minimum magnetic flux density (T) for $H = 800\,Am^{-1}$	Minimum stacking factor
		1.5 T	1.7 T		
089-27-N 5	0.27	0.89	1.40	1.75	0.950
097-30-N 5	0.30	0.97	1.50	1.75	0.955
111-35-N 5	0.35	1.11	1.65	1.75	0.960

b) Grade	Thickness (mm)	Maximum specific total loss ($W\,kg^{-1}$) at 1.7 T peak induction	Minimum magnetic flux density (T) for $H = 800\,Am^{-1}$	Minimum stacking factor
130-27-S 5	0.27	1.30	1.78	0.950
140-30-S 5	0.30	1.40	1.78	0.955
155-35-S 5	0.35	1.55	1.78	0.960

c) Grade	Thickness (mm)	Maximum specific total loss ($W\,kg^{-1}$) at 1.7 T peak induction	Minimum magnetic flux density (T) for $H = 800\,Am^{-1}$	Minimum stacking factor
111-30-P 5	0.30	1.11	1.85	0.955
117-30-P 5	0.30	1.17	1.85	0.955
125-35-P 5	0.35	1.25	1.85	0.960
135-35-P 5	0.35	1.35	1.85	0.960

Table 4.3-18 Basic properties of grain-oriented Fe–3.2 wt% Si alloys [3.7]

Property	Value
Density	$7.65 \times 10^3\,kg\,m^{-3}$
Thermal concuctivity	$16.3\,W\,°C^{-1}\,kg\,m^{-3}$
Electrical resistivity	$48 \times 10^{-8}\,\Omega\,m$
Young's modulus	
Single crystals	
[100] direction	120 GPa
[110] direction	216 GPa
[111] direction	295 GPa
(110)[001] texture	
Rolling direction (RD)	122 GPa
45° to RD	236 GPa
90° to RD	200 GPa
Yield strength	
(110)[001] texture	
Rolling direction	324 MPa
Tensile strength	
(110)[001] texture	
Rolling direction	345 MPa
Saturation induction	2.0 T

Table 4.3-18 Basic properties of grain-oriented Fe–3.2 wt% Si alloys [3.7], cont.

Property	Value
Curie temperature	745 °C
Magnetocrystalline anisotropy	3.6×10^4 J m^{-3}
Magnetostriction constants	
λ_{100}	23×10^{-6}
λ_{111}	-4×10^{-6}

Rapidly Solidified Fe–Si Alloys

The Fe–6.5 wt% Si alloy exhibits good high-frequency soft magnetic properties due to a favorable combination of low values of the saturation magnetostriction λ_s, as well as the low values of the magnetocrystalline anisotropy energy K_1, and a high electrical resistivity. But as mentioned above, Fe–Si alloys which contain more than about 4 wt% Si are brittle and thin sheets cannot be manufactured by rolling. Therefore, Fe–6.5 wt% Si sheets and ribbons are manufactured via two different routes by which the adverse mechanical properties are circumvented: a continuous "siliconizing" process in commercial scale production and a rapid quenching process.

In the siliconizing process Fe–3 wt% Si sheet reacts with a Si-containing gas at 1200 °C. The sheet is held at 1200 °C in order to increase and homogenize the Si content by diffusion. After this treatment the ductility of Fe–6.5 wt% Si sheets amounts to about 5% elongation to fracture.

By rapid quenching from the melt, the formation of the B2 and $D0_3$ type ordered structures, based on conventional cooling after casting of Fe–6.5 wt% Si alloys, may be suppressed. Thus, the ensuing material brittleness can be overcome. The ribbons formed by the rapid quenching process are about 20 to 60 μm thick and are ductile, with a microcrystalline structure. By means of an annealing treatment above 1000 °C fol-

Table 4.3-19 Physical and magnetic properties of rapidly quenched Fe–6.5 wt% Si alloys [3.7]

Property	Value
Density	7.48×10^3 kg m^{-3}
Thermal conductivity (31 °C)	4.5 cal m^{-1} °C^{-1} s^{-1}
Specific heat (31 °C)	128 cal °C^{-1} kg^{-1}
Coefficient of thermal expansion (150 °C)	11.6×10^{-6} °C^{-1}
Electrical resistivity	82×10^{-8} Ω m
Tensile strength (rapidly-quenched ribbons 60 μm thick)	630 MPa
Saturation magnetization	1.8 T
Curie temperature	700 °C
Saturation magnetostriction	0.6×10^{-6}

Table 4.3-20 Magnetic properties of 30–40 μm thick, rapidly-quenched ribbons of Fe–6.5 wt% Si, in the as-quenched state and after 24 h annealing at various temperatures [3.7]

Annealing temperature (°C)	H_c (A m^{-1})	H_{10} (T)	B_r/B_{10}	μ_{max}/μ_0
As-quenched	112	1.25	0.70	3100
500	100	1.27	0.93	4300
700	72	1.32	0.90	5400
800	45	1.31	0.92	9400
850	37.5	1.28	0.94	10 000
900	33	1.30	0.77	12 500
1000	21	1.31	0.83	17 000
1100	18	1.30	0.84	18 000
1200	20	1.32	0.87	22 000

lowed by rapid cooling to restrain D0$_3$ type ordering, large-grain-sized, recrystalllized (100) ⟨0vw⟩ textured material with good ductility and good soft magnetic properties is obtained, as characterized in Tables 4.3-19 and 4.3-20. Figures 4.3-8 and 4.3-9 show the loss behavior as a function of magnetizing frequency and ribbon thickness.

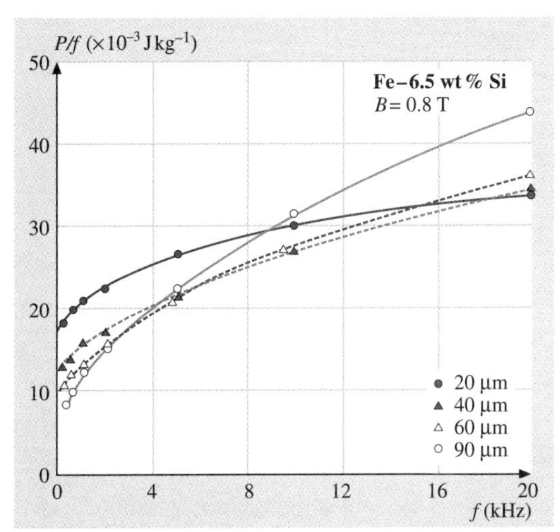

Fig. 4.3-8 Power loss per cycle vs. magnetizing frequency for rapidly quenched Fe–6.5 wt% Si ribbons of various thicknesses [3.7]

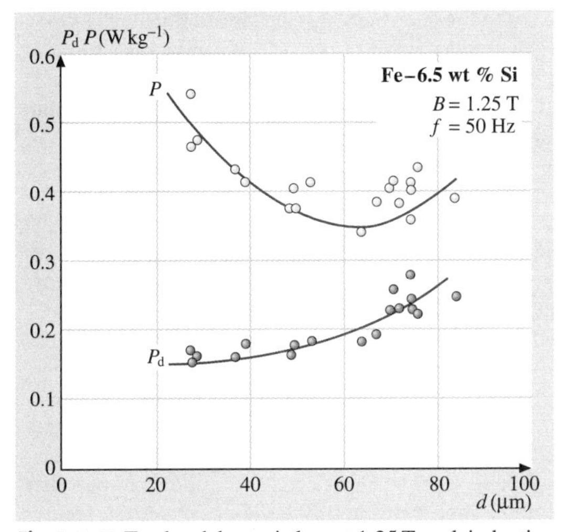

Fig. 4.3-9 Total and dynamic loss at 1.25 T peak induction and 50 Hz vs. ribbon thickness for rapidly quenched Fe–6.5 wt% Si ribbons characterized by a strong (100) [0uv] grain texture induced by vacuum annealing [3.7]

4.3.2.4 Nickel–Iron-Based Alloys

The fcc phase in the Ni–Fe alloy system and the formation of the ordered Ni_3Fe phase provide a wide range of structural and magnetic properties for developing soft magnetic materials with specific characteristics for different applications. The phase diagram is shown in Sect. 3.1.5. Before amorphous and nanocrystalline soft magnetic alloys were introduced, the Ni–Fe materials

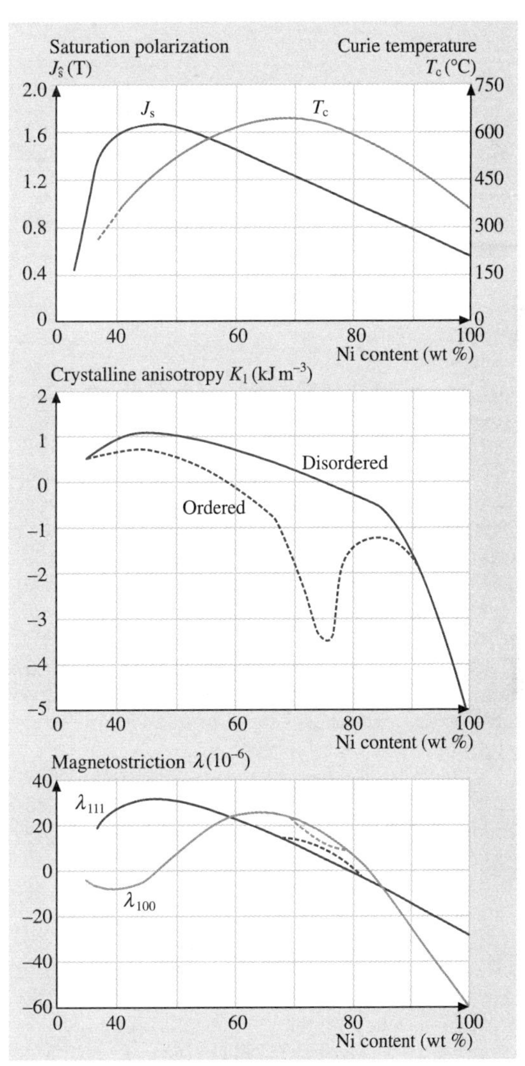

Fig. 4.3-10 The dependence of the intrinsic magnetic parameters I_s, T_c, K_1, λ of Ni–Fe alloys on the Ni concentration [3.12]

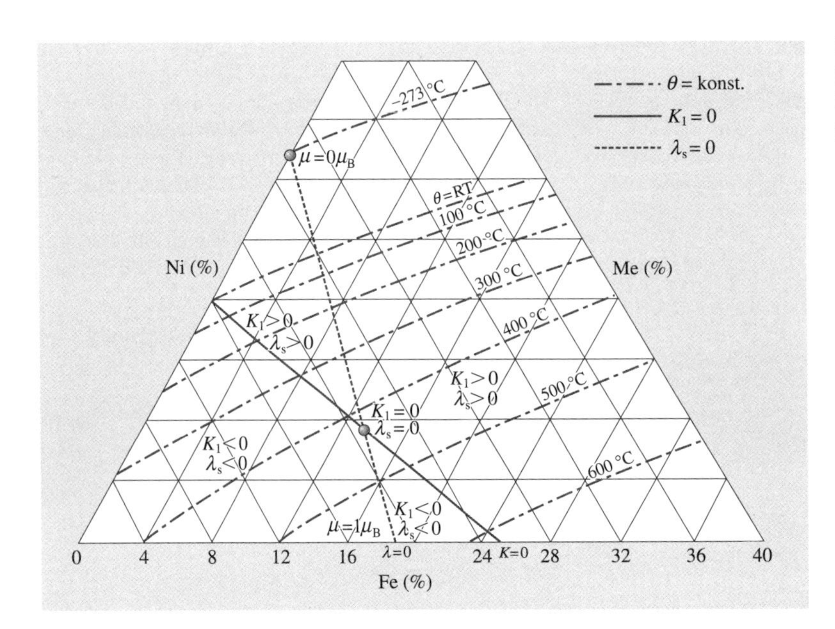

Fig. 4.3-11 Zero lines for $K_1 = 0$ and $\lambda_s = 0$ in the Ni–Fe–Me system [3.14, 15]

containing about 72 to 83 wt% Ni with additions of Mo, Cu, and/or Cr were the magnetically softest materials available.

Based on the low magnetocrystalline anisotropy K_1 and low saturation magnetostriction λ_s, the alloys containing about 80 wt% Ni where K_1 and λ_s pass through zero attain the lowest coercivity $H_c \approx 0.5\,\mathrm{A\,m^{-1}}$ and the highest initial permeability $\mu_i \approx 200\,000$. Figure 4.3-10 shows the variation of the decisive intrinsic magnetic parameters I_s, T_c, K_1, λ_{100}, and λ_{111} of binary Ni–Fe alloys with the Ni concentration. The strong effect of structural ordering on the magnitude of K_1 should be noted because it permits control of this intrinsic magnetic property by heat treatment.

In the binary Ni–Fe system, $K_1 = 0$ at about 76 wt% Ni and $\lambda_s = 0$ at about 81 wt% Ni. Small additions of Cu lower the Ni content for which $\lambda_s = 0$ while Mo additions increase the Ni content for $K_1 = 0$. Thus different alloy compositions around 78 wt% Ni are available which have optimal soft magnetic properties. General relations of the effect of alloying elements in Ni–Fe-based alloys on K_1, λ_s, and on the permeability have been developed in [3.14, 15]. Figure 4.3-11 shows the position of the lines for $K_1 = 0$ and $\lambda_s = 0$ in the disordered Ni–Fe–Me system, where Me = Cu, Cr, Mo, W, and V. High permeability regions in the Ni–Fe–Me system with a different valence of Me are delineated in Fig. 4.3-12.

The main fields of application of high permeability Ni–Fe alloys are fault-current circuit breakers, LF and HF transformers, chokes, magnetic shielding, and high sensitivity relays.

It should be noted that annealing treatments in a magnetic field of specified direction induce atomic rearrangements which provide an additional anisotropy termed uniaxial anisotropy K_u. It can be used to modify the field dependence of magnetic induction in such a way that the hysteresis loop takes drastically different forms, as shown in Fig. 4.3-13.

The extremes of a steep loop (Z type) and a skewed loop (F type) are obtained by field annealing with the direction of the field (during annealing) longitudinal and transverse to the operating field of the product, respectively. Materials with Z and F loops produced by magnetic field annealing are used in magnetic amplifiers, switching and storing cores, as well as for pulse and instrument transformers and chokes.

Combined with small alloy variations, primary treatments and field annealing treatments, a wide variety of annealed states can be realized to vary the induction behavior. The field dependence of the permeability of some high-permeability Ni–Fe alloys (designations according to Vacuumschmelze GmbH) are shown in Fig. 4.3-14 [3.16].

Alloys of Ni–Fe in the range of 54–68 wt% Ni combine relatively high permeability with high saturation

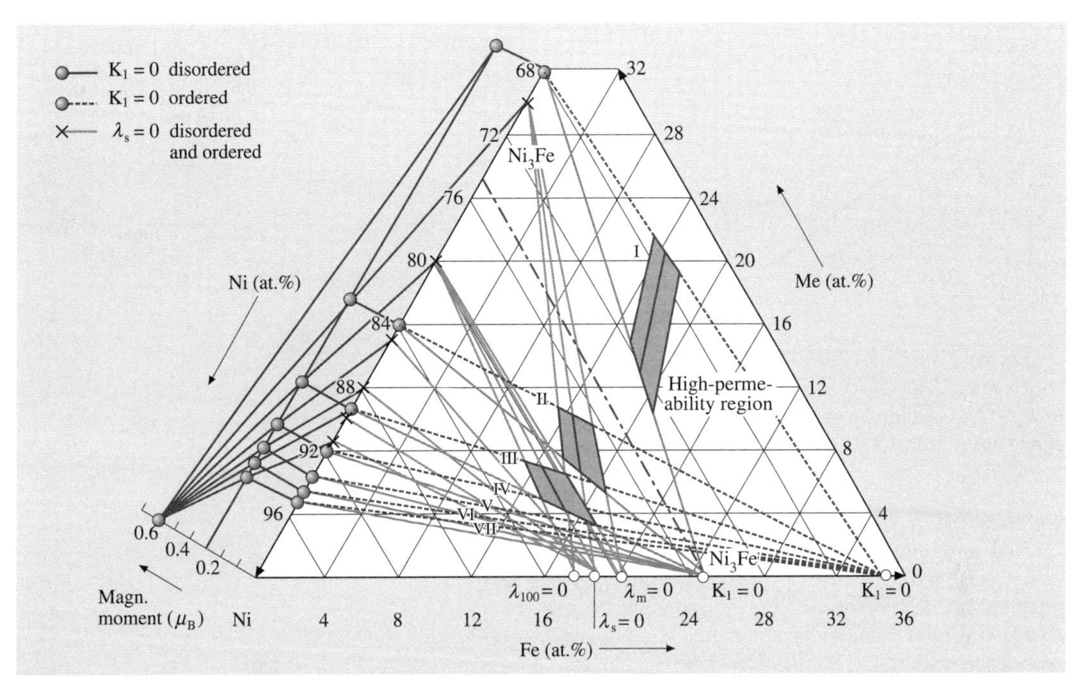

Fig. 4.3-12 Schematic representation of high-permeability regions in the ternary system Ni–Fe–Me by means of the zero curves K_1, $\lambda_{100} = 0$ and $\lambda_{111} = 0$ for additives with valence I–VIII [3.15]

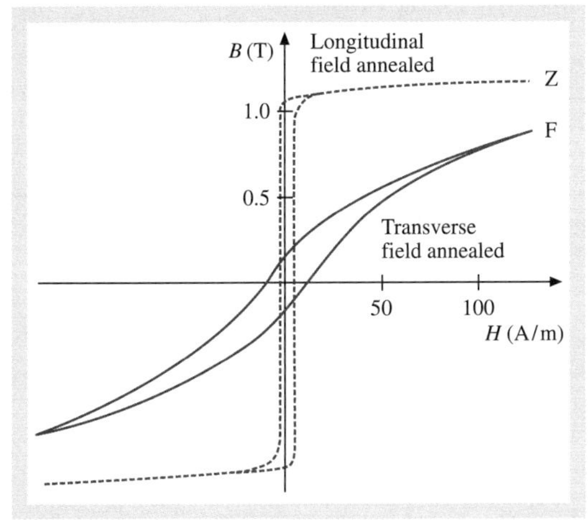

Fig. 4.3-13 Alloy of the 54–68 wt% NiFe-group with Z- and F-loop [3.12]

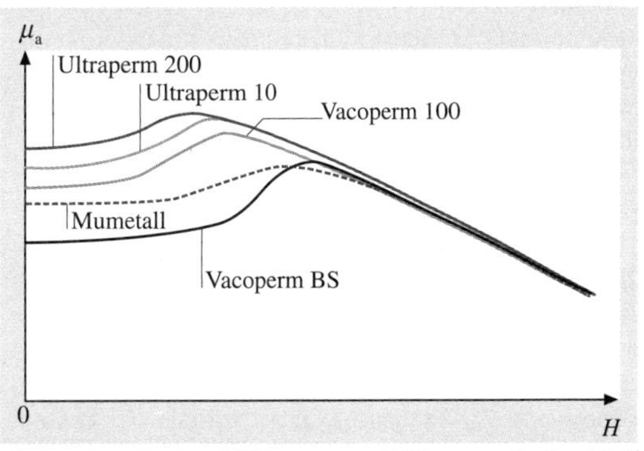

Fig. 4.3-14 Permeability versus field strength for high permeability high Ni-content Ni–Fe-alloys of Vacuumschmelze Hanau

polarization. Magnetic field annealing of these alloys provides a particularly high uniaxial anisotropy with ensuing Z and F type loops [3.12].

Alloys containing 45 to 50 wt% Ni reach maximum saturation polarization of about 1.6 T. Under suitable rolling and annealing conditions a cubic texture with an ensuing rectangular hysteresis loop and further loop variants over a wide range can be realized. The microstructure may vary from fine grained to coarsely grained.

Alloys containing 35 to 40 wt% Ni show a small but constant permeability, $\mu_r = 2000–8000$, over a wide range of magnetic field strength. Moreover,

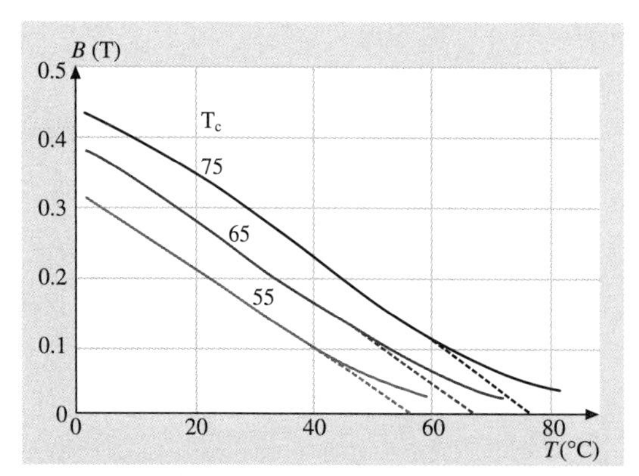

Fig. 4.3-15 Induction vs. temperature of Fe–Ni alloys with approximately 30% Ni as a function of Curie temperature [3.17]

they have the highest resistivity of all Ni–Fe alloys and a saturation polarization between 1.3 and 1.5 T.

The Curie temperature of the alloys with about 30 wt% Ni is near room temperature. Accordingly, the magnetization is strongly temperature-dependent in this vicinity (see Fig. 4.3-15). By slight variation of the Ni content (a composition increase of 0.1 wt% Ni gives rise to an increase of T_c by 10 K) T_c can be varied between 30 °C and 120 °C. These alloys are used mainly for temperature compensation in permanent magnet systems, measuring systems, and temperature sensitive switches [3.16].

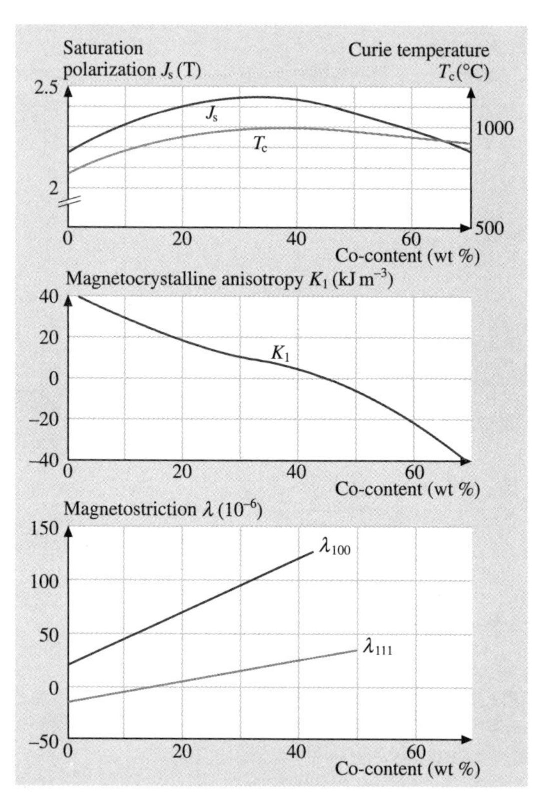

Fig. 4.3-16 The dependence of intrinsic magnetic parameters I_s, T_c, K_1 and λ of Co–Fe alloys on the Co content [3.12]

4.3.2.5 Iron–Cobalt Alloys

Of all known magnetic materials, Fe–Co alloys with about 35 wt% Co have the highest saturation polarization $I_s = 2.4\,\mathrm{T}$ at room temperature and the highest Curie temperature of nearly 950 °C. The intrinsic magnetic properties I_s, T_c, K_1, and λ_{hkl} as a function of Co content are shown in Fig. 4.3-16.

Since K_1 and λ_s have minima at different Co contents, different compositions for different applications have been developed. A Fe–49 wt% Co–2 wt% V composition is commonly used. The V addition reduces the brittleness by retarding the structural ordering transformation, improves the rolling behavior, and increases the electrical resistivity. The workability of Co–Fe alloys is difficult altogether. Two further alloy variants containing 35 wt% Co and 27 wt% Co, respectively, are of technical interest. They are applied where highest flux density is required, e.g., in magnet yokes, pole shoes, and magnetic lenses. The high T_c makes Fe–Co based alloys applicable as high temperature magnet material.

4.3.2.6 Amorphous Metallic Alloys

By rapid quenching of a suitable alloy from the melt at a cooling rate of about 10^{-5}–10^{-6} K/s, an amorphous metallic state will be produced where crystallization is suppressed. Commonly, casting through a slit nozzle onto a rotating copper wheel is used to form a ribbon-shaped product. The thickness of the ribbons is typically between 20 and 40 μm.

From the magnetic point of view amorphous alloys have several advantages compared to crystalline alloys: they have no magnetocrystalline anisotropy, they combine high magnetic softness with high mechanical hardness and yield strength, their low ribbon thickness and their high electrical resistivity (100–150 μΩ cm)

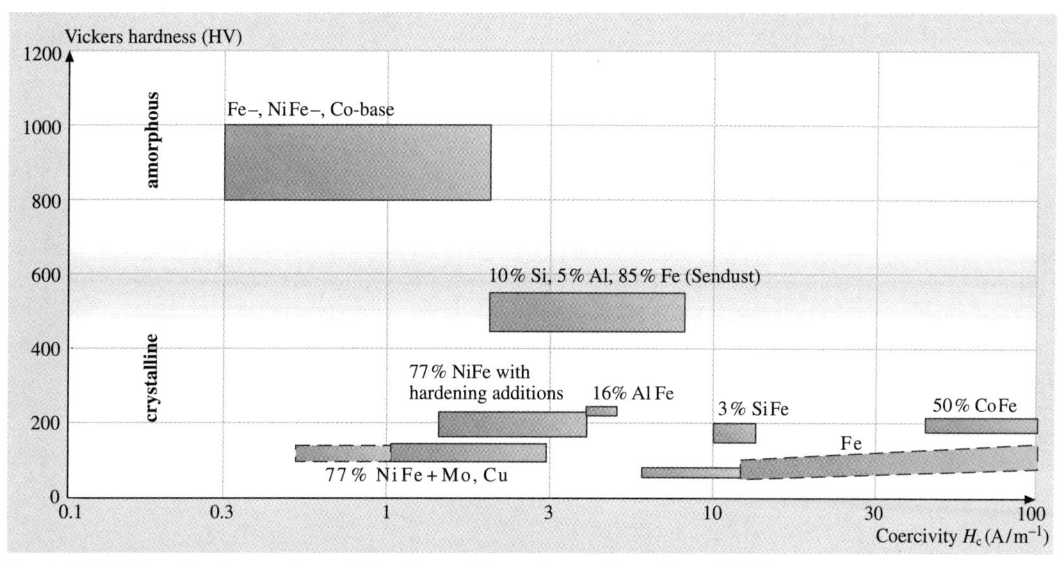

Fig. 4.3-17 Vickers hardness and coercivity of crystalline and amorphous alloys [3.12]

provide excellent soft magnetic material properties for high frequency applications, in particular low losses. A plot of Vickers hardness (HV) vs. coercivity (H_c) for crystalline and amorphous alloys is shown in Fig. 4.3-17 and indicates that a particularly favorable combination for soft magnetic amorphous alloys applies: being magnetically soft and mechanically hard.

The soft magnetic properties of amorphous alloys depend essentially on alloy composition, focusing on a low saturation magnetostriction λ_s, high glass forming ability required for ribbon preparation at technically accessible cooling conditions, and annealing treatments which provide structural stability and field-induced anisotropy K_u.

The soft magnetic amorphous alloys are based on the ferromagnetic elements Fe, Co, and Ni with additions of metalloid elements, the so-called glass forming elements Si, B, C, and P. The most stable alloys contain about 80 at.% transition metal (TM) and 20 at.% metalloid (M) components.

Depending on their base metal they exhibit characteristic differences of technical significance. Accordingly they are classified into three groups: Fe-based alloys, Co-based alloys, and Ni-based alloys. The characteristic variation of their intrinsic magnetic properties saturation polarization I_s, saturation magnetostriction λ_s, and the maximum field induced magnetic anisotropy energy K_u, as functions of alloy concentration is shown in Figures 4.3-18 to 4.3-20.

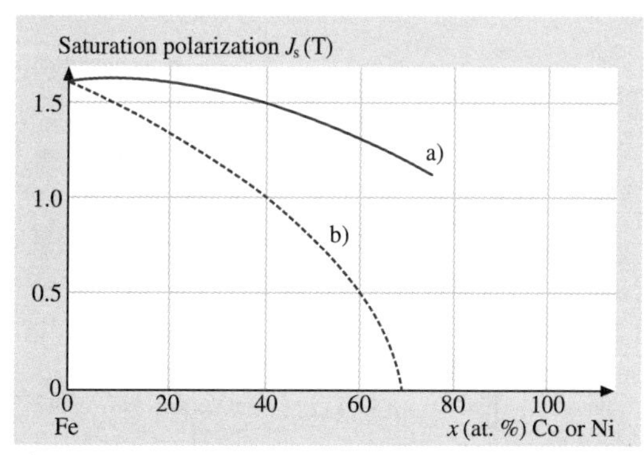

Fig. 4.3-18a,b Saturation polarization J_s of Fe-based amorphous alloys depending on Co and Ni content: **(a)** $Fe_{80-x}Co_xB_{20}$ (*O'Handley*, 1977) [3.18]. **(b)** $Fe_{80-x}Ni_xB_{20}$ (*Hilzinger*, 1980) [3.12, 19]

Iron-Based Amorphous Alloys

Of all amorphous magnetic alloys, the iron-rich alloys on the basis $Fe_{\sim 80}(Si, B)_{\sim 20}$ have the highest saturation polarization of 1.5–1.8 T. Because of their relatively high saturation magnetostriction (λ_s) of around 30×10^{-6}, their use as soft magnetic material is limited. The application is focused on transformers at low and

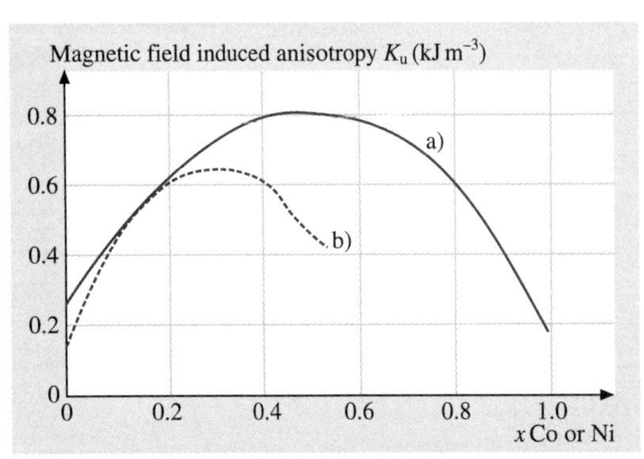

Fig. 4.3-19a,b Field induced anisotropy K_u of FeCo-based and FeNi-based amorphous alloys: **(a)** $(Fe_{1-x}Co_x)_{77}Si_{10}B_{13}$ (*Miyazaki* et al., 1972) [3.20]. **(b)** $(Fe_{1-x}Ni_x)_{80}B_{20}$ (*Fujimori* et al., 1976) [3.12,21]

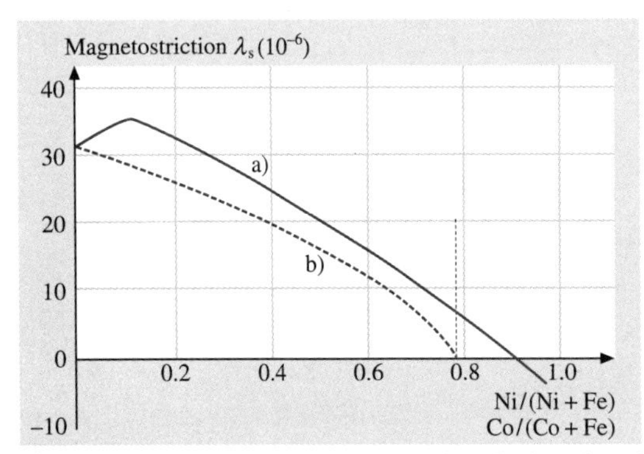

Fig. 4.3-20a,b Magnetostriction λ_s of FeCo-based and FeNi-based amorphous alloys depending on Co and Ni Content: **(a)** $(FeCo)_{80}B_{20}$ [3.18]. **(b)** $(FeNi)_{80}B_{20}$ [3.12, 18]

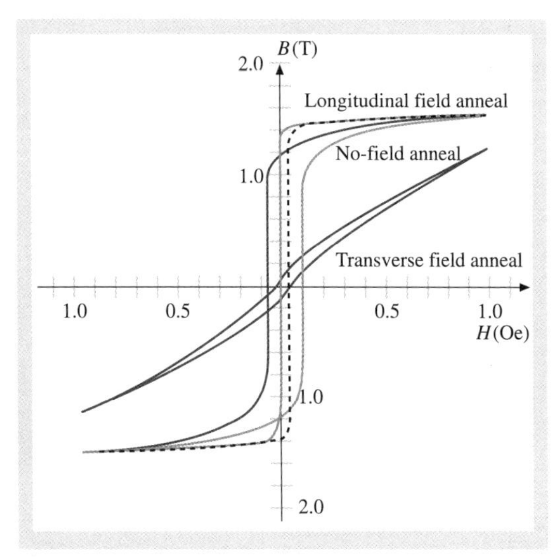

Fig. 4.3-21 Typical dc hysteresis loop of Fe-rich METGLAS alloy SA 1 [3.22]

medium frequencies in electric power distribution systems.

Compared to grain-oriented silicon steels the iron-rich amorphous alloys show appreciably lower coercivity and consequently lower total losses.

The physical and magnetic properties of a characteristic commercial Fe-rich Metglas amorphous alloy are shown in Fig. 4.3-21 and Table 4.3-21 [3.22].

Cobalt-Based Amorphous Alloys

In the $(Fe_xCo_{1-x})_{\sim 80}B_{\sim 20}$ system the saturation magnetostriction λ_s passes through zero. Along with a proper selection of alloy composition this behavior gives rise to a particularly low coercivity, the highest permeability of all amorphous magnetic alloys, low stress sensitivity, and extremely low total losses.

The saturation polarization ranging from 0.55 to 1.0 T is lower than in Fe-rich amorphous alloys but comparable to the $\sim$ 80 wt% Ni crystalline permalloy materials. By applying magnetic field annealing, well-controlled Z-type and F-type loops can be realized.

Nickel-Based Amorphous Alloys

A typical composition of this amorphous alloy group is $Fe_{40}Ni_{40}(Si, B)_{20}$, with a saturation polarization I_s of 0.8 T and a saturation magnetostriction λ_s of 10×10^{-6}. The latter is causing magnetoelastic anisotropy which can be applied to design magnetoelastic sensors where a change in the state of applied stress causes a change in permeability and loop shape, respectively. Upon annealing, Fe–Ni-based alloys will have an R-(round) type

Table 4.3-21 Physical and magnetic properties of Fe-rich METGLAS alloy SA 1 [3.22]

Property	Value
Ribbon Thickness (μm)	25
Density (g/cm^3)	7.19
Thermal Expansion (ppm/°C)	7.6
Crystallization Temperature (°C)	550
Curie Temperature (°C)	415
Countinuous Service Temperature (°C)	155
Tensile Strength (MN/m^2)	1–1.7 k
Elastic Modulus (GN/m^2)	100–110
Vicker's Hardness (50 g load)	860
Saturation Flux Density (Tesla)	1.56
Permeability (depending on gap size)	Variable
Saturation Magnetostriction (ppm)	27
Electrical Resistivity (μΩ cm)	137

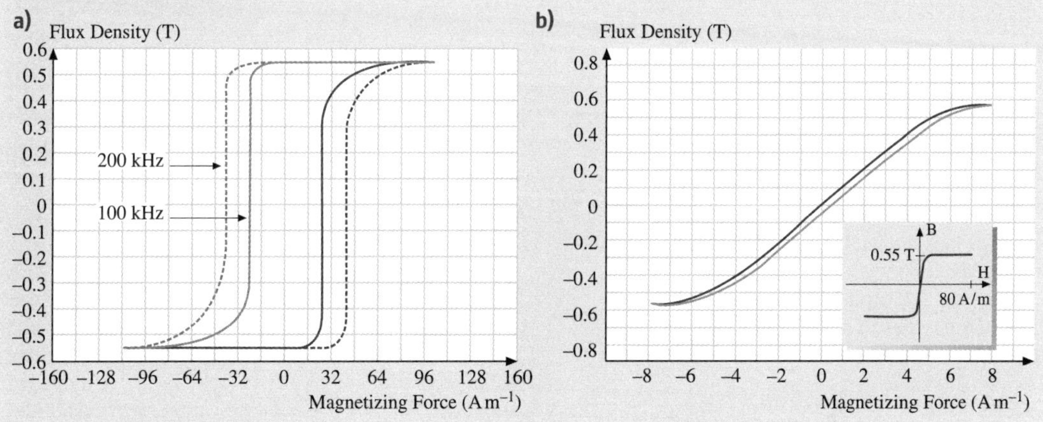

Fig. 4.3-22a,b Typical dc hysteresis loop of Co-based METGLAS alloy: (**a**) alloy 2714 AS (Z-loop) and (**b**) alloy 2714 AF (F-loop) [3.22]

Table 4.3-22 Physical and magnetic properties of the cobalt-based amorphous alloy METGLAS alloy 2714 AF [3.22]

Property	Value
Ribbon Thickness (μm)	18
Density (g/cm^3)	7.59
Thermal Expansion (ppm/°C)	12.7
Crystallization Temperature (°C)	560
Curie Temperature (°C)	225
Countinuous Service Temperature (°C)	90
Tensile Strength (MN/m^2)	1–1.7 k
Elastic Modulus (GN/m^2)	100–110
Vicker's Hardness (50 g load)	960
Saturation Flux Density (Tesla)	0.55
Permeability (μ@1 kHz, 2.0 mA/cm)	90 000±20%
Saturation Magnetostriction (ppm)	$\ll 1$
Electrical Resistivity (μΩ cm)	142

Table 4.3-23 Survey of soft magnetic amorphous and nanocrystalline alloys. Some amorphous alloy of METGLAS (Allied Signal Inc., Morristown/NJ) and VITROVAC (Vacuumschmelze GmbH, Hanau, Germany) have been selected from commercially available alloys [3.12]

Composition	Typical properties						
	Saturation polarization in (T)	Curie temperature in (°C)	Saturation magnetostriction in 10^{-6}	Coercivity (dc) in ($A\,m^{-1}$)	Permeability[a] at $H = 4\,mA\,m^{-1}$ $\times 10^3$	Density in ($g\,cm^{-3}$)	Specific electrical resistivity[b] in ($\Omega\,mm^2\,m^{-1}$)
Amorphorus							
Fe-based:							
$Fe_{78}Si_9B_{13}$	1.55	415	27	3	8	7.18	1.37
$Fe_{67}Co_{18}Si_1B_{14}$	1.80	~550[c]	35	5	1.5	7.56	1.23
FeNi-based:							
$Fe_{39}Ni_{39}Mo_2Si_{12}B_8$	0.8	260	+8	2	20	7.4	1.35
Co-based:							
$Fe_{67}Fe_4Mo_1Si_{17}B_{11}$	0.55	210	< 0.2	0.3	100	7.7	1.35
$Fe_{74}Fe_2Mn_4Si_{11}B_9$	1.0	480[c]	< 0.2	1.0	2	7.85	1.15
Nanocristalline							
$Fe_{73.5}Cu_1Nb_3Si_{13.5}B_9$	1.25	600	+2	1	100	7.35	1.35

[a] Materials with round (R) or flat loops (F), $f = 50\,Hz$
[b] $1\,\Omega\,mm^2/m = 10^{-4}\,\Omega\,cm$
[c] Extrapolated values ($T_c > T_x$, T_x: crystallization temperature)

hysteresis loop associated with high initial permeability, or an F-type loop with low losses.

Table 4.3-23 gives a survey of the magnetic and physical properties of several soft magnetic amorphous alloys

4.3.2.7 Nanocrystalline Soft Magnetic Alloys

Nanocrystalline soft magnetic alloys are a rather recent class of soft magnetic materials with excellent magnetic properties such as low losses, high permeability, high saturation polarization up to 1.3 T, and near-zero magnetostriction. The decisive structural feature of this alloy type is its ultra-fine microstructure of bcc α-Fe–Si nanocrystals, with grain sizes of 10–15 nm which are embedded in an amorphous residual phase. Originally, this group of materials was discovered in the alloy system Fe–Si–B–Cu–Nb with the composition $Fe_{73.5}Si_{15.5}B_7Cu_1Nb_3$. This material is prepared by rapid quenching like an amorphous Fe–Si–B alloy with a subsequent annealing treatment and comparatively high temperature in the range of 500 to 600 °C which leads to partial crystallization.

The evolution of the nanocrystalline state during annealing occurs by partial crystallization into randomly oriented, ultrafine bcc α-Fe–Si grains that are 10–15 nm in diameter. The residual amorphous matrix phase forms a boundary layer that is 1–2 nm thick. This particular nano-scaled microstructure is the basis for ferromagnetically-coupled exchange interaction of and through these phases, developing excellent soft magnetic properties: $\mu_a \approx 10^5$, $H_c < 1\,A\,m^{-1}$. Annealing above 600 °C gives rise to the precipitation of the borides Fe_2B and/or Fe_3B with grain sizes of 50–100 nm. At higher annealing temperatures, grain coarsening arises. Both of these microstructural changes are leading to a deterioration of the soft magnetic properties.

The influence of the annealing temperature on grain size, H_c, and μ_i of a nanocrystalline type alloy is shown in Fig. 4.3-23 [3.23].

The small additions of Cu and Nb favor the formation of the nanocrystalline structure. Copper is thought to increase the rate of nucleation of α-Fe–Si grains by a preceding cluster formation, and Nb is supposed to lower the growth rate because of its partitioning effect and decrease of diffusivity in the amorphous phase. Figure 4.3-24 illustrates the formation of the nanocrystalline structure schematically.

It is useful to note the influence of the atomic diameter of alloying additions on the grain size of the α-Fe–Si phase starting from the classical alloy composition $Fe_{73.5}Si_{15.5}B_7Cu_1Nb_3$. This effect is shown for

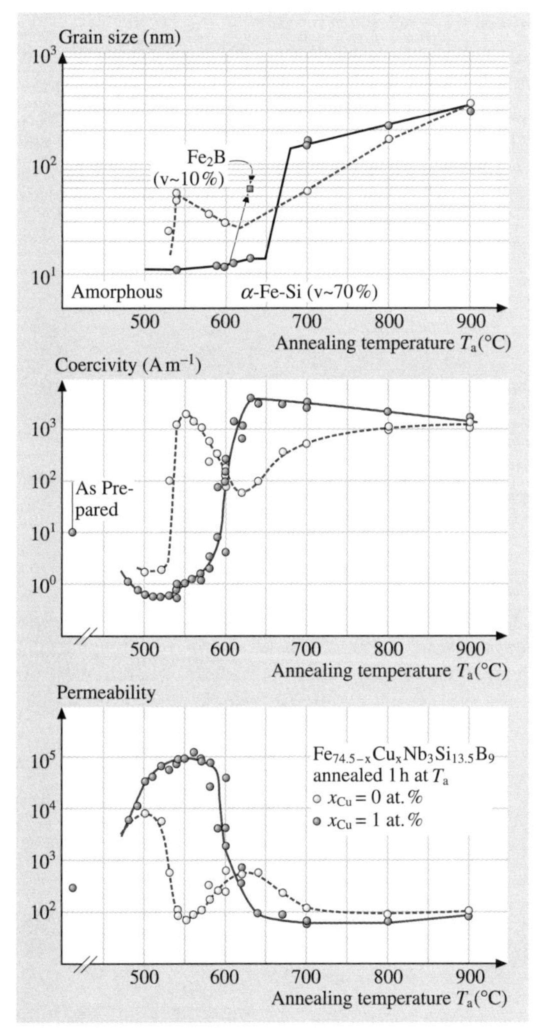

Fig. 4.3-23 Average grain size, coercivity and initial permeability of a nanocrystalline soft magnetic alloy as a function of the annealing temperature [3.23]

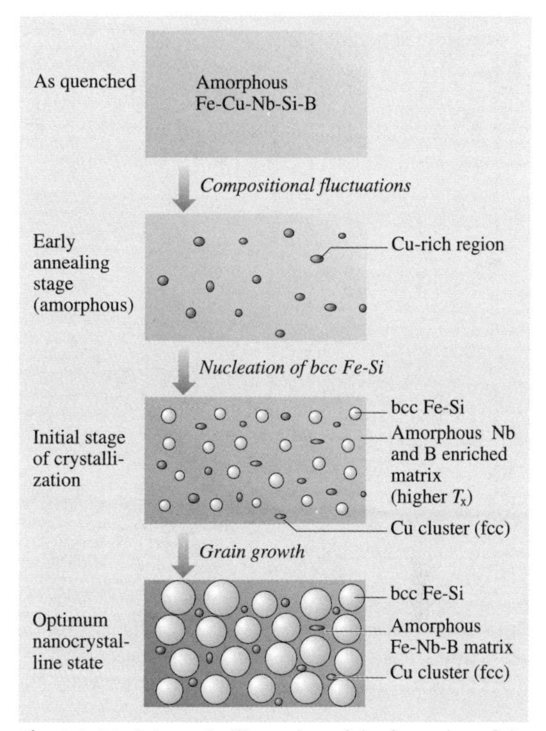

Fig. 4.3-24 Schematic illustration of the formation of the nanocrystalline structure in Fe–Cu–Nb–Si–B alloys, based on atom probe analysis results and transmission electron microscopy observations by *Hono* et al. [3.23, 24]

partial substitution of Nb by V, Mo, W, and Ta (2 at.% each) in Fig. 4.3-25. The larger the atomic diameter, the smaller the resulting grain size. The elements Nb and Ta have the same atomic diameter. Furthermore, the smaller the atomic diameter of the alloying element, the sooner the crystallization of the α-Fe-Si phase begins.

One of the decisive requirements for excellent soft magnetic properties is the absence of magnetostriction. Amorphous Fe–Si–B–Cu–Nb alloys have a saturation

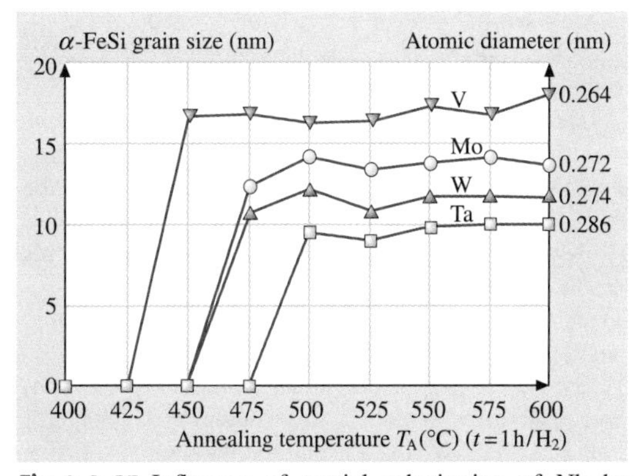

Fig. 4.3-25 Influence of partial substitution of Nb by the refractory elements R = *V*, Mo, W and Ta on the α-FeSi grain size during annealing of the alloy $Fe_{73.5}Si_{15.5}B_7Cu_1Nb_3R_2$ [3.25]

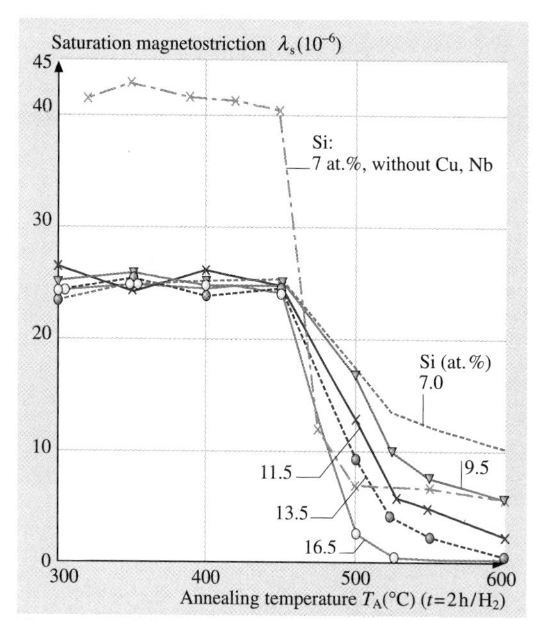

Fig. 4.3-26 Influence of the Si-content on the saturation magnetostriction λ_s during annealing of the nanocrystalline alloys $Fe_{(75-73.5)}Si_{(7-16.5)}B_{(14-6)}Cu_1Nb_3$ [3.26]

magnetostriction $\lambda_s \approx 24\times10^{-6}$, with a magnetoelastic anisotropy energy $K_\sigma \approx 50\,J\,m^{-3}$. With partial crystallization of the α-Fe–Si phase during annealing, λ_s varies significantly. At higher Si contents it decreases strongly and passes through zero at about 16 at.% Si, as shown in Fig. 4.3-26.

This behavior is caused by the compensation of the negative saturation magnetostriction λ_s of the α-Fe–Si phase $\lambda_s \approx -8\times10^{-6}$ and the positve values of λ_s of the residual amorphous phase of $\lambda_s \approx +24\times10^{-6}$.

The superposition of the local magnetostrictive strains to an effective zero requires a large crystalline volume fraction of about 70 vol.% to compensate the high positive value of the amorphous residual phase of about 30 vol.%. By anealing at about 550 °C this relation can be realized. The second requirement for superior soft magnetic properties is a small or vanishing magnetocrystalline anisotropy energy K_1.

By developing a particular variant of the random anisotropy model, it was shown [3.27] that for grain diameters D smaller than the magnetic exchange length L_0, the averaged anisotropy energy density $\langle K\rangle$ is given by

$$\langle K\rangle \approx v_{cr}^2 K_1 (D L_0^{-1})^6 = v_{cr}^2 D^6 K_1^4 A^{-3} ,$$

where v_{cr} = crystallized volume, K_1 = intrinsic crystal anisotropy energy of α-Fe–Si, $L_0 = \sqrt{A K_1^{-1}}$, and A = exchange stiffness constant. A schematic representation of this model is given in Fig. 4.3-27. The basic effect of decreasing the grain size consists of local averaging of the magnetocrystalline anisotropy energy K_1 for D = 10–15 nm at L_0 = 30–50 nm (about equal to

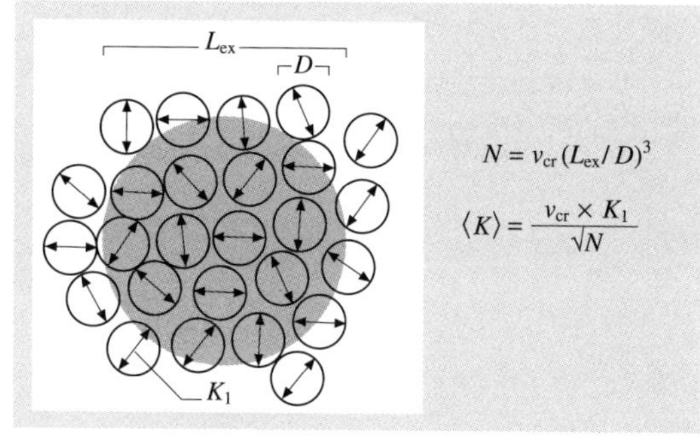

Fig. 4.3-27 Schematic representation of the random anisotropy model for grains embedded in an ideally soft ferromagnetic matrix. The *double arrows* indicate the randomly fluctuating anisotropy axis; the *dark* area represents the ferromagnetic correlation volume determined by the exchange length $L_{ex} = (A/\langle K\rangle)^{1/2}$ [3.23]

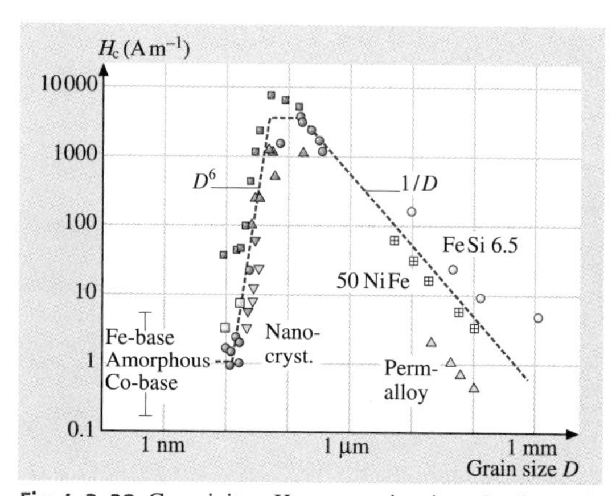

Fig. 4.3-28 Coercivity, H_c, vs. grain size, D, for various soft magnetic metallic alloys [3.27]: Fe–Nb–Si–B (*solid up triangles*) [3.28], Fe–Cu–Nb–Si–B (*solid circles*) [3.29, 30], Fe–Cu–V–Si–B (*solid* and *open down triangles*) [3.31], Fe–Zr–B (*open squares*) [3.32], Fe–Co–Zr (*open diamonds*) [3.33], NiFe-alloys (+ *center squares* and *open up triangles*) [3.34], and Fe–Si(6.5 wt%) (*open circles*) [3.35]

Table 4.3-24 Typical values of grain size D, saturation magnetization J_s, saturation magnetostriction λ_s, coercivity H_c, initial permeability μ_i, electrical resistivity ρ, core losses P_{Fe} at 0.2 T, 100 kHz and ribbon thickness t for nanocrystalline, amorphous, and crystalline soft magnetic ribbons

Alloy	D (nm)	J_s (T)	λ_s (10^{-6})	H_c $(\mathrm{A\,m^{-1}})$	μ_i (1 kHz)	ρ ($\mu\Omega$ cm)	P_{Fe} $(\mathrm{W\,kg^{-1}})$	t (μm)	Ref
$Fe_{73.5}Cu_1NB_3Si_{13.5}B_9$	13	1.24	2.1	0.5	100 000	118	38	18	a
$Fe_{73.5}Cu_1NB_3Si_{15.5}B_7$	14	1.23	~0	0.4	110 000	115	35	21	b
$Fe_{84}NB_7B_9$	9	1.49	0.1	8	22 000	58	76	22	c
$Fe_{86}Cu_1Zr_7B_6$	10	1.52	~0	3.2	48 000	56	116	20	c
$Fe_{91}Zr_7B_3$	17	1.63	−1.1	5.6	22 000	44	80	18	c
$Co_{68}Fe_4(MoSiB)_{28}$	amorphous	0.55	~0	0.3	150 000	135	35	23	b
$Co_{72}(FeMn)_5(MoSiB)_{23}$	amorphous	0.8	~0	0.5	3000	130	40	23	b
$Fe_{76}(SiB)_{24}$	amorphous	1.45	32	3	8000	135	50	23	b
80 % Ni–Fe (permalloys)	~100 000	0.75	< 1	0.5	100 000[d]	55	>90[e]	50	b
50–60 % Ni–Fe	~100 000	1.55	25	5	40 000[d]	45	>200[e]	70	b

[a] [3.36]
[b] Typical commercial grades for low remanence hysteresis loops, Vacuumschmelze GmbH 1990, 1993
[c] [3.37, 38]
[d] 50 Hz-values
[e] Lower bound due to eddy currents

the domain wall thickness), which leads to the extreme variation of $\langle K\rangle$ with the sixth power of grain size. These relations were confirmed experimentally and result in an anomalous variation of the coercivity with grain size, as shown in Fig. 4.3-28.

Another type of nanocrystalline soft magnetic materials is based on Fe–Zr–B–Cu alloys [3.37, 38]. A typical composition is $Fe_{86}Zr_7B_6Cu_1$. As Zr provides high glass-forming ability, the total content of glass-forming elements can be set to $\ll 20$ at.%. As a consequence the Fe content is higher, which implies higher saturation polarization. The nanocrystalline microstructure consists of a crystalline α-Fe phase with grain sizes of about 10 nm embedded in an amorphous residual phase. After annealing at 600 °C, an optimum combination of magnetic properties is obtained: $I_s \geq 1.5$ T, $H_c \approx 3\,\mathrm{A\,m^{-1}}$, $\lambda_s \approx 0$. Because of the high reactivity of Zr with oxygen the preparation of this type of alloy is difficult. The production of these materials on an industrial scale has not yet succeeded.

Table 4.3-24 shows the magnetic and physical properties of some commercially-available nanocrystalline alloys for comparison to amorphous and Ni–Fe-based crystalline soft magnetic alloys. Magnetic field annealing allows the shape of the hysteresis loops of

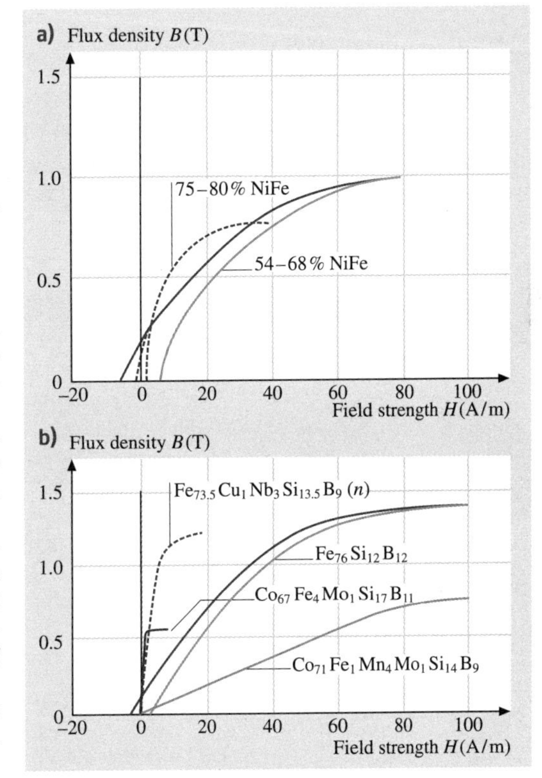

Fig. 4.3-29a,b Soft magnetic alloys with flat hysteresis loops: **(a)** Crystalline. **(b)** Amorphous and nanocrystalline (curve indicated by n) [3.12]

nanocrystalline soft magnetic alloys to be varied according to the demands of the users. Accordingly, different shapes of hystesis loops (Z, F, or R type) may be achieved. For comparison several characteristic hysteresis loops of crystalline, amorphous, and nanocrystalline soft magnetic alloys are shown in Figs. 4.3-29a,b and 4.3-30a,b.

A survey of the field dependence of the amplitude permeability of various crystalline, amorphous, and nanocrystalline soft magnetic alloys is given in Fig. 4.3-31 [3.12]. Figure 4.3-32 [3.23] represents the frequency behavior of the permeability $|\mu|$ of different soft magnetic materials for comparison.

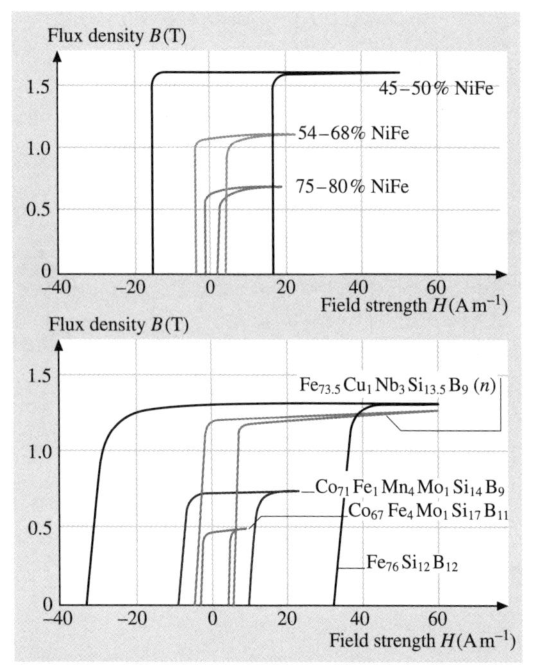

Fig. 4.3-30a,b Soft magnetic alloys with rectangular hysteresis loops: **(a)** Crystalline. **(b)** amorphous and nanocrystalline (curve indicated by n) [3.12]

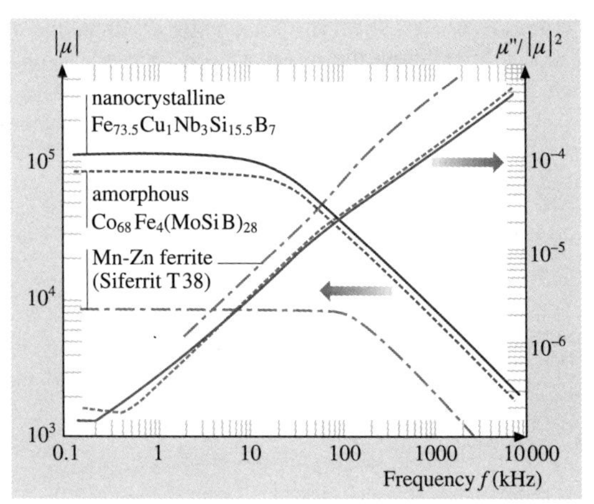

Fig. 4.3-32 Frequency dependence of permeability, $|\mu|$, and the relative loss factor, $\mu''/|\mu|^2$, for nanocrystalline $Fe_{73.5}Cu_1Nb_3Si_{15.5}B_7$ and comparable, low remanence soft magnetic materials used for common mode choke cores [3.23]

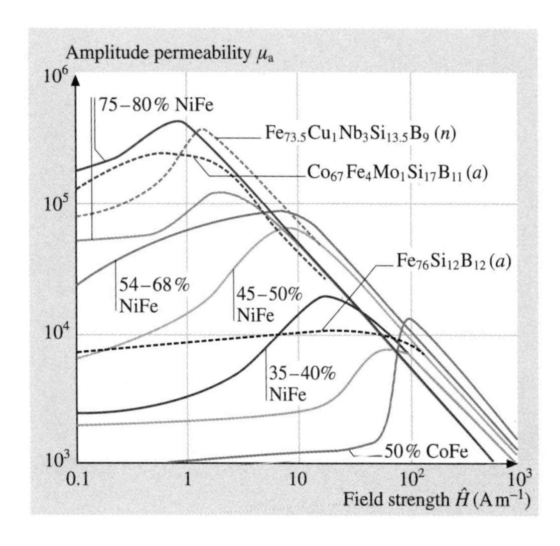

Fig. 4.3-31 Amplitude permeability–field strength curves of soft magnetic alloys ($f = 50$ Hz): amorphous (a); nanocrystalline (n) [3.12]

4.3.2.8 Invar and Elinvar Alloys

The term invar alloys is used for some groups of alloys characterized by having temperature-invariant properties, either temperature-independent volume (invar) or temperature-independent elastic properties (elinvar) in a limited temperature range. A comprehensive survey of the physics and applications of invar alloys is given in [3.17].

Invar Alloys

With the discovery of an Fe–36 wt% Ni alloy with an uncommonly low thermal expansion coefficient (TEC) around room temperature and called "Invar" by Guillaume in the 1890s, the history of invar and elinvar type alloys began. Below the magnetic transition temperature, Curie temperature T_C or Néel temperature T_N of ferromagnetic or antiferromagnetic materials, a spontaneous volume magnetostriction ω_S sets in. With some alloy compositions, ω_S is comparable in magnitude to the linear thermal expansion but opposite in sign. As a result the coefficient of linear thermal expansion may become low and even zero.

The linear (α) and volumetric (β) thermal expansion coefficients (TECs) are defined as:

$$\alpha = (1/l)(\Delta l/\Delta T)_P \quad [\mathrm{K}^{-1}]$$

and

$$\beta = (1/V)(\Delta V/\Delta T)_P \quad [\mathrm{K}^{-1}],$$

with l = length, V = volume, and T = temperature. If the alloys are isotropic the volumetric thermal expansion coefficient is equal to three times the linear TEC:

$$\beta = 3\alpha .$$

The spontaneous volume magnetostriction ω_S is in a first approximation:

$$\omega_S = \kappa C M_S^2 ,$$

with κ = compressibility, C = magnetovolume coupling constant, and M_S = spontaneous magnetization. Figure 4.3-33 [3.42] illustrates schematically the temperature-dependent behavior of thermal expansion, ω_S and M_S, which give rise to a small linear expansion coefficient.

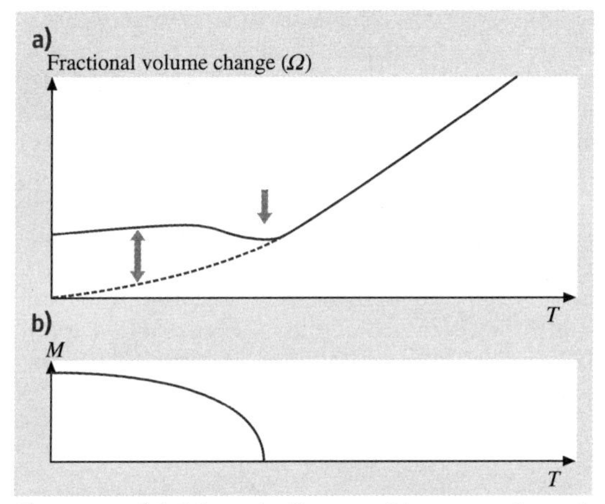

Fig. 4.3-33 **(a)** Schematic diagram of invar-type thermal-expansion anomaly. The *dashed curve* indicates thermal expansion for hypothetical paramagnetic state. The difference between the two curves corresponds to the spontaneous volume magnetostriction, ω_S. **(b)** Temperature dependence of the spontaneous magnetization

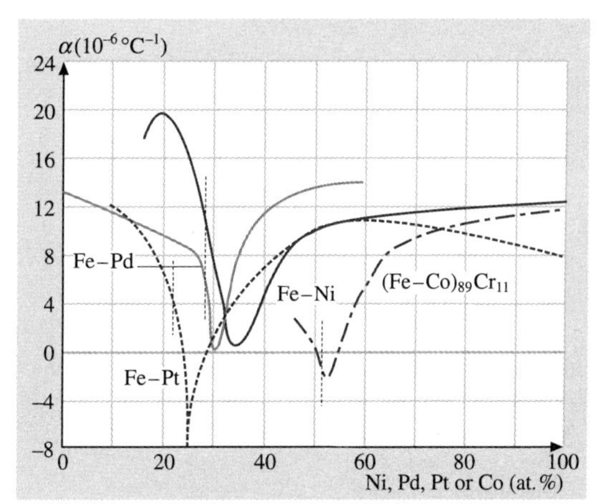

Fig. 4.3-34 Temperature coefficient of linear thermal expansion α at room temperature as a function of the composition in typical invar alloy systems: Fe−Ni [3.39], Fe−Pt [3.40], Fe−Pd [3.40], and Fe−Co−Cr [3.39]. *Vertical dotted lines* show boundary between bcc and fcc phases

Crystalline invar alloys are essentially based on 4 binary alloy systems: Fe−Ni, Fe−Pt, Fe−Pd, and Fe−Co, containing a few percent of Cr. Figure 4.3-34 [3.17] shows the thermal expansion coefficient of these invar type alloy systems. The invar alloy in each system is fcc

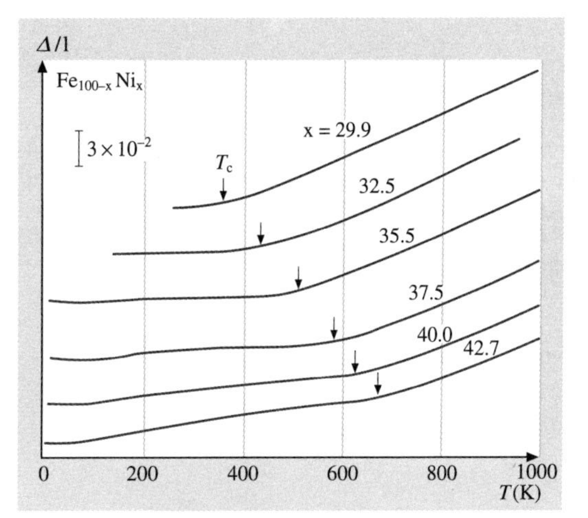

Fig. 4.3-35 Thermal expansion curves of Fe−Ni alloys annealed at 1323 K for 5 days [3.41]

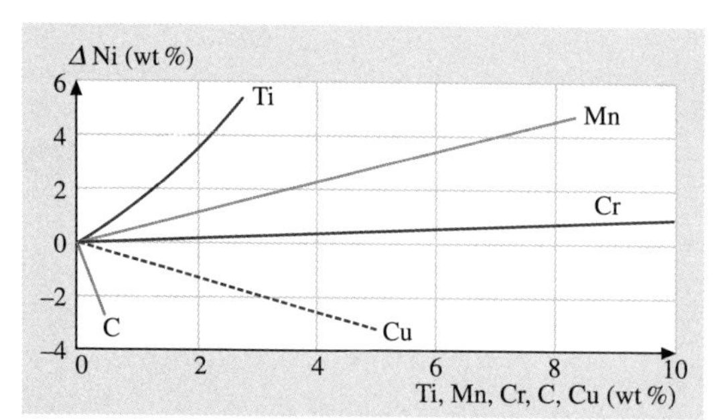

Fig. 4.3-36 Displacement of the composition corresponding to the minimum thermal expansion coefficient of Fe–Ni alloys by the addition of Ti, Mn, Cr, Cu, and C [3.17]

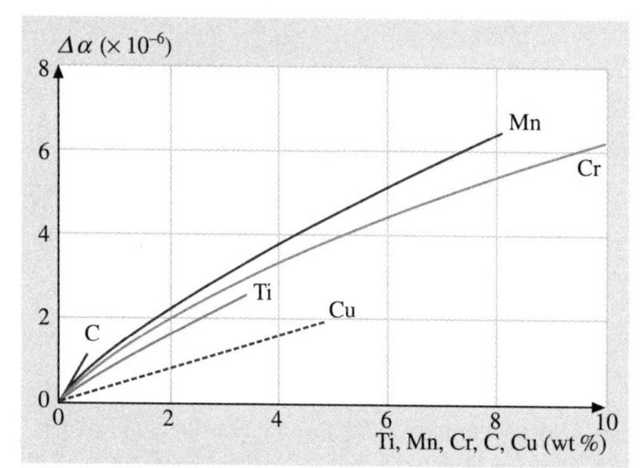

Fig. 4.3-37 Increase of the minimum value of thermal expansion coefficient of Fe–Ni alloys by the addition of Ti, Mn, Cr, Cu, and C [3.17]

and its composition is near the boundary between the bcc and fcc phase fields.

Fe–Ni-Based Invar Alloys. An iron alloy containing 34 to 36.5 wt% Ni is well known as a commercial invar material. The materials C < 0.12 wt%, Mn < 0.50 wt%, and Si < 0.50 wt% are generally added for metallurgical purposes. Figure 4.3-35 [3.41] shows the thermal expansion curves of some Fe–Ni alloys.

The thermal expansion of invar alloys is affected significantly by the addition of third elements, by cold working and by thermal treatment as shown in Fig. 4.3-36 [3.17], Fig. 4.3-37 [3.17] and Fig. 4.3-38 [3.41].

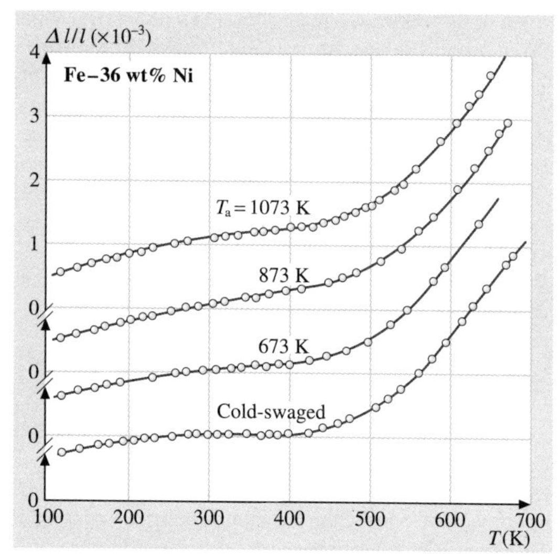

Fig. 4.3-38 Effect of thermal annealing at a temperature T_a and of mechanical treatment on the thermal expansion of Fe–36 at.% Ni invar alloy [3.41]

Invar is an austenitic alloy and cannot be hardened by heat treatment. The effect of heat treatment on the TEC α depends on the method of cooling after annealing. Air cooling or water quenching from the annealing temperature results in a reduction of α but at the same time α becomes unstable. In order to stabilize the material, annealing at low temperature and slow cooling to room temperature are necessary.

Following heat treatments for an optimum and stable magnitude of α are recommended: 830 °C, 1/2 h, water quenching; 315 °C, 1 h, air cooling; 95 °C, 20 h, air cooling. Cold working reduces α. But before use in high precision instruments a stress-relief anneal at 320 to 370 °C for 1 h followed by air cooling is required.

The mechanical properties of some invar-type alloys are listed in Table 4.3-25 [3.42].

Table 4.3-25 Thermal expansion coefficient, α and mechanical properties of invar alloys for practical use [3.42]

	Composition (wt%)	α at RT ($10^{-6}\,K^{-1}$)	Vickers hardness	Tensile stress (MPa)
Invar	Fe–36Ni	<2	150–200	500–750
Super Invar [a]	Fe–32Ni–4Co	<0.5	150–200	500–800
Stainless Invar [b]	Fe–54Co–9Cr	<0.5		
High strength Invar	Fe–Ni–Mo–C [c]	<4		1250
	Fe–Ni–Co–Ti [d]	<4	300–400	1100–1400

[a] Masumoto, 1931
[b] Masumoto, 1934
[c] Yokota et al., 1982
[d] Yahagi et al., 1980

Table 4.3-26 Thermal expansion coeffcient of invar 36 and free-cut invar [3.17]

Temperature	α ($\times 10^{-6}\,K^{-1}$)			
	As annealed		As cold-drawn	
(°C)	Invar 36	Free-cut Invar 36	Invar 36	Free-cut Invar 36
25–100	11.18	1.60	0.655	0.89
25–200	1.72	2.91	0.956	1.62
25–300	4.92	5.99	2.73	3.33
25–350	6.60	7.56	3.67	4.20
25–400	7.82	8.88	4.34	4.93
25–450	8.82	9.80	4.90	5.45
25–500	9.72	10.66	5.40	5.92
25–600	11.35	12.00	6.31	6.67
25–700	12.70	12.90	7.06	7.17
25–800	13.45	13.60	7.48	7.56
25–900	13.85	14.60	7.70	8.12

Table 4.3-27 Some physical properties of invar alloys [3.17]

Property	Value	
Melting point	1.425 °C	2600 °F
Density	$8\,g\,cm^{-3}$ (8.0–8.13)	$500\,lb\,ft^{-3}$
Thermal electromotive force to copper (0–96 °C)	9.8 μV/K	
Specific resistance (Annealed)	$82\times10^{-6}\,\Omega\,cm$ (81–88)	495 Ω circ. mil/ft
Temperature coefficient of electric resistivity	$1.21\times10^{-3}\,K^{-1}$	$0.67\times10^{-3}\,°F^{-1}$
Specific heat	0.123 cal/g K (25–100 °C)	0.123 Btu/lb °F (77–212 °F)
Thermal conductivity	0.0262 cal/sec cm K (22–100 °C)	72.6 Btu in./hr ft^2 °F (68–212 °F)
Curie temperature	277 °C (277–280 °C)	530 °F
Inflection temperature	191 °C	375 °F
Modulus of elasticity (in tension)	15.0×10^{10} Pa	$21.4\times10^{6}\,lb\,in.^{-2}$
Temperature coefficient of elastic modulus	$+50\times10^{-5}\,K^{-1}$ (16–50 °C)	$+27\times10^{-5}\,°F^{-1}$ (60–122 °F)
Modulus of rigidity	5.7×10^{10} Pa	$8.1\times10^{6}\,lb\,in.^{-2}$
Temperature coefficient of rigidity modulus	$+58\times10^{-5}\,K^{-1}$	$+30\times10^{-5}\,°F^{-1}$
Posson's ratio	0.290	

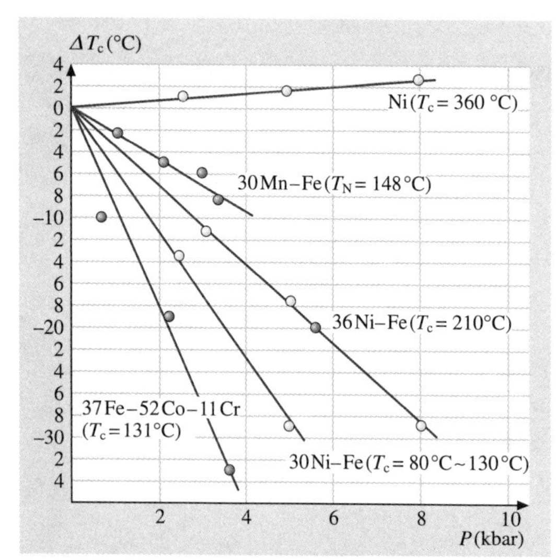

Fig. 4.3-39 The displacement of the Curie point ΔT_c vs. pressure in Fe–Ni invar alloys [3.43] and in Fe–Co–Cr invar alloy [3.44]. The results for Ni [3.43] and for a 30Mn–Fe alloy [3.44] are also shown for comparison. Numerical values in parentheses show the Curie or Néel point at atmospheric pressure

Invar-type alloys show large effects of pressure on magnetization and on the Curie temperature, which suggests a high sensitivity to the interatomic spacing, as shown in Fig. 4.3-39 [3.17].

Thermal expansion coefficients of invar 36 and free-cut invar 36 (containing S and P, or Se) between 25 and 900 °C are listed in Table 4.3-26 [3.17]. Some physical properties of Invar alloys are given in Table 4.3-27 [3.17].

Low thermal expansion coefficients are observed at ternary and quaternary Fe-alloy systems, too. The composition Fe–32 wt% Ni–4 wt% Co was the starting point of superinvar, whose TEC α is in the order of 10^{-7} K^{-1}. The thermal expansion curves of different variants are shown in Fig. 4.3-40 [3.17].

In order to improve the corrosion resistance of invar alloys, "stainless invar" was developed. The basic composition is Fe–54 wt% Co–9.5 wt% Cr. Stainless invar has the bcc structure at room temperature in the equilibrium state. As an invar material, it is used after quenching from a high temperature to retain the fcc structure [3.17].

To improve the mechanical properties two types of high strength invar materials were developed: a work-hardening type based on Fe–Ni–Mo–C, and

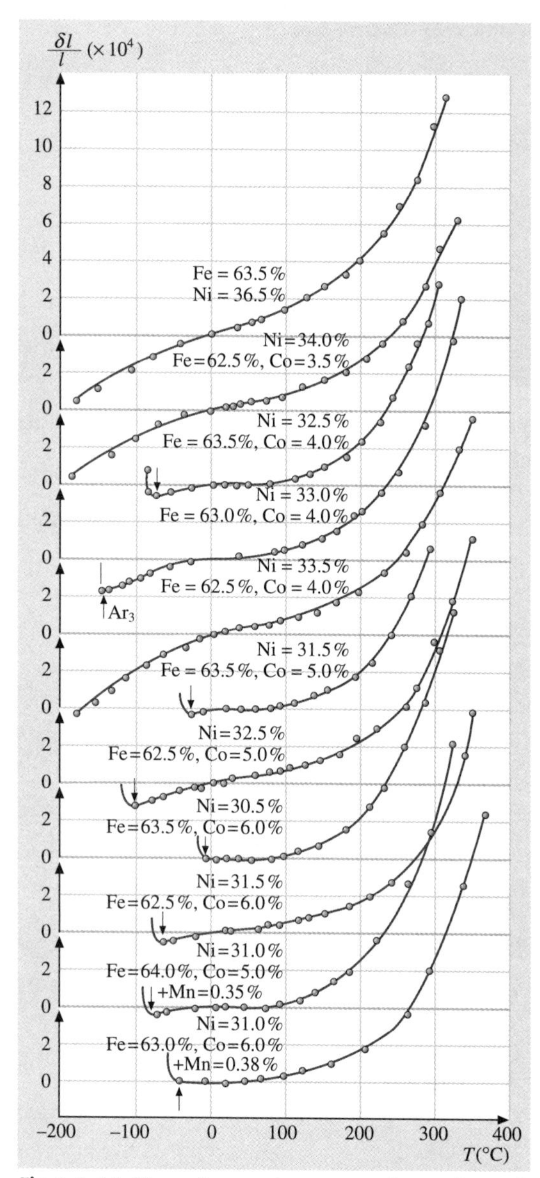

Fig. 4.3-40 Thermal expansion curves of super invar alloys [3.17]

a precipitation-hardening type Fe–Ni–Co–Ti alloy [3.17].

Fe-Pt-Based Invar Alloys. Among the ordered phases of the Fe–Pt system (Fig. 4.3-41 [3.17]), the Fe_3Pt phase shows the invar type thermal expansion anomaly. In

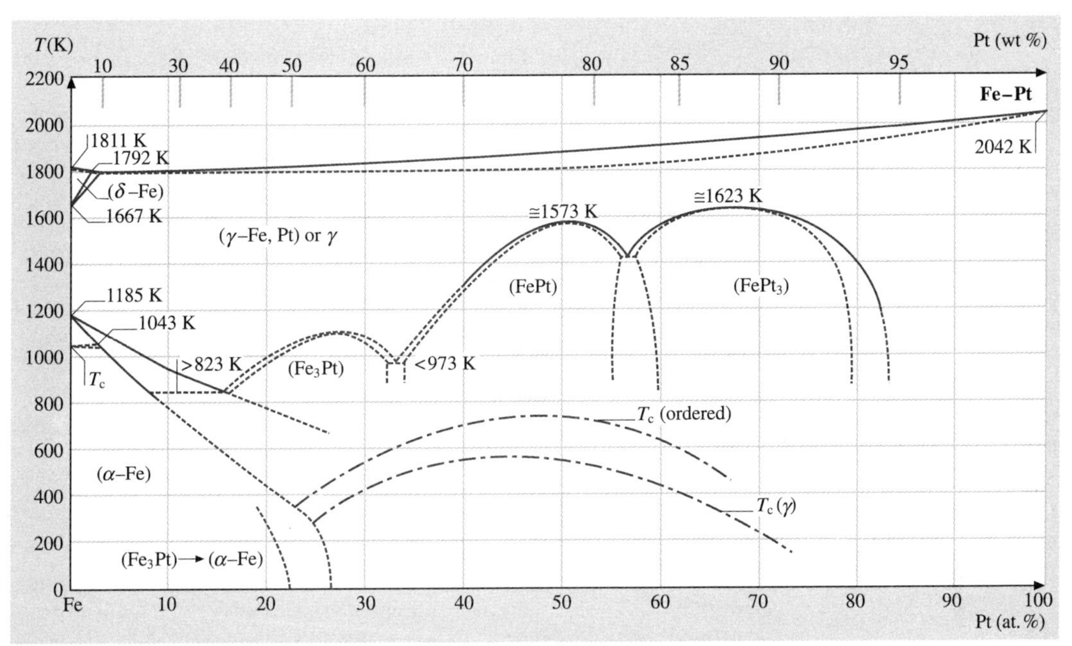

Fig. 4.3-41 Fe–Pt phase diagram. *Dashed-dotted lines*: Curie temperature T_c

order to obtain a well ordered state, a long annealing time is necessary (600 °C, 160 h). Disordered fcc alloys which show invar anomalies, too, may be obtained by rapid quenching from above the order–disorder transformation temperature. The Curie temperature of the disordered state is lower than that of the ordered state. A high negative value of α is observed just below the Curie temperature, particularly for disordered alloys. Annealed alloys containing 52 to 54 wt% Pt have small TEC and those containing 52.5 to 53.5 wt% Pt show negative values of α, Fig. 4.3-42 [3.17].

Fe–Pd-Based Invar Alloys. Alloys of Fe–Pd (see Fig. 4.3-43) containing 28 to 31 at.% Pd show invar characteristics, as seen in Fig. 4.3-44 [3.41]. In order to obtain invar behavior, the alloys are quenched from the high temperature γ phase field such that phase transformations at lower temperatures are suppressed. As shown in Fig. 4.3-45 [3.41], the thermal expansion is strongly decreased by cold deformation, i. e., by disordering, lattice defects, and internal stresses. After cold working, an instability of the invar property is observed.

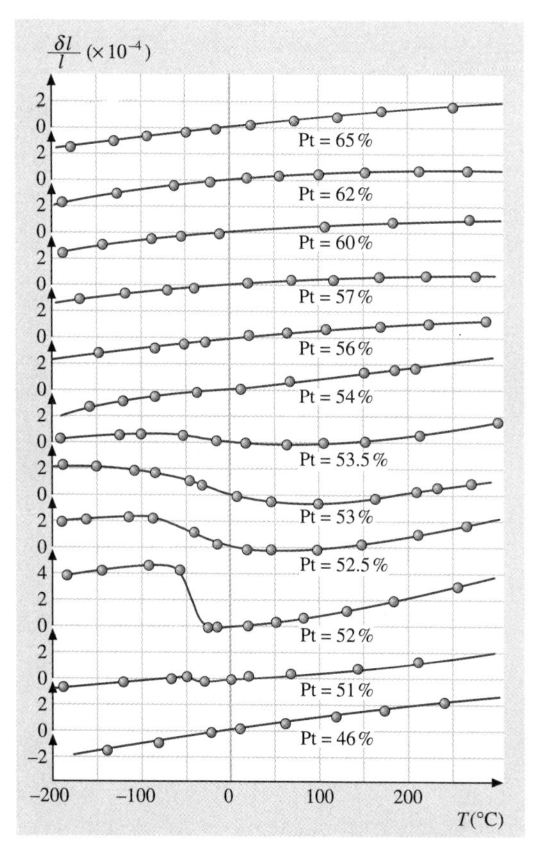

Fig. 4.3-42 Thermal expansivity curves of Fe−Pt alloys [3.17]

Other Alloy Systems with Invar Behavior. Beyond the invar-type alloys mentioned, many other Fe-based alloy systems show invar behavior, for instance, Fe−Ni−Cr, Fe−Ni−V, Fe−Pt−Re, Fe−Ni−Pt, Fe−Ni−Pd, Fe−Pt−Ir, Fe−Cu−Ni, and Fe−Mn−Ni. Some amorphous melt/quenched alloy systems show invar characteristics too, e.g., $Fe_{83}B_{17}$, $Fe_{85}P_{15}$, $Fe_{79}Si_9B_{12}$. Similarly, amorphous alloys prepared by sputtering show invar characteristics: $Fe_{75}Zr_{25}$, $Fe_{72}Hf_{28}$. Some antiferromagnetic Mn- and Cr-based alloys also exhibit a remarkable anomaly of thermal expansion due to magnetic ordering: $Pd_{64.5}Mn_{35.5}$, $Mn_{77}Ge_{23}$, $Cr_{92.5}Fe_{4.3}Mn_{0.5}$, $Cr_{96.5}Si_3Mn_{0.5}$.

Elinvar Alloys

Elinvar or constant-elastic-modulus alloys are based on work by *Guillaume* [3.45] and *Chevenard* [3.46]. They show a nearly temperature-independent behavior of the Young's modulus (E). For technical applications, elinvar alloys are designed to show the anomaly in the range of their operating temperature, usually close to room temperature. Since the resonance frequency f_0 of an oscillating body is related to its E modulus by $f_0 \sim \sqrt{E/\varrho}$ (ϱ =density), the thermoelastic properties of elinvar alloys are utilized for components in oscillating systems of the precision instrument industry. In these applications highly constant resonance frequencies are required. Typical examples are: resonators in magnetomechanical filters; balance springs in watches, tuning forks; helical springs in spring balances or seismographs, as well as in pressure or load cells. The condition for temperature compensation of the Young's modulus E is:

$$2\Delta f/f\Delta T = \Delta E/E\Delta T + \alpha \approx 0\,,$$

with T = temperature and α = linear thermal expansion coefficient. Elinvar characteristics also refer to the temperature independence of the shear modulus G, which is related to the E modulus via

$$3/E = 1/G + 1/3B$$

and

$$E = 2(1+\nu)G = 3(1-\nu)B\,,$$

with B = bulk modulus and ν = Poisson's ratio.

In ordinary metallic materials, E decreases with increasing temperature according to the variation of the elastic constants with temperature according to the anharmonicity in the phonon energy term. The temperature compensation of the E-modulus in ferromagnetic and antiferromagnetic elinvar alloys is caused by an anomaly in its temperature behavior (ΔE effect). As a consequence in ferromagnetic alloys, the E modulus in the demagnetized state is different from that in the magnetized state. The ΔE effect with ferromagnetic elinvar alloys consists of three parts [3.47]:

$$\Delta E = \Delta E_\lambda + \Delta E_\omega + \Delta E_A\,.$$

A schematic description of the components of the ΔE effect in ferromagnetic elinvar alloys is given in Fig. 4.3-46.

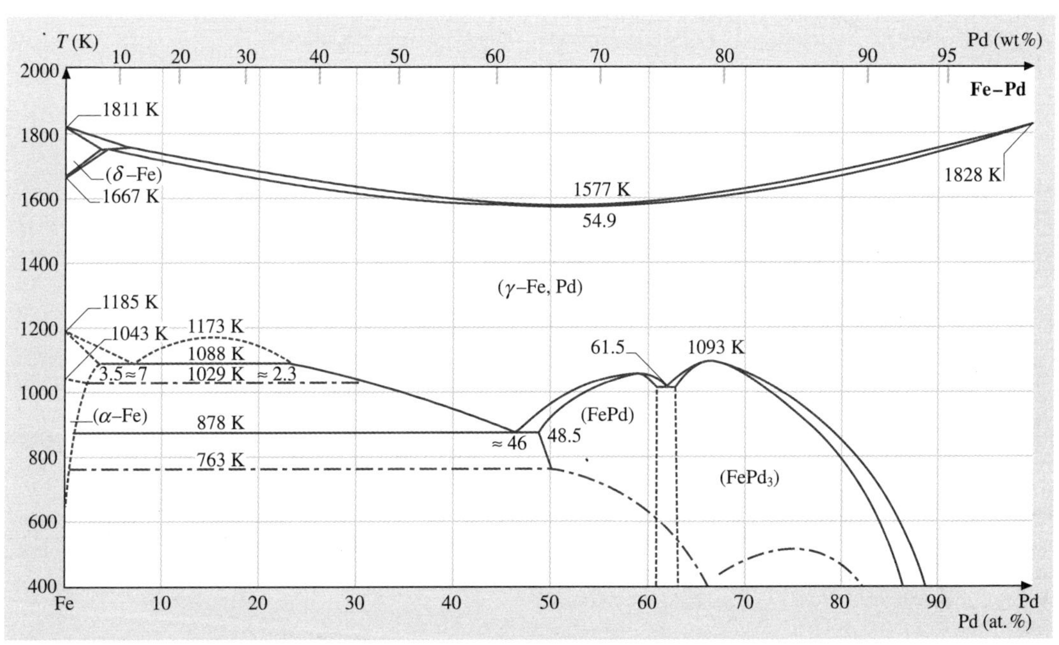

Fig. 4.3-43 Fe–Pd phase diagram. *Dashed-dotted lines*: Curie temperature T_c

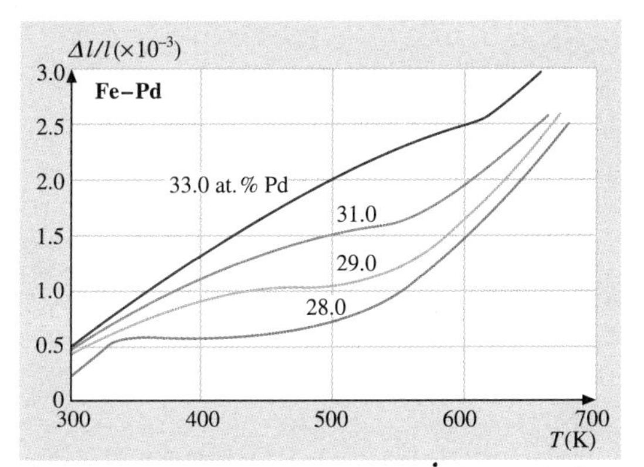

Fig. 4.3-44 Thermal expansion curves of Fe–Pd alloys rapidly cooled from high temperature [3.41]

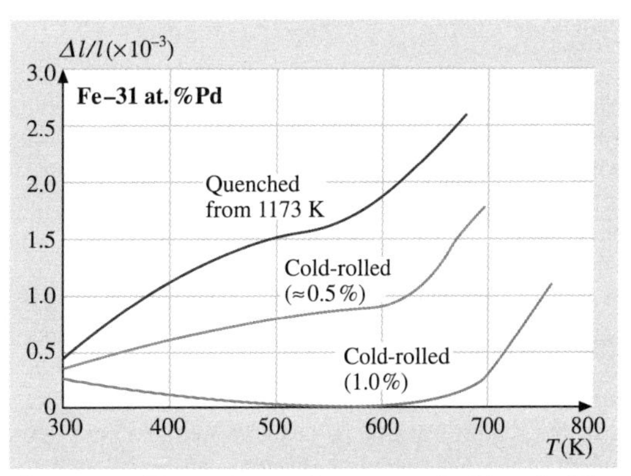

Fig. 4.3-45 Thermal expansion curves of cold-worked Fe–31 at.% Pd alloy. The rolling ratio is given by percentage

These components are attributable to the following relations:

$$\Delta E_\lambda = -2/5 E^2 \lambda_s / \sigma_i \,,$$

where ΔE_λ is attributed to the shape magnetostriction (λ_s) that changes the direction of the spontaneous magnetization, owing to domain wall motion and rotation processes as a consequence of the influence of mechanical stresses (σ_i) or magnetic fields on the unsaturated state of material [3.48]. The value ΔE_ω is caused by forced volume magnetostriction (ω) as a consequence of changes of the interatomic distances induced by stress or very high magnetic fields, which lead to a change of the magnetic interaction [3.49]:

$$\Delta E_\omega = -1/9 E^2 [(\partial\omega/\partial H)^2 / \partial J/\partial H] \,,$$

where $\partial\omega/\partial H =$ forced volume magnetostriction and $\partial J/\partial H =$ para-susceptibility. The value ΔE_A takes the

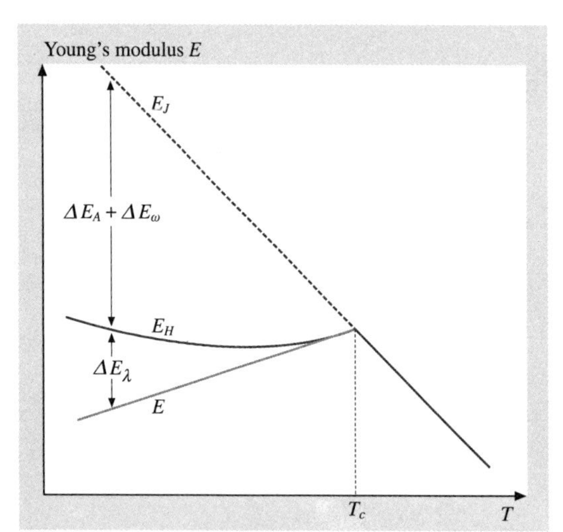

Fig. 4.3-46 Young's modulus E as a function of temperature of ferromagnetic elinvar alloys: $E = E$ modulus in the absence of a magnetic field, $E_H = E$ modulus measured in a magnetic field H, $E_J = E$ modulus at constant polarization J

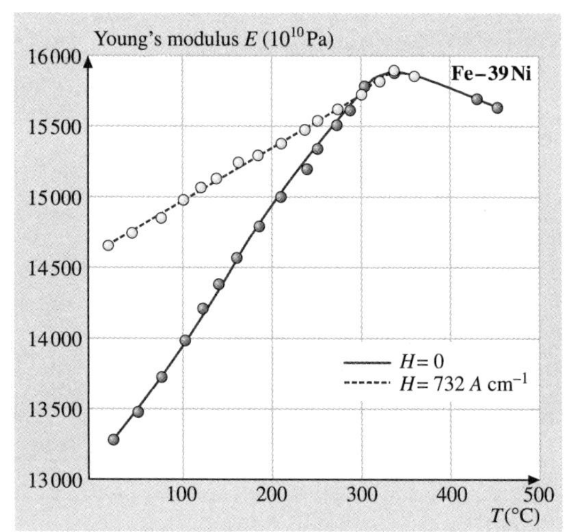

Fig. 4.3-47 E modulus and its dependence on the temperature of an annealed Fe–39 wt% Ni alloy, with and without influence of a magnetic field H [3.51]

role of the exchange energy into account [3.50]:

$$\Delta E_A \sim -\omega_s \sim J_s^2 ,$$

where $\omega_s =$ volume magnetostriction and $J_s =$ saturation polarization. It originates from the spontaneous volume magnetostriction ω_s as a function of the change of exchange energy with temperature due to the variation of magnetic ordering up to the Curie temperature.

Ferromagnetic Elinvar Alloys. The development of ferromagnetic elinvar alloys is based on affecting the shape magnetostriction λ_s by control of internal stresses resulting from deformation and/or precipitation hardening. But this requires a zero or negative temperature coefficient of the E_H modulus. Based on an Fe–39 wt% Ni alloy, it can be shown how this behavior is achievable. Alloys of Fe–Ni with 36 to 45 wt% Ni have a positive sign of the temperature coefficient of the E_H modulus (see Fig. 4.3-47 [3.51]).

By addition of 7 wt% Cr this coefficient is reduced to zero or to slightly negative values. By a deformation it can be influenced furthermore while the absolute value

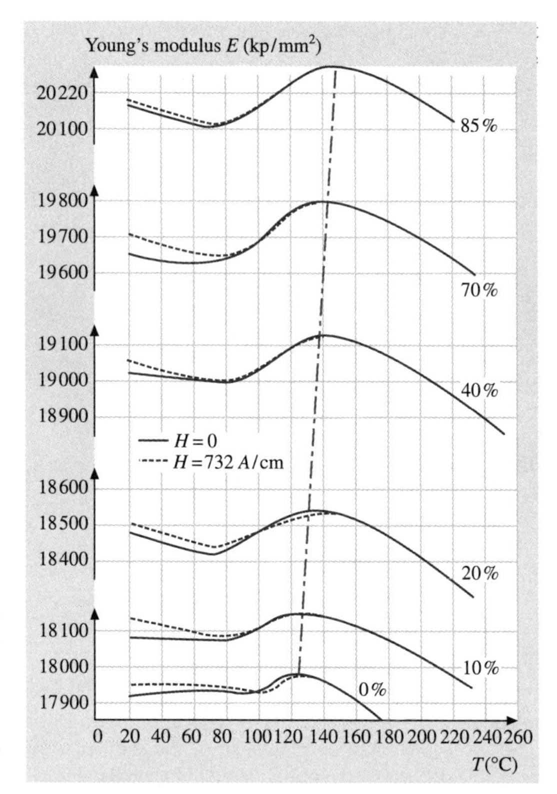

Fig. 4.3-48 Young's modulus E of an Fe–39Ni–7Cr–0.8Be–1.0Ti (wt%) alloy as a function of temperature, degree of cold deformation η (%) with and without a magnetic field H [3.51]

and 1.0 wt% Ti on the $E = f(T)$ characteristic. Cold deformation causes the Curie temperature to rise.

The final processing steps consist of solution annealing at 1150 °C, water quenching, cold deformation, and a final precipitation annealing at about 600 °C. Accordingly commercial elinvar alloys are produced, such as those listed in Tables 4.3-28, 4.3-29, and 4.3-30.

Apart from the well-tried Fe–Ni–Cr–(Be, Ti)-based elinvar-type alloys, Fe–Ni–Mo-based alloys have gained technical application. Addition of Mo improves the elastic properties, lowers the Curie temperature, and increases the resistance to corrosion. While in Europe and the USA, Fe–Ni-based elinvar-type alloys mainly were developed, in Japan Co–Fe-based elinvar alloys were discovered. Ternary Co–Fe–Cr alloys and quaternary alloys containing Ni (Co-elinvar) attained significant technical relevance. Distinguishing marks worth mentioning include: higher Young's modulus than Fe–Ni-based alloys, corrosion resistance, wide range of temperature compensating of E modulus, and easy hardening by cold-working.

Antiferromagnetic Elinvar Alloys. In antiferromagnetic alloys no domains are formed and no ΔE_λ effect occurs. With antiferromagnetic ordering the ΔE_A effect can only be exploited in combination with a pronounced cubic to tetragonal lattice distortion associated with antiferromagnetic ordering. This requires methods to develop temperature compensating elastic behavior which are different from those for ferromagnetic thermoelastic materials. The development of antiferromagnetic alloys with Elinvar properties has been concentrated on Mn–Cu, Mn–Ni, and Fe–Mn base alloys. In [3.52] the distinct anomalies of Young's modulus at the Néel

Table 4.3-28 Compositions in (wt%) of Fe–Ni-based elinvar alloys [3.17]

	Bal. Fe															
	C	Ni	Cr	Ti	Mo	W	Mn	Si	Al	Be	Nb	Cu	V	Co	P	S
Durinal	0.1	42		2.1			2		2							
Elinvar Extra	0.04	43	5	2.75			0.6	0.5	0.3					0.35		
	0.6	42	5.5	2.5			0.5	0.5	0.6							
Elinvar New	1	35	5			2	1									
Elinvar Original		36	12													
	0.5–2	33–35	21–5			1–3	0.5–2	0.5–2								
	0.71	33.5	8.4			2.98	2.4	0.33							0.018	0.01
Iso-elastic		36	8		0.5	(other small constituents)										
	0.1	36	7.5		0.5		0.6	0.5				0.2				
Isoval	0.6	30			2.2	3.2	0.15	0.2			3.8		4.2			
Métélinvar	0.6	40	6		1.5	3	2									
Ni-Span C	0.03	42.2	5.3	2.5			0.4	0.4	0.4			0.05			<0.04	<0.04
Ni-Span C 902	<0.06	42	5.2	2.3			<0.8	<0.1	0.5							
Nivarox CT	0.02	37	8	1			0.8	0.2		0.8						
Nivarox CTC	0.2	38	8	1						1						
Nivarox M	0.03	31			6		0.7	0.1		0.7						
Nivarox M30		30			9					1						
Nivarox M40		40			9					1						
Nivarox W		36		1						1						
Sumi-Span 1		36	9													
Sumi-Span 2	0.4	38	11													
Sumi-Span 3		42.5	5.5	2.4												
Thermelast 4009		40			9					0.5						
Thermelast 5409		40			9					0.5						
Vibralloy		39			9											
		40			10											
YNiC	0.03	41–43	5.1–5.5	2.2–3					0.5–0.6							

Table 4.3-29 Fe–Ni-based commercial elinvar-type alloys and the corresponding values of: density d, melting point T_m, Curie temperature T_c, electrical resistivity ρ, thermal expansion coefficient α, Young's modulus E, its temperature coefficient e, and shear modulus G [3.17]

	d ($g\,cm^{-3}$)	T_m (°C)	T_c (°C)	ρ ($\mu\Omega$ cm)	α ($10^{-6}\,K^{-1}$)	E (10^{10} Pa)	e ($10^{-5}\,K^{-1}$)	G (10^{10} Pa)
Durinal			90				−1.0–1.0	
Elinvar		1420–1450	−100		8	7.8–8.3	−0.3–0.3	
Elinvar Extra [a]	8.15				6.5	18.9	0	6.9
Iso-elastic				88	6.7	18.0	−3.3–2.5	6.4
Métélinvar			260, 295				0	
Ni-Span C	8.15	1450–1480		80	7.1	18.9	−1.7–1.7	
Ni-Span C 902 [a]	8.14	1460–1480	160–190	100–120	8.1	17.7–19.6		6.9–7.4
Nivarox CT	8.3		80	97	7.5	18.6	−2.5–2.5	
Sumi-Span 1	8.15		−140	100	−10	18.1	0–2.5	
Sumi-Span 2	8.08		−190	105	−10	18.1	−1.5–0 [b]	
Sumi-Span 3 [a]	8.05		−190	110	≈8	19.2	−1.0–1.0	7.8
Thermelast 4002 [a]	8.3			100	8.5	18.6		6.4
Thermelast 4005 [a]	8.3			100	8.3	17.2		6.4
Thermelast 5405 [a]	8.3			100	8.0	18.6		6.4
Thermelast 5429 [a]	8.3			100	8.0	18.6		6.4
Vibralloy [a]	8.3		300		8	17.4	0	
YNiC	8.15		90–180		8.1	19.6	−1.8–1.5	6.5–6.8

[a] Properties in the fully aged state
[b] Temperature coefficient of the frenquency of proper vibration

Table 4.3-30 Co–Fe-based elinvar-type alloys (annealed state). Composition, thermal expansion coefficient α, Young's modulus E and shear modulus G, and their respective temperature coefficients, e and g [3.17]

	Composition in wt%								α [a]	E [b]	e [c]	G [b]	g [c]
	Co	Fe	Cr	V	Mo	W	Mn	Ni	($10^{-6}\,K^{-1}$)	(10^{10} Pa)	($10^{-5}\,K^{-1}$)	(10^{10} Pa)	($10^{-5}\,K^{-1}$)
Co-elinvar	60.0	30.0	10.0						5.1	17.07		6.91	−0.2
	51.5	38.5	10.0						8.7	18.84	−1.0	7.55	
	47.3	34.5	9.1					9.1	7.8			6.61	0.2
	27.7	39.2	10.0					23.1	8.1			6.48	−0.3
	26.7	50.8	5.8					16.7	7.8			4.99	0.3
	17.9	42.8	10.7					28.6	8.3			6.74	0.9
Elcolloy	40.0	35.0	5.0			5.0		15.0	5.0				−0.2
	35.0	36.0	5.0		4.0	4.0		16.0	9.0				0.5
Mangelinvar	38.0	37.0					15.0	10.0	9.7	18.0	−1.0		
Moelinvar	50.0	32.5			17.5				9.6			7.36	−0.2
	45.0	35.0			10.0			10.0	8.5			6.15	0.7
	20.0	40.0			20.0			20.0	8.4			7.70	0.9
	10.0	45.0			15.0			30.0	9.8			7.85	−0.4
Tungelinvar	50.0	28.5				21.5			7.4			6.45	−0.7
	39.0	32.0				19.0		10.0	7.8			8.13	0.4

Table 4.3-30 Co–Fe-based elinvar-type alloys (annealed state). Composition, thermal expansion coefficient α, cont.

	Composition in wt%								α [a]	E [b]	e [c]	G [b]	g [c]
	Co	Fe	Cr	V	Mo	W	Mn	Ni	(10^{-6} K^{-1})	(10^{10} Pa)	(10^{-5} K^{-1})	(10^{10} Pa)	(10^{-5} K^{-1})
Velinvar	60.0	30.0		10.0					8.1			6.53	0.0
	50.0	31.8		8.2				9.1	11.0			6.70	1.0
	37.5	35.5		7.0				20.0	11.1			6.45	−0.6
	22.5	43.5		4.0				30.0	12.3			5.66	−0.3

[a] For the temperature range 10–50 °C
[b] At 20 °C
[c] For the temperature range 0–50 °C

temperature of Mn–Ni- and Mn–Cu-based alloys were modified by additions of Cr, Fe, Ni, Mo, and W as well as by suitable technological treatment in such a way that useful thermoelastic coefficients could be reached. However, the mechanical workability, the sensitivity of elastic properties to the degree of cold-working, the unsufficient spring properties as well the low corrosion resistance and high mechanical damping could not satisfy the conditions of technical application. Figures 4.3-49 and 4.3-50 [3.17] indicate the temperature dependence of E of some Mn–Ni- and Mn–Cu-based alloys. Table 4.3-31 [3.17] lists the compositions and the thermoelastic and mechanical properties of this group of alloys.

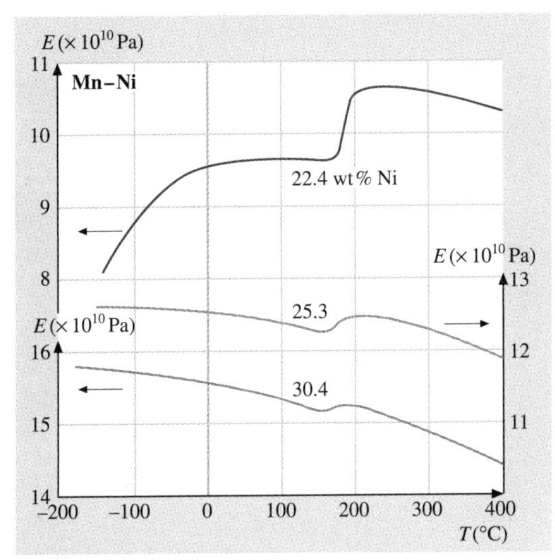

Fig. 4.3-49 Mn–Ni binary alloys. Young's modulus E vs. temperature for alloys annealed at 1223 K for 1 h after cold-working [3.17]

By optimization of the chemical composition and the processing technology, an Fe–24Mn–8Cr–7Ni–0.8Be

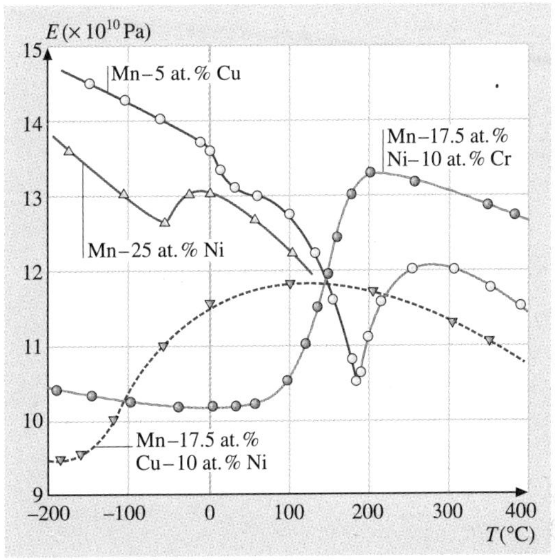

Fig. 4.3-50 Temperature dependence of Young's modulus E [3.17] for various Mn-based alloys annealed at 1223 K for 1 h and then quenched

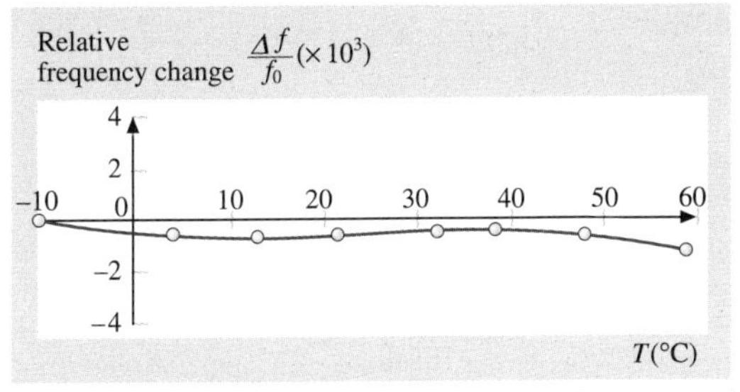

Fig. 4.3-51 Temperature dependence of the frequency of a screw spring [3.53]

Table 4.3-31 Nonferromagnetic Mn-based elinvar-type alloys. Composition, thermal expansion coefficient α, Young's modulus E and shear modulus G, and their respective temperature coefficients, e and g and hardness HV [3.17]

Composition in wt%									α [a]	E	e [a]	G	g [a]	HV
Mn	Cu	Ni	Cr	Fe	Co	Mo	W	Ge	($10^{-6}\,K^{-1}$)	(10^{10} Pa)	($10^{-5}\,K^{-1}$)	(10^{10} Pa)	($10^{-5}\,K^{-1}$)	
87		10		3					23.7	12.3	1.25	5.18	1.05	121
82		15					3		23.0	12.1	0.55	5.07	0.78	149
80		16			4				21.1	12.2	−0.13	4.63	−0.75	150
80		9				11			20.3	11.9	0.05	5.00	1.10	380
80								20	12.3	9.0	1.5			255
79		21								9.8	−2.5	3.60	−2.7	235
67	20	13							22.4	14.4	0.21	4.55	0.29	125
59		16	25						21.6	16.2	0.85	5.21	0.83	250
49	41			10					22.4	13.5	−0.97	5.53	−0.20	250
44	55		1						22.1	13.2	0.11	5.03	0.08	145
43	57								23.6	11.2	0.3	4.2	−0.9	131
43	55				2				22.9	8.50	−1.11	4.02	−2.57	135
42	55					3			23.0	15.2	2.30	6.77	1.88	149
39	56						5		23.2	12.0	−0.25	5.05	−0.56	140

[a] For the temperature range 0–40 °C
[b] At 20 °C

Table 4.3-32 Properties of an antiferromagnetic Fe−24Mn−8Cr−7Ni−0.8Be elinvar alloy [3.53]

Property	Value
Young's modulus E	165–195 GPa
Thermoelastic coefficient TKE	(1–10) MK^{-1}
Compensation range of E	0–50 °C
Shear modulus G	74–82 GPa
Coefficient of thermal expansion α (20–100 °C)	13 MK^{-1}
Tensile strength σ_B	1200–1800 MPa
Yield point σ_s	1100–1650 MPa
Elongation δ	12–2 %
Vicker's Hardness HV 10	420–540 HV
Quality factor Q	20 000–10 000
Specific electrical resistance ρ	80 μΩ cm
Density γ	7.6 g cm^{-3}
Electrochemical breahdown potential ϵ_D	−0.25 V
Melting temperature T_s	1450–1480 °C

antiferromagnetic elinvar alloy was developed which fulfills the complex requirements for an antiferromagnetic, corrosion resistant, temperature compensating thermoelastic spring material for applications near room temperature, Fig. 4.3-51 [3.53]. The thermoelastic, mechanical and physical properties are summarized in Table 4.3-32 [3.53].

Other Nonmagnetic Elinvar-Type Alloys. In general, the elastic constants decrease with increasing tem-

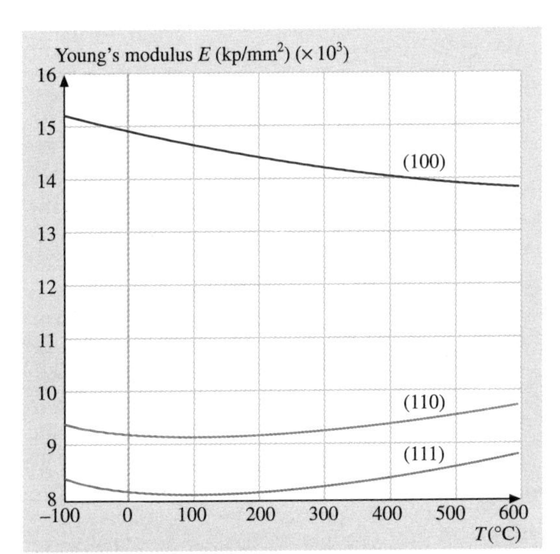

Fig. 4.3-52 Thermoelastic behavior of single crystalline Nb in the major lattice directions [3.54]

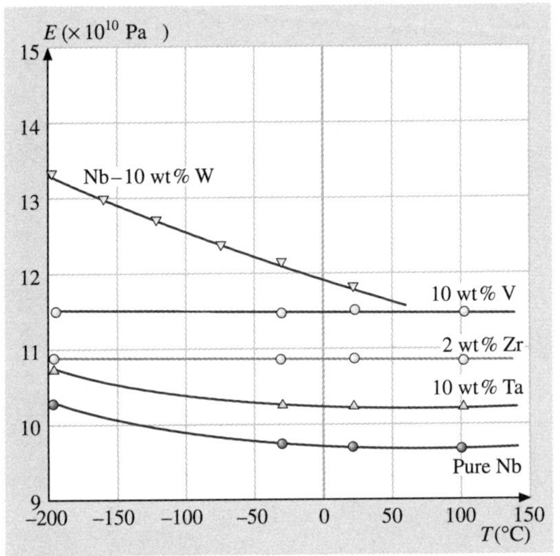

Fig. 4.3-53 Young's modulus E vs. temperature. Samples annealed at 1400 °C for 4 h [3.41] of pure Nb and Nb-based binary alloys

perature. The temperature dependence of the elastic properties of Nb shows highly anisotropic anomalies (see Fig. 4.3-52) [3.54]. In a randomly oriented polycrystal, Nb becomes an elinvar-type material, Fig. 4.3-53 [3.41]. This plot of Young's modulus vs. temperature also shows the influence of alloying elements.

Elinvar behavior is, also, found in concentrated Nb–Zr and Nb–Ti alloys. Furthermore, some amorphous alloys, e.g., in the Fe–B-, Fe–P-, and Fe–Si–B-based systems, show elinvar behavior. Examples for amorphous Fe–B alloys are shown in Fig. 4.3-54.

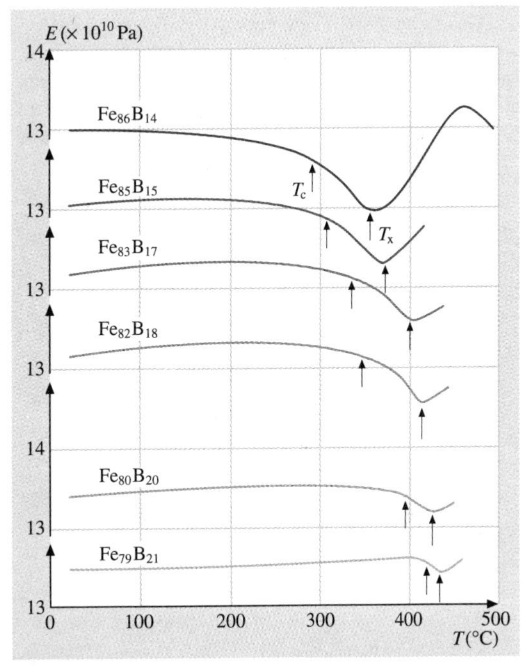

Fig. 4.3-54 Fe–B alloys. Temperature dependence of Young's modulus E for amorphous alloys annealed at 200 °C for 2 h. T_x and T_c show the crystallization and the Curie temperature, respectively [3.41]

4.3.3 Hard Magnetic Alloys

Permanent or hard magnetic materials comprise traditionally some special steels but consist essentially of multiphase alloys and intermetallic and ceramic compounds today. Relatively few magnetic alloys and compounds fulfill the requirement of combining high magnetic efficiency and competitive cost: Fe–Ni–Al–Co (Alnico), Fe–Cr–Co, Mn–Al–C, hard ferrites, and rare-earth transition metal compounds of the Co–Sm, Fe–Nd–B and Fe–Sm–N alloy systems. Surveys may be found in handbooks [3.1, 3, 55] and data collections [3.10, 56]. A survey of permanent magnetic materials is given in Table 4.3-33. The metallic hard magnets are treated in this chapter, the oxidic hard magnetic materials are dealt with in Sect. 4.3.4.

All of the hard magnetic materials are based on choosing a base alloy with a sufficiently high saturation magnetization M_s and a high magnetocrystalline anisotropy constant K_1, and tailoring the microstructure to exploit this crystal anisotropy. In some cases, shape anisotropy is generated in addition. This microstructural control is achieved by

1. Inducing a texture by processing in such a way that a macroscopic direction in the material, e.g., the rolling direction of a sheet or the pressing direction in a sintered material, is an easy direction, and processing at 90°, i. e., the transverse direction, is a hard direction for magnetization. This is the basic magnetic hardening mechanism of hard magnetic steels which have lost their importance in present technology; but it is the basis of producing the more recent high energy magnets made from intermetallic compounds such as Co_5Sm and ferrites.
2. Inducing a two-phase microstructure by coherent precipitation or decomposition and promoting, by suitable magnetic field annealing procedures, the alignment of the elongated precipitates in one direction of easy magnetization, i. e., inducing both

Table 4.3-33 Survey commercially used permanent magnetic materials. Survey

Material	Fe, Co content (wt%)	B_r^a (T)	$_JH_c$ (kA m^{-1})	$(BH)_{max}$ (kJ m^{-3})	appr. T_c (°C)	T_{max}^b (°C)	Process[c]
Dense magnets							
3.5Cr steel	94–95	0.95	5	2.3	745		C[d]
6W steel	92–93	0.95	6	2.6	760		C
36Co steel	90–91	0.95	19	7.4	890		C
Alnico	67–74	0.52–1.4	40–135	13–69	810–900	450–550	C, P[d]
Fe–Cr–Co	65–73	1.1–1.4	40–65	25–55	670	500	C
hard ferrite	58–63	0.37–0.45	160–400	26–40	460	250	P
Pt-Co	23.3	0.64	430	73	480	350	C, P
MnAlC	–	0.55	250	44	500	300	P
Co_5Sm	63–65	0.85–1	>1600	140–200	730	250	P
$TM_{17}Sm_2$	61–68	0.95–1.15	480–2000	190–220	810	330–550	P
Fe–Nd–B	66–72	1.05–1.5	950–2700	240–400	320	60–180	P
Bonded magnets							
hard ferrite	58–63	0.1–0.31	180–300	2–18		140	P
$TM_{17}Sm_2$	61–68			70–120		150	P
$Fe_{14}Nd_2B$	70–72	0.47–0.69	600–1200	35–80		80–110	P
$Fe_{17}Sm_2N_3$		0.77	650	105		100	P

[a] B_r values are for magnets operated at load lines $B/H \gg 1$

[b] The maximum operating temperature of bonded magnets is determined by the organic binder used

[c] Magnets are manufactured either by a casting/heat treatment technique or by a powder metallurgical process. Powder metallurgy is applied for small magnets where small and intricate shapes to precise tolerances are required

[d] C: magnets produced by cast and heat treatment; P: magnets produced by means of powder metallurgical techniques

magnetocrystalline and shape anisotropy. This is termed magnetic shape anisotropy and is employed in the Alnico and Fe−Co−Cr hard magnetic materials.
3. Inducing a fine grained microstructure with a magnetically insulating phase at the grain boundaries so that the grains are magnetically decoupled and, as a consequence, the nucleation of reverse magnetization is requiring an extremely high nucleation energy. This is applied, for example, to $TM_{17}Sm_2$, $Fe_{14}Nd_2B$ and bonded magnets.

In some of the hard magnetic materials two of these variants of microstructural design are combined.

4.3.3.1 Fe−Co−Cr

Hard magnetic materials made of ternary Fe−Co−Cr alloys are based on the high atomic moment of Fe−Co alloys and the miscibility gap occurring when Cr is added. Intrinsic magnetic properties are compiled in [3.6]. Extensive magnetic materials data are found in [3.56]. Figure 4.3-55 shows the relevant metastable phase relations in the ternary equilibrium diagram. If an alloy is homogenized in the solid solution range above the solvus surface with $T_{max} > 700\,°C$ first and annealed in the miscibility gap subsequently, coherent decomposition occurs, which results in a two-phase microstructure on the nanometer scale. The α_1 phase is rich in Fe and Co and ferromagnetic while the α_2 phase is rich in Cr and antiferromagnetic. This two-phase microstructure on the nm scale has hard magnetic properties which can be varied by adjusting the alloy composition and heat treatment. The term spinodal decomposition is frequently applied to all kinds of coherent decomposition, e.g., in [3.56]. But it is used correctly only if referring to a special mode of compositional evolution associated with particular kinetics in the initial stage of decomposition within the spinodal of a miscibility gap.

Three groups of materials have been developed, differing essentially in the Co content (< 5, 10–15, 23–25 wt% Co), while the Cr content ranges from 22 to 40 wt% Cr. Table 4.3-34 lists data obtained by varying composition, mode of manufacturing, and heat treatment systematically for the group characterized by < 5 wt% Co as an example. The variation of the magnetic properties is determined by the intrinsic properties of the decomposed phases α_1 and α_2 and their microstructural array.

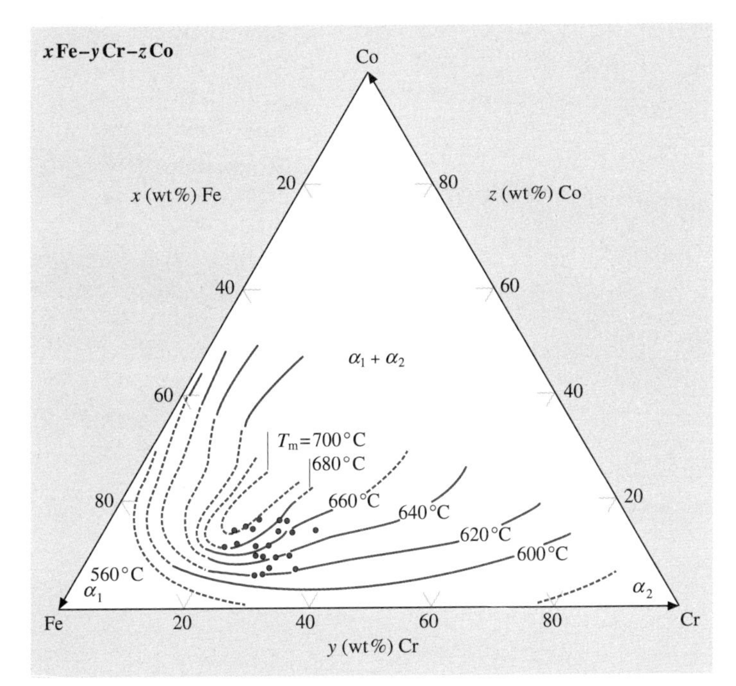

Fig. 4.3-55 The miscibility gap ($\alpha_1 + \alpha_2$) of the bcc α-phase in the Fe−Co−Cr phase diagram [3.56]

Table 4.3-34 Survey of magnetic properties of Fe–Cr–Co (≤ 5 wt% Co) alloys in relation to the composition, mode of manufacturing and heat treatment [3.56]

Alloy components (wt%)			Mode of manufacturing and heat treatment	Magnetic properties					
				B_r		$_BH_c$		$(BH)_{max}$	
Cr	Co	others		(T)	(kG)	(kA m^{-1})	(kOe)	kJ m^{-3}	(MGOe)
33	2	1 Hf	H(700, 15):MCL > T_s, R, 500):FCL	1.25	12.5	16.2	0.203	14.0	1.76
32	3		As above	1.29	12.9	35.9	0.449	32.4	4.08
32	4	0.5 Ti	As above	1.26	12.6	42.7	0.534	40.1	5.06
28	5		As above	1.38	13.8	29.0	0.362	27.9	3.52
30	5		As above	1.34	13.4	42.2	0.528	42.1	5.31
33	5		As above	1.22	12.2	40.8	0.51	36.3	4.58
35	5		As above	1.15	11.5	37.0	0.462	29.3	3.69
30	5	0.1 B	As above	1.31	13.1	42.0	0.525	40.2	5.07
30	5	0.25 B	As above	1.29	12.9	39.8	0.498	34.8	4.39
30	5	0.1 C	As above	1.31	13.1	42.2	0.527	38.8	4.89
30	5	0.8 Ge	As above	1.32	13.2	24.8	0.31	39.1	4.93
30	5	0.25 Ti	As above	1.34	13.4	27.2	0.34	40.2	5.07
30	5	0.5 Ti	As above	1.30	13.0	41.4	0.518	40.1	5.06
30	5	1.5 Ti	As above	1.27	12.7	40.8	0.51	38.1	4.81
30	5	0.25 Hf	As above	1.32	13.2	43.0	0.537	41.2	5.2
30	5	0.5 Hf	As above	1.29	12.9	43.9	0.549	41.4	5.22
30	5	1 Hf	As above	1.30	13.0	43.0	0.537	40.4	5.1
30	5	3 Hf	As above	1.24	12.4	41.5	0.519	34.8	4.39
23	2	1 Hf	MCL(> T_s, 550): CCL (550, 500)	1.24	12.4	36.8	0.46	34.1	4.3
32	3		As above	1.25	12.5	40.0	0.5	34.1	4.3
30	5		As above	1.34	13.4	42.4	0.53	42.0	5.3
32	4	0.5 Ti	As above	1.26	12.6	42.8	0.535	40.4	5.1
28	5	4 Ni	As above	1.27	12.7	29.6	0.37	30.1	3.8
28	7		As above	1.25	12.5	40.8	0.51	41.2	5.2
27	9		MCL: CCL (as above)	1.30	13.0	46.4	0.58	49.2	6.2
33	5		CL(680, 40 K/h):HW(D:67%):H (600): CCL (15–4 K/h, 500)[a]	1.15	11.5	24.8	0.31	19.0	2.4
33	7	2 Cu	As above	1.19	11.9	38.8	0.485	26.2	3.3
33	7		As above	1.18	11.8	42.0	0.525	33.3	4.2
33	9		As above	1.24	12.4	46.4	0.58	32.5	4.1
28	7		CL(> T_s, 60 ($T_s = 645\,°C$))	0.97	9.7	26.4	0.33	11.1	1.4
31	5		(Sintered 1400 °C, 4 h; H_2)[b]:WQ: H (700, 30): FCL (700, 640):MCL (640, 0.9 K/h, 500)	1.23	12.3	40.0	0.5	34.9	4.4

[a] For the deformation aging process the initially aged state corresponds to an overaged state

[b] Sintering ST

Commercial materials are characterized by the fact that they can be quenched from temperatures above the miscibility gap first, which results in mechanical properties amenable to forming by conventional processes such as rolling, stamping, drilling. The final annealing treatment in the miscibility gap results in the magnetically hard state. This is associated with a drastic decrease in ductility. If the annealing treatment is carried out in a magnetic field, the final product has an anisotropic behavior. Table 4.3-35 shows the property range of commercial Fe–Cr–Co materials.

4.3.3.2 Fe–Co–V

Magnetic materials based on the Fe–Co–V alloy system were the first ductile magnets. The intrinsic magnetic properties may be found in [3.6] while extensive magnetic materials data are treated in [3.56]. The optimum magnetic behavior is obtained for alloy compositions around Fe–55 wt% Co–10 wt% V. As the isothermal sections of the Fe–Co–V phase diagram Fig. 4.3-56a and Fig. 4.3-56b show, this alloy is mainly in the fcc γ-phase (austenite) state at 900 °C,

Table 4.3-35 Commercial Fe–Cr–Co magnetic materials

Composition nominal wt%	Variant	Remanence (T)	Coercivity ($kA\,m^{-1}$)	Energy density ($kJ\,m^{-3}$)	Curie temperature (°C)	Maximum application temperature (°C)	Hardness HV	Commercial designation [a]
Fe–27Cr–11Co–Mo	isotropic	0.85–0.95	36–42	13	640	480	480	12/160
Fe–28Cr–16Co–Mo	isotropic	0.80–0.90	39–45	15	640	480	480	16/160
Fe–27Cr–11Co–Mo	anisotropic	1.15–1.25	47–55	35	640	480	480	12/500
Fe–28Cr–16Co–Mo	anisotropic	1.10–1.20	53–61	37	640	480	480	16/550

[a] Designation of CROVAC® by Vacuumschmelze, Hanau, Germany

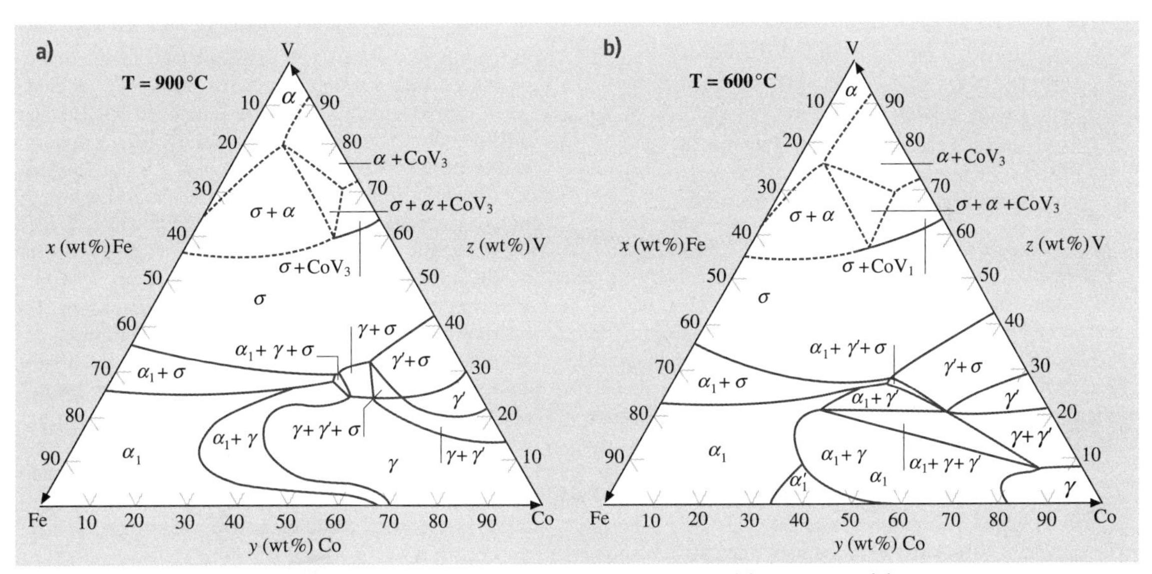

Fig. 4.3-56a,b Isothermal sections of the Fe–Co–V phase diagram at 900 °C (**a**) and 600 °C (**b**). α: bcc disordered; α_1: bcc ordered (CsCl type); γ: fcc disordered (austenite); γ': fcc ordered (Au_3Cu type) [3.56]

Table 4.3-36 Commercial Fe–Co–V-based magnetic materials

Composition nominal (wt%)	Remanence (T)	Coercivity (kA m^{-1})	Energy density (kJ m^{-3})	Curie temperature (°C)	Maximum application temperature (°C)	Hardness (HV) As rolled	Hardness (HV) Heat treated	Alloy code[a]
34Fe–52Co–13V	0.80–0.90	25–30	12	700	500	480	900	35U
34Fe–53Co–8.5V–3.5Cr	1.00–1.10	30–35	20	700	500	520	950	93

[a] Designation of MAGNETOFLEX® by Vacuumschmelze, Hanau, Germany

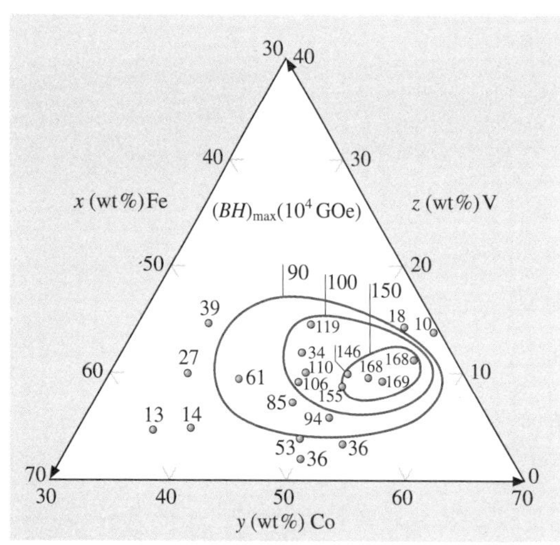

Fig. 4.3-57 Contour map of $(BH)_{max}$ of Fe–Co–V alloys in the optimum annealed state. It is obtained by annealing in the temperature range $T_a = 555–750\,°C$ in the magnetically preferred direction of the anisotropic sample [3.56]

while it decomposes into the bcc ordered $\alpha_1 + \gamma$ state upon annealing at lower temperature such as 600 °C. In combinations of heat treatment with plastic deformation (also serving to form the product, e.g., wire) an optimum anisotropic hard magnetic state can be realized.

If quenched from the γ-phase state, the alloy can be deformed. By a judicious choice of annealing temperatures in the range of 555 to 750 °C the maximum energy product as a function of alloy composition, as shown in Fig. 4.3-57, may be obtained. This annealing treatment is associated with a drastic increase in hardness, as indicated in Table 4.3-36, and a concomitant loss in ductility.

Based on these interrelations of phase equilibria and thermomechanical treatments as well as by optimization through further alloying additions, commercial magnetic materials such as those listed with their properties in Table 4.3-36 have been developed.

4.3.3.3 Fe–Ni–Al–Co, Alnico

The term Alnico refers to two-phase hard magnetic materials based on the Fe–Ni–Al system (Fig. 4.3-58). The intrinsic magnetic properties may be found in [3.6], while extensive magnetic materials data are treated in [3.56]. Table 4.3-37 lists some of the magnetic properties of Alnico type magnets. The magnetically optimized microstructure consists essentially of elongated ferromagnetic Fe-rich precipitates (α_1-phase, bcc disordered) in a non-magnetic matrix of NiAl (α_2-phase, bcc ordered, CsCl type). The remanence B_r is increased significantly by adding Co, which leads to the formation of precipitates rich in Fe–Co. The coercivity H_c is optimized by adding Ti and Cu. The two-phase state is obtained by a homogenization at about 1300 °C, followed by annealing treatments which lead to decomposition into structurally coherent phases on the nanometer scale. The particles are aligned preferentially along the ⟨100⟩ directions of the bcc lattice. This decomposition microstructure is the essential microstructural feature. Higher remanence and coercivity prevails in chill-cast magnets with a columnar microstructure and ⟨100⟩ fiber texture, providing additional magnetocrystalline anisotropy. More extensive treatments and data may be found in [3.10, 56, 57].

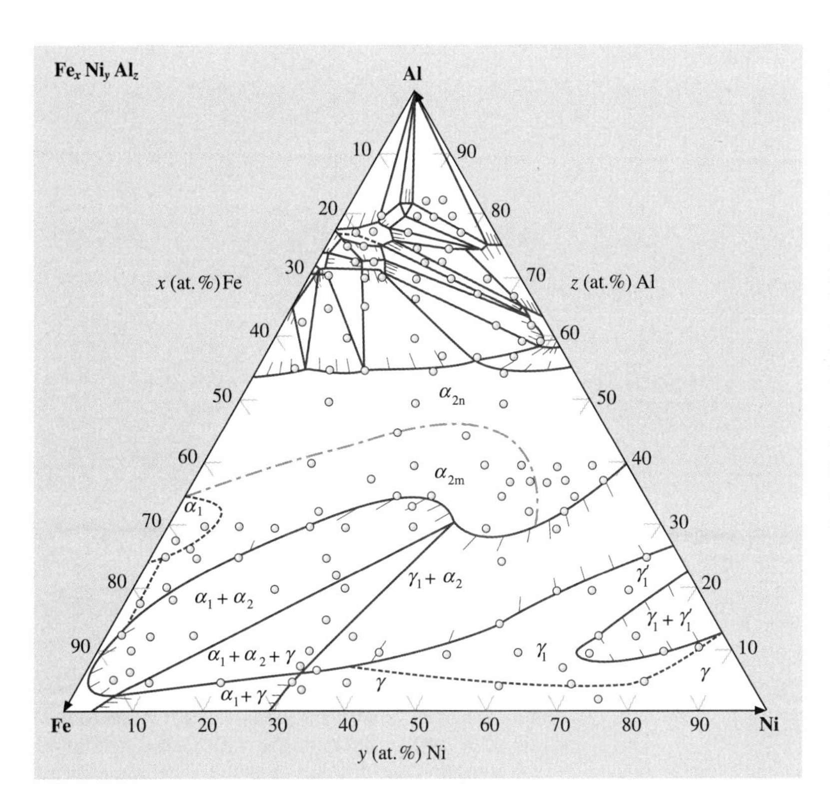

Fig. 4.3-58 Effective $Fe_xNi_yAl_z$ phase diagram after cooling from the melt at 10 K/h. *Broken lines* indicate superlattice phase boundaries; the *point-dash line* the magnetic phase boundary in the Ni(Al,Fe) phase field. α_1: bcc; α'_1: Fe_3Al-type superlattice phase; α_2: (Fe,Ni)Al-type superlattice phase; γ: fcc; γ_1: Ni_3Al-type superlattice phase; γ'_1 : as γ_1 but with the larger lattice spacing. The indices m and n indicate magnetic and non-magnetic phases, respectively

Part 4 | 3.3

Table 4.3-37 Magnetic properties of Alnico type magnets

Designation according to DIN 17410[a]	Remanence B_r		Coercivity $_BH_c$		$_JH_c$		Energy density $(BH)_{max}$		Alloy code[b]
	(mT)	(G)	(kA m^{-1})	(Oe)	(kA m^{-1})	(Oe)	(kJ m^{-3})	(MGOe)	
AlNiCo 9/5i	550	5500	54	679	57	716	10.3	1.3	130
AlNiCo 12/6i	650	6500	57	716	60	754	13.5	1.7	160
AlNiCo 19/11i	640	6400	105	1319	115	1433	22.3	2.8	260
AlNiCo 15/6a	750	7500	60	54	62	57	16.7	2.1	190
AlNiCo 28/6a	1100	11 000	64	804	65	817	31.8	4.0	400
AlNiCo 39/12a	880	8800	115	1445	119	1495	43.8	5.5	450
AlNiCo 37/5a	1240	12 400	51	641	51	641	41.4	5.2	500
AlNiCo 39/15a	740	7400	150	1855	160	2011	43.8	5.5	1800

[a] i = isotropic; a = anisotropic

[b] Commercial designations of Koerzit® by WIDIA Magnettechnik

Table 4.3-38 Intrinsic properties of $Fe_{14}RE_2B$ at $T = 300$ K

RE	K_1 (10^7 erg cm^{-3})	$4\pi M_s$ (T)	$H_a = 2K_1/M_s$ (MA m^{-1})	T_c (K)
Ce	1.7	1.28 (4)	3.0	430 (6)
Pr	4.5	1.59 (8)	6.3	563 (3)
Nd	4.8 (3)	1.68 (9)	5.7	590 (5)
Sm	plane	1.55 (10)		618
Gd	1.0	0.94 (2)	2.3	665 (4)
Tb		0.72 (5)	11.1 [a]	639
Dy	4.1	0.75 (5)	12.6	597 (5)
Ho		0.84 (12)	5.7 [a]	576
Er	plane	1.02 (12)		556 (3)
TM		1.25 (2)		543 (2)
Lu		–	2.1 [a]	538
La		–		543
Y	1.1(1)	1.40(5)	2.2	566 (2)

[a] Data taken from [3.58]. H_a value obtained directly by extrapolation of the magnetization curves for the easy and the hard direction. The average value is given where more than one reference is available; the number in parentheses indicates the standard deviation in the last figure

4.3.3.4 Fe–Nd–B

The most powerful permanent magnets presently available consist essentially of the tetragonal $Fe_{14}Nd_2B$ phase. The intrinsic magnetic properties may be found in [3.6] while extensive magnetic materials treatments and data may be found in [3.1, 10, 56]. Two different production routes are used to prepare dense anisotropic magnets: conventional powder metallurgy and a rapid quenching process to produce flake-shaped powder particles with a nanocrystalline microstructure as a starting material. The flakes are then processed further into dense isotropic or anisotropic magnets by means of a combination of cold pre-forming, hot pressing, and hot deformation steps.

The $Fe_{14}RE_2B$ phase is formed with all rare earth (RE) elements with the exception of Eu. Their intrinsic properties have been investigated extensively. They are listed in Table 4.3-38. Neodymium shows the highest permanent magnet potential based on its combination of high values of K_1 and M_S.

Conventional Powder Metallurgical Processing

Figure 4.3-59 shows the approximate phase relations of Fe–Nd–B at room temperature. According to the phase diagram of the Fe–Nd–B system the $Fe_{14}Nd_2B$ phase forms at 1180 °C. In powder metallurgical processing of the magnets, sintering at 1050 °C leads to the formation of $Fe_{14}Nd_2B$ in equilibrium with a Nd-rich liquid and with the Fe_4NdB_4 boride phase. The liquid phase solidifies below the ternary eutectic at 630 °C. The resulting non-magnetic Nd-rich solid phase spreads along

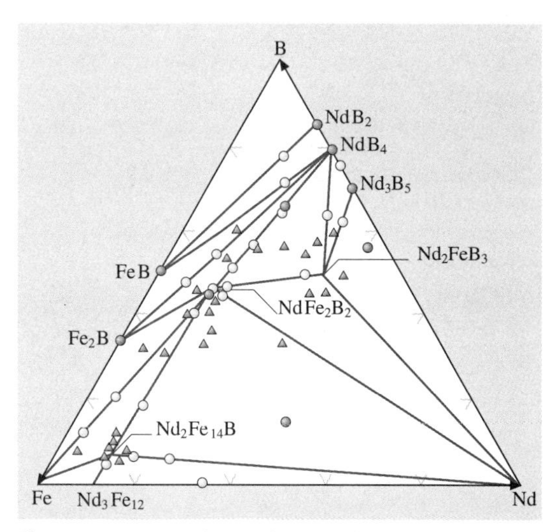

Fig. 4.3-59 Approximate phase relations of Fe–Nd–B at room temperature

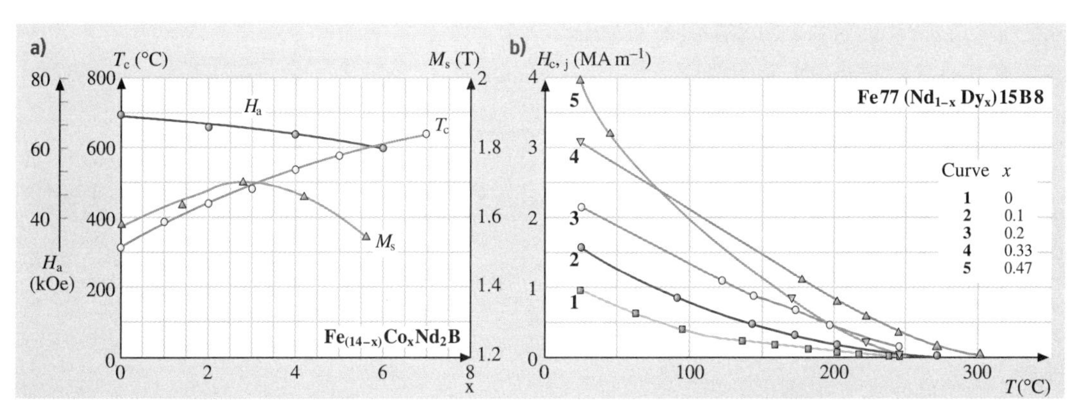

Fig. 4.3-60a,b Influence of substitutional elements on $Fe_{14}Nd_2B$ type magnets. **(a)** Influence of the Co content on the intrinsic properties at room temperature. Data for the magnetization M_s, for the anisotropy field H_a, and for the Curie temperature T_c. **(b)** Influence of the Dy content on the temperature dependence of the coercivity $_JH_c$ of sintered magnets. There is an approximately linear increase of $_JH_c$ at room temperature. The temperature dependence increases with increasing Dy content

the grain boundaries and provides the magnetic decoupling of the $Fe_{14}Nd_2B$ grains, thus providing the basic coercivity of the sintered magnet.

Additions of Dy and Al are increasing the coercivity. Dysprosium enters the RE sites in the $Fe_{14}Nd_2B$ structure, increasing the magnetocrystalline anisotropy but decreasing the magnetic remanence B_r. At compositions of > 2 at.% Al, the anisotropy field H_a decreases linearly at a rate of $0.13\,\mathrm{MA\,m^{-1}}$ per at.% Al. Nevertheless the coercivity increases significantly due to an optimization of the microstructure: Al is enriched in the Nd grain boundary phase which is spreading more uniformly around the magnetic grains, thus leading to better decoupling of exchange interactions. This is a basic condition for the increase in coercivity.

As indicated by Fig. 4.3-60a, Co addition leads to a strong increase of the Curie temperature. However, the anisotropy field H_a is reduced by Co, and the decrease of the coercive field is even larger than expected from this decrease in H_a. On the other hand, there is only a small increase in the magnetic saturation with a maximum at 20 at.% Fe substituted by Co. Accordingly the alloying is limited to 20 at.% Co.

The vulnerability of RE compounds to corrosion is a problem. The corrosion behavior of Fe–Nd–B magnets has been improved by adding elements which influence the electrochemical properties of the Nd-rich grain boundary phase. Additions of small amounts of more noble elements such as Cu, Co, Ga, Nb, and V result in the formation of compounds which replace the highly corrosive Nd-rich phase. Table 4.3-39 lists some of the elements used for manufacturing Fe–Nd–B magnets.

A multitude of grades of Fe–Nd–B magnets is produced by varying chemical composition and processing, such as the press technique applied in order to satisfy the different specifications required for the different fields of application. A maximum remanence is needed, for instance, for disc drive systems in personal computers and for background field magnets in magnetic resonance imaging systems. On the other hand, straight line demagnetization curves up to operating temperatures of 150 °C are specified for application in highly dynamic motors. This requires very high coercive fields at room temperature. Magnetic remanence values of $B_r > 1.4\,\mathrm{T}$ as well as $_JH_c$ values of $> 2500\,\mathrm{kA/m}$ can be achieved. However, high B_r values are attainable only with lowering the $_JH_c$ value and the operating temperature, and vice versa. Possible combinations of B_r and $_JH_c$ for a given manufacturing process (pressing technique), can be rep-

Table 4.3-39 Elements used for manufacturing Fe–Nd–B magnets

Element	Fe	Nd	B	Dy	Co	Al	Ga	Nb, V
wt%	balance	15–33	0.8–1	0–15	0–15	0.5–2	0–2	0–4

Table 4.3-40 Physical properties of sintered Nd–Fe–B magnets[a]

Density	Curie temperature	Electrical resistivity	Specific heat	Thermal conductivity	Thermal expansion coefficient		Young's modulus	Flexural strength	Compression strength	Vickers Hardness
					‖-c axis	⊥-c axis				
($g\,cm^{-3}$)	(K)	($\Omega\,mm^2\,m^{-1}$)	($J\,kg^{-1}\,K^{-1}$)	($W\,m^{-1}\,K^{-1}$)	($10^{-6}\,K^{-1}$)	($10^{-6}\,K^{-1}$)	($kN\,mm^{-2}$)	($N\,mm^{-2}$)	($N\,mm^{-2}$)	
7.5 (0.05)	580–605	1.50 (0.10)	430 (10)	9	4.4 (0.6)	−1	155 (5)	260 (10)	930 (170)	580 (10)

[a] All values are for 300 K. The values are the averages taken from the companies brochures of: VAC Vacuumschmelze, Hanau, Germany; MS Magnetfabrik Schramberg, Schramberg, Germany; Ugimag Inc, Valparaiso, USA; Neorem Magnets Oy, Ulvila, Finland; TDK Corporation, Tokyo, Japan; Hitachi Metals Ltd, Tokyo, Japan. The numbers in parentheses indicate the maximum deviation

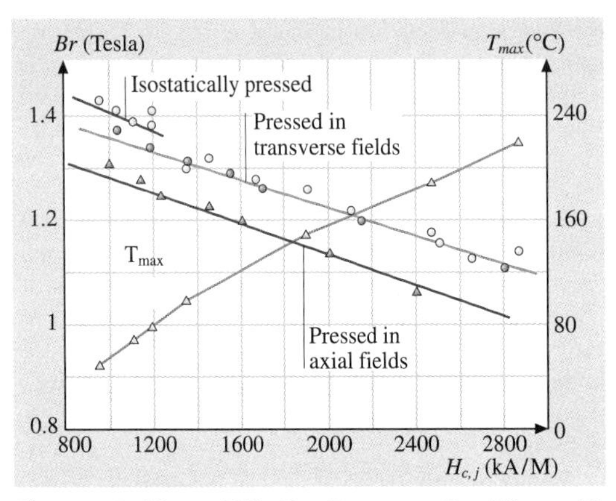

Fig. 4.3-61 Sintered Nd–Fe–B magnets. Possible combinations of B_r and ${}_JH_c$

resented by a straight line, as shown in Fig. 4.3-61. The physical properties of sintered Nd–Fe–B magnets are given in Table 4.3-40.

Magnets Processed by the Rapid Quenching/Hot Working Technique

This alternative technology uses isotropic or amorphous material processed by a rapid quench technique involving melt spinning the molten alloy through a nozzle on to a rotating wheel. Flakes that are about 30 μm thick are obtained. Their microstructure shows typically an average grain size of 50 to 60 nm. The material is subjected to controlled deformation at elevated temperatures, which gives rise to a texture oriented perpendicular to the direction of mass flow during deformation. Hot working processes applied are the die-upset method and indirect extrusion. The major steps of the process are: alloy preparation, melt spinning, cold forming, hot working, coating, and magnetizing. The first step of the hot working procedure is pressing to 100% density at temperatures between 700 and 800 °C. Isotropic magnets are obtained. The isotropic parts are deformed at about 800 °C by using the die-upset technique. Constant strain rates have to be applied. Typically, the strain rate is about 0.1 sec^{-1}. The strain rate and the degree of deformation determine the alignment factor of the final magnet. A maximum value for the magnetic remanence B_r of about 1.35 T may be achieved under economically reasonable conditions. The relationship between the degree of deformation and B_r is shown in Fig. 4.3-62. Properties of commercially availabe magnets are included in Table 4.3-41. The application of the magnets is limited to special fields where complicated shapes would otherwise require expensive machining

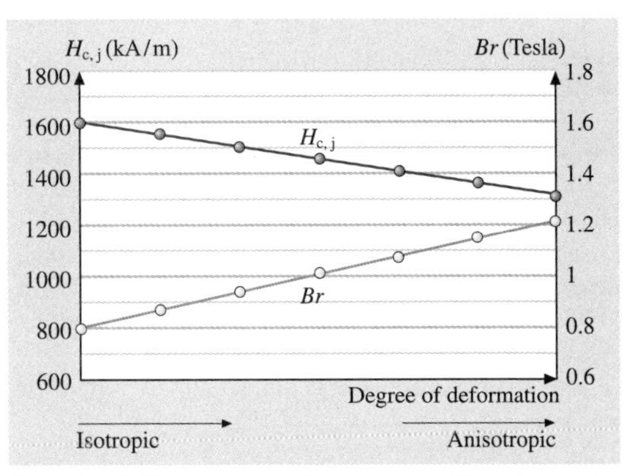

Fig. 4.3-62 Dependence of remanence B_r and coercivity ${}_JH_r$ of dense magnets, manufactured by the hot pressing/hot working technique, on the degree of deformation. Data provided and authorized for publication by Magnequench Int., Tübingen

Table 4.3-41 Magnetic properties of $Fe_{14}Nd_2B$-based magnetic materials at room temperature. Typical values of commercially available magnets

Remanence (T)	Coercivity (kA m^{-1})		Energy density (kJ m^{-3})	Temperature coefficients (% K^{-1})		Max. operation temperature[a] (°C)	Product code[b]
B_r	$_BH_c$	$_JH_c$	$(BH)_{max}$	$TC(B_r)$	$TC(_JH_c)$	T_{max}	
Sinter Route							
1.47	915	955	415	−0.115	−0.77	50	VD722HR
1.44	1115	1195	400	−0.115	−0.73	70	VD745HR
1.35	1040	1430	350	−0.095	−0.65	110	VD633HR
1.30	980	1035	325	−0.115	−0.80	70	VD335HR
1.18	915	2465	270	−0.085	−0.55	190	VD677HR
1.43	915	9550	395	−0.115	−0.77	50	VD722TP
1.41	1090	1195	385	−0.115	−0.73	70	VD745TP
1.32	1020	1430	335	−0.095	−0.65	110	VD633TP
1.25	965	1195	300	−0.115	−0.75	70	VD335TP
1.14	885	2865	250	−0.080	−0.51	220	VD688TP
1.32	965	1115	335	−0.115	−0.73	80	VD510AP
1.26	965	1510	305	−0.095	−0.64	120	VD633AP
1.22	900	1195	285	−0.115	−0.75	80	VD335AP
1.08	830	2865	225	−0.080	−0.51	230	VD688AP
Hot Working							
0.83	575	1400	120	−0.10	−0.50	180	MQ2-E15
1.28	907	995	302	−0.10	−0.60	125	MQ3-E38
1.25	915	1313	287	−0.09	−0.60	150	MQ3-F36
1.31	979	1274	334	−0.09	−0.60	150	MQ3-F42
1.16	876	1592	255	−0.09	−0.06	200	MQ3-G32SH

[a] Maximum operating temperature is defined by a straight demagnetization line up to an operation point of the magnet of $B/\mu_0 H = -2$
[b] VD = Vacodym, trade name of Vacuumschmelze GmbH for NdFeB based magnets; HR grades = isostatically pressed, TP grades = transverse field pressed, AP grades = axial field pressed. MQ2, MQ3 = trade names of Magnequench International Inc. for hot pressed - hot deformed magnets

work, as for instance screwed arcs for non-cogging motors.

Typical technical alloys and their properties are listed in Table 4.3-41. Characteristic demagnetization curves are shown in Fig. 4.3-63.

4.3.3.5 Co–Sm

Numerous magnetic, binary, rare earth (RE) transition metal (TM) compounds exist, of which the Co_5RE and $Co_{17}RE_2$ phases form the basis for materials with excellent permanent magnetic properties. They combine high saturation magnetization M_s with high crystal anisotropy K_1 and high Curie temperature T_c. The intrinsic magnetic properties of the Co_5RE and $Co_{17}RE_2$ phases are summarized in Tables 4.3-42 and 4.3-43, respectively (Co_5RE and $Co_{17}RE_2$ are, also, referred to as 5/1 and 17/2 phases). It is obvious that Co_5Sm and $Co_{17}Sm_2$ have the best potential for manufacturing permanent magnetic materials.

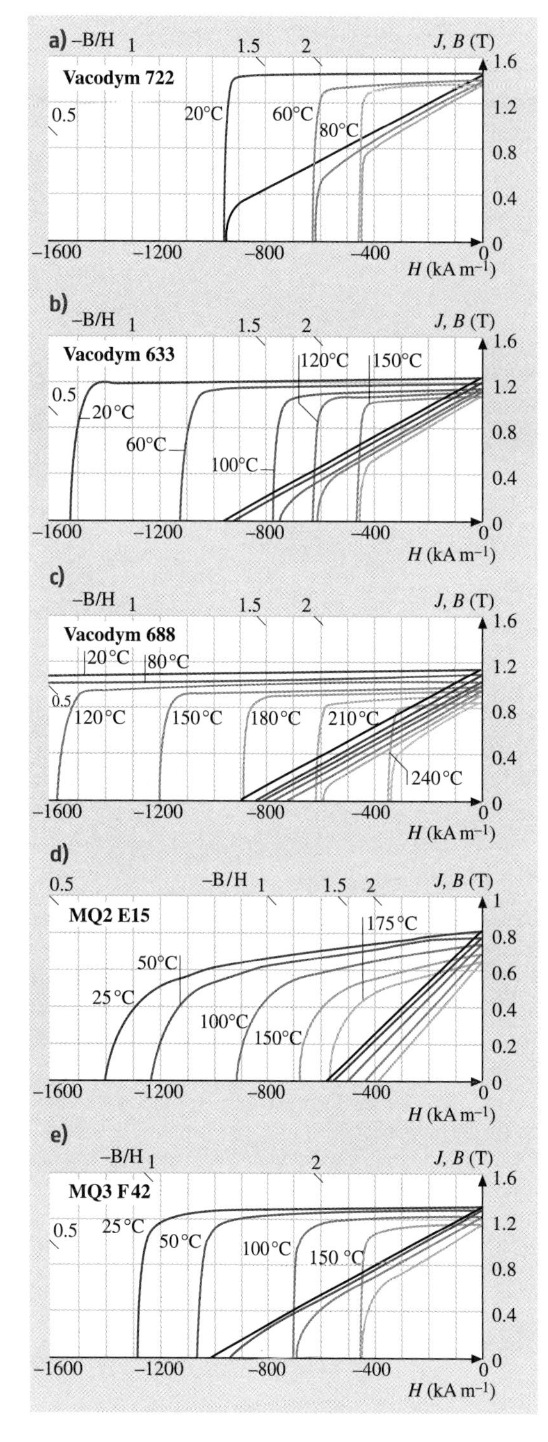

Fig. 4.3-63a–e $Fe_{14}Nd_2B$-based commercial magnets [3.10]. **(a)** Top grade magnet with highest remanence B_r. Isostatically pressed, designed to meet exceptional requirements for maximum energy density at operating temperatures up to 60 °C. **(b)** Magnet, axially pressed, with an optimum combination of high coercivity and energy product **(c)** Magnet, axially pressed, with a very high coercivity; exceptionally well suited for use in highly dynamic servo motor applications. **(d)** Isotropic dense magnet made from rapidly quenched powder by means of the hot pressing technique. **(e)** Magnet made from rapidly quenched powder by means of the hot pressing and hot deforming technique; especially well suited for complicated shaped magnets, such as scewed arcs in special motor applications

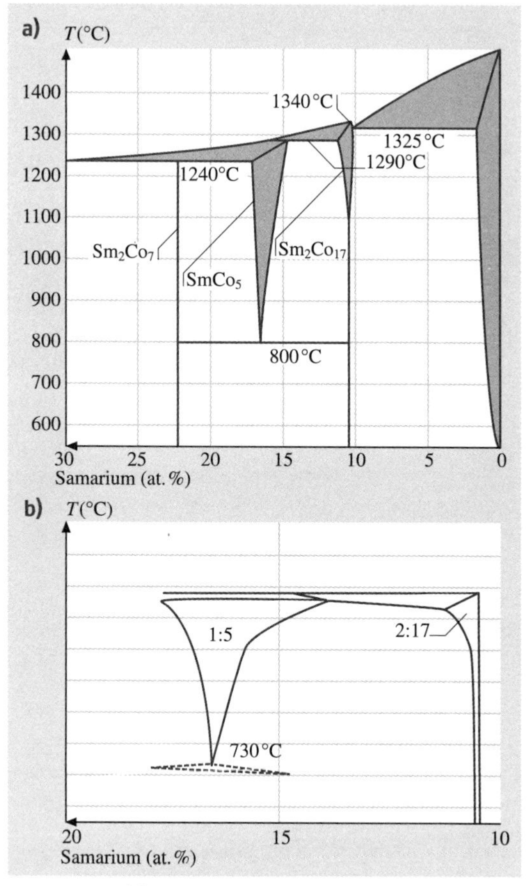

Fig. 4.3-64 **(a)** Part of the Co–Sm phase diagram. **(b)** Section of the Co–Sm–Cu phase diagram at 10 at.% Cu [3.10]

Table 4.3-42 Co_5RE alloys. Room temperature intrinsic magnetic properties, the crystal anisotropy constant K_1, the magnetic saturation M_s, the anisotropy field $H_a = 2K_1/M_s$, and the Curie temperature, T_c [3.10]

RE[a]	K_1^b (10^7 erg cm^{-3})	$4\pi M_s^b$ (T)	H_a (MA m^{-1})	T_c^b (K)
Ce	5.3 (6)	0.83	13	660 (20)
Pr	8.1 (9)	1.29	12.5	900 (10)
Nd	0.7 (5)	1.34	1.1	910 (2)
Sm	17.2 (9)	1.12 (3)	31	990 (10)
Gd	4.6 (5)	0.35 (3)	26	1035
Tb	NC			973
Dy	NC			961
Ho	3.6 (3)	0.52	13.8	996
Er	4.2 (4)	0.4	14.6	983
La	5.9 (5)	0.91	10.6	834
Y	5.2 (4)	1.11	9.3	980

[a] No data are available for the RE elements Tm, Yb, and Lu

[b] The mean value is given where more than one reference is available; the number in parentheses indicates the standard deviation in the last figure. NC: non-collinear spin structure

Table 4.3-43 Room temperature intrinsic magnetic properties, of $Co_{17}RE_2$ alloys [3.10]

RE[a]	K_1^b (10^7 erg cm^{-3})	$4\pi M_s^b$ (T)	H_a (MA m^{-1})	T_c^b (K)
Ce	−0.6	1.15		1068 (10)
Pr	−0.6	1.38		1160 (10)
Nd	−1.1	1.39		1160 (3)
Sm	3.3 (1)	1.25 (3)	5.3 (1)	1196 (1)
Gd	−0.5	0.73		1212 (6)
Tb	−3.3 (6)	0.67		1189 (4)
Dy	−2.6 (5)	0.69		1173 (8)
Ho	−1.0 (2)	0.85		1177 (1)
Er	0.41 (3)	1.05 (5)	0.75	1175 (15)
TM	0.50 (5)	1.21	0.83	1179 (2)
Yb	−0.38	1.35		1180
Lu	−0.20 (5)	1.41		1202 (6)
Y	−0.34 (2)	1.27		1199 (13)

[a] La does not form the 17/2 phase. The 17/2 phases with Sm, Er, and Tm show uniaxial anisotropy while all others have easy plane anisotropy

[b] The mean value is given where more than one reference is available; the number in parentheses indicates the standard deviation in the last figure

Phase Equilibria of Co–RE Systems

The phase diagrams of Co–RE systems are very similar. Figure 4.3-64 shows the magnetically relevant part of the Co–Sm phase diagram. The features to note are: Co and $Co_{17}Sm_2$ form a eutectic, while Co_5Sm forms as a result of a peritectic reaction between $Co_{17}Sm_2$ and liquid. Both Co_5Sm and $Co_{17}Sm_2$ show a significant range of homogeneity at elevated temperatures. The compound Co_5Sm is unstable at room temperature and decomposes via an eutectoid reaction into Co_7Sm_2 and $Co_{17}Sm_2$.

Iron and Cu are two important substitutional elements for Co–RE alloys with respect to manufacturing permanent magnets. In Co_5RE compounds, Fe may substitute up to 5 at.% Co while complete solubility occurs in $(Co_{1-x}Fe_x)_{17}RE_2$. Copper is essential in $Co_{17}Sm_2$ type alloys. The different solubility of Cu in 5/1 and 17/2 is used to form precipitates in the 17/2 phase, resulting in a microstructure which provides high coercivity.

Co–Sm-Based Permanent Magnets

Powder Metallurgical Processing. Cobalt–Samarium magnets are produced using powder metallurgical techniques. Alloys are prepared by an inductive melting process or a Ca reduction process. The starting alloys are crushed and pulverized to single crystalline particles 3–4 μm in diameter. The powders are compacted in a magnetic field to obtain anisotropic magnets: by uniaxial compaction in magnetic fields, either parallel or transverse to the direction of the applied force; or by isostatic compaction of powders in elastic bags after subjecting the filled bags to a pulsed field. The magnetically aligned green compacts are sintered in an inert atmosphere to achieve an optimum combination of high density and high coercive field.

5/1 Type Magnets. Binary Co_5Sm is the basis of 5/1 type magnets. Table 4.3-44 lists some of the magnetic properties at room temperature and Table 4.3-45 lists the physical properties of sintered Co_5Sm magnets. Partial substitution of Sm by Pr increases B_s while still yielding sufficiently high ${}_JH_c$. The microstructure of Co_5Sm magnets consists of single domain grains. Magnetization reversal starts by nucleation of domains in a demagnetizing field. The domain wall moves easily through the particle.

The application of permanent magnets in measuring devices or in devices in aircraft or space systems requires a small temperature coefficient (TC) of B_r. The combination of Co_5Sm which has a negative TC, with Co_5Gd, which has a positive TC, yields magnets with reduced temperature dependence of B_r, reaching about

Table 4.3-44 Co_5Sm-based magnetic materials. Magnetic properties at room temperature, typical values [3.10]

B_r (T)	${}_BH_c$ ($kA\,m^{-1}$)	${}_JH_c$ ($kA\,m^{-1}$)	$(BH)_{max}$ ($kJ\,m^{-3}$)	Press mode [a]	Material	Producer code [b]
1.01	755	1500	200	Iso	Co_5Sm	Vacomax 200
0.95	720	1800	180	TR	Co_5Sm	Vacomax 170
1.0	775	2400	200	Iso	Co_5Sm	Recoma 25
0.94	730	2400	175	TR	Co_5Sm	Recoma 22
0.9	700	2400	160	A	Co_5Sm	Recoma 20
0.73	570	>2400	105	A	$Co_5Sm_{0.8}Gd_{0.2}$	EEC 1.5TC-13
0.61	480	>2400	70	A	$Co_5Sm_{06}Gd_{0.4}$	EEC 1.5TC-9

[a] Iso: isostatically pressed; TR: uniaxially pressed in transverse oriented aligning fields; A: uniaxially pressed in axially oriented aligning fields
[b] Vacomax: Trademark of Vacuumschmelze GmbH, Germany; Recoma: Trademark of Ugimag AG, Switzerland; EEC: Trademark of EEC Electron Energy Corporation, USA

Table 4.3-45 Physical properties of sintered Co_5Sm magnets [a] [3.10]

Density ($g\,cm^{-3}$)	Curie temperature (K)	Electrical resistivity ($\Omega\,mm^2\,m^{-1}$)	Specific heat ($J\,kg^{-1}\,K^{-1}$)	Thermal conductivity ($W\,m^{-1}\,K^{-1}$)	Thermal expansion coefficient $\parallel$ c axis ($10^{-6}\,K^{-1}$)	Thermal expansion coefficient $\perp$ c axis ($10^{-6}\,K^{-1}$)	Young's modulus ($kN\,mm^{-2}$)	Flexural strength ($N\,mm^{-2}$)	Compression strength ($N\,mm^{-2}$)	Vickers Hardness
8.40 (0.10)	990 (10)	0.53 (0.03)	372 (3)	11.5 (1.0)	6.0 (1.5)	12.5 (0.5)	150 (40)	125 (35)	900 (300)	580 (50)

[a] All values for 300 K. Average values are taken from the companies brochures of: VAC Vacuumschmelze, Hanau, Germany; MS Magnetfabrik Schramberg, Schramberg, Germany; Ugimag AG, Lupfig, Switzerland; EEC Electron Energy Corporation, Landisville, USA; TDK Corporation, Tokyo Japan; Hitachi Metals Ltd, Tokyo, Japan. The numbers in parentheses indicate the standard deviation

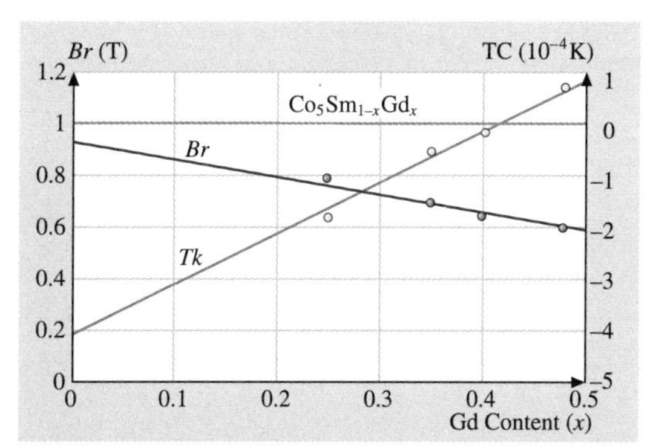

Fig. 4.3-65 Temperature coefficient TC of $Co_5Sm_{1-x}Gd_x$ magnets in the temperature range between 20 and 200 °C. The coercivity of these ternary magnets is comparable to that of binary Co_5Sm magnets due to the high anisotropy field of Co_5Gd $H_a = 24\,MA\,m^{-1}$ while for Co_5Sm, $H_a = 31\,MA\,m^{-1}$ [3.10]

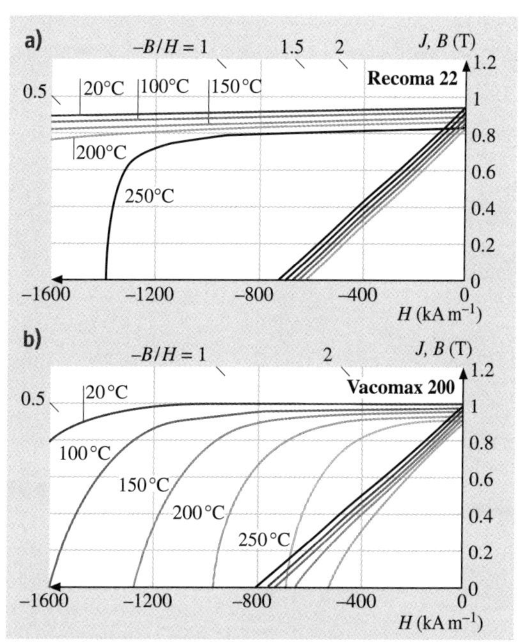

Fig. 4.3-66a,b Co_5Sm magnets. Demagnetization curves of typical commercially available magnets [3.10]. **(a)** Magnet with high intrinsic coercivity $_JH_c$ up to 300 °C, uniaxially pressed in a transverse aligning field; Recoma: trademark of Ugimag AG, Switzerland. **(b)** Magnet with highest energy density obtained by cutting from isostatically pressed block; Vacomax: trademark of Vacuumschmelze GmbH, Germany

zero for $Co_5Sm_{0.6}Gd_{0.4}$ between room temperature and 200 °C (Fig. 4.3-65).

Tables 4.3-44 and 4.3-45 list the magnetic properties of typical Co_5Sm based magnets and their physical properties, respectively. Figures 4.3-66a,b show characteristic demagnetisation curves.

17/2 Type Magnets. The permanent magnetic potential of the binary phase $Co_{17}Sm_2$ is increased by partially substituting Co by other transition metals. The general chemical composition of commercial magnets corresponds to $(Co_{bal}Fe_vCu_yZr_x)_zSm$. Copper is the essential addition. Its solubility in the 17/2 phase is strongly temperature-dependent, and this is used for precipitation hardening of 17/2 magnets.

The influence of Fe on the intrinsic magnetic properties is summarized in Fig. 4.3-67. Both Zr and Hf increase coercivity. Figures 4.3-67a,b show the combined effects of Fe, Cu, and Zr additions and the Co/Sm ratio on the temperature dependence of coercivity $_JH_c$. The magnets are sintered between 1200 and 1220 °C. A single phase with $Zn_{17}Th_2$ structure is obtained by homogenization at temperatures between 1160 and 1190 °C. After rapid cooling the magnets are finally annealed between 800 and 850 °C, followed by cooling to 400 °C. The microstructure leading to high coercivity $_JH_c$ consists of 17/2 matrix grains, a 5/1 boundary phase enriched in Cu, and platelet-shaped precipitates enriched in Fe and Zr. The coercivity of these magnets

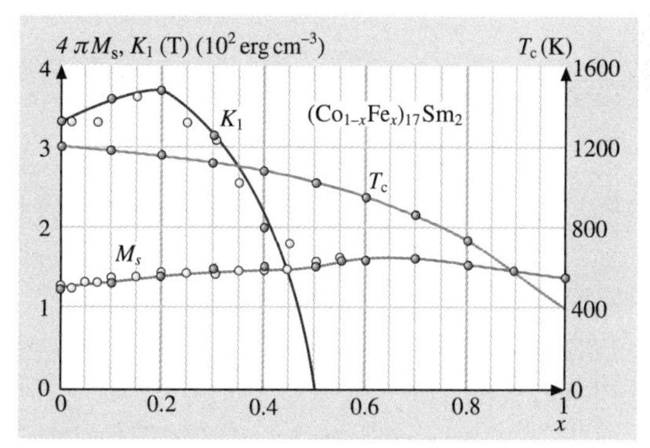

Fig. 4.3-67 $(Co_{1-x}Fe_x)_{17}Sm_2$. Dependence K_1, $4\pi M_s$, and T_c on the Fe content x [3.10]

is based on pinning of the Bloch walls at the 5/1 grain boundary phase.

Table 4.3-46 Chemical composition of commercial high energy 17/2 magnets [3.10]

Element	Sm [a]	Co	Cu	Fe	Zr
wt%	25–27	balance	4.5–8	14–20	1.5–3

[a] The Sm content includes the fraction of Sm which is present as Sm_2O_3, typically 2.5 wt%

Table 4.3-47 Annealing treatments for 17/2 type magnets [3.10]

	Isothermal ageing	Cooling rate to 400 °C	$_JH_c$ (kA m^{-1})	Magnetizing field H_m (kA m^{-1})
high $_JH_c$	850 °C/10 hours	1 K/min	> 2000	5000
low $_JH_c$	800 °C/30 min.	5 K/min	500–800	1500

Table 4.3-48 High energy $TM_{17}Sm_2$ magnets for applications up to 300 °C, typical values [3.10]

B_r (T)	$_BH_c$ (kA m^{-1})	$_JH_c$ (kA m^{-1})	$(BH)_{max}$ (kJ m^{-3})	Press mode [a]	Material	Producer code [b]
1.14	670	800	247	Iso	$TM_{17}Sm_2$	Hicorex -30CH
1.12	730	800	240	Iso	$TM_{17}Sm_2$	Vacomax 240HR
1.05	720	800	210	A	$TM_{17}Sm_2$	Vacomax 240
1.10	820	2070	225	Iso	$TM_{17}Sm_2$	Vacomax 225HR
1.07	800	2000	215	TR	$TM_{17}Sm_2$	Recoma 28
1.04	760	2070	205	A	$TM_{17}Sm_2$	Vacomax 225
0.9	654	> 2000	150	A	$TM_{17}(Sm_{0.8}, Gd_{0.2})_2$	EEC2:17TC-18
0.8	575	> 2000	115	A	$TM_{17}(Sm_{0.6}, Gd_{0.4})_2$	EEC2:17TC-15

[a] Iso: isostatically pressed; TR: uniaxially pressed in transverse oriented aligning fields; A: uniaxially pressed in parallel oriented aligning fields

[b] Hicorex: Trademark of Hitachi Metals Ltd., Japan; Vacomax: Trademark of Vacuumschmelze GmbH, Germany; Recoma: Trademark of Ugimag AG, Switzerland; EEC: Trademark of EEC Electron Energy Corporation, USA

Table 4.3-49 Composition of 17/2 magnets for operating up to 550 °C [3.10]

Element	Sm	Co	Cu	Fe	Zr
wt%	26.5–29	balance	6–12	7	2.5–6

Table 4.3-50 Physical properties of sintered $TM_{17}Sm_2$ magnets [a] [3.10]

Density	Curie temperature	Electrical resistivity	Specific heat	Thermal conductivity	Thermal expansion coefficient		Young's modulus	Flexural strength	Compression strength	Vickers hardness
					‖-c axis	⊥-c axis				
(g cm^{-3})	(K)	(Ω mm^2 m^{-1})	(J kg^{-1} K^{-1})	(W m^{-1} K^{-1})	(10^{-6} K^{-1})	(10^{-6} K^{-1})	(kN mm^{-2})	(kN mm^{-2})	(kN mm^{-2})	
8.38 (0.08)	1079 (8)	0.86 (0.05)	355 (30)	10.5 (1.5)	8.4 (1)	11.5 (0.5)	160 (40)	130 (15)	760 (150)	600 (50)

[a] All values are for 300 K. The values are the average of values taken from company brochures

High Energy Magnets for Applications at Temperatures up to 300 °C. The compositional parameters v, y, x, and z of the 17/2 type magnets are normally optimized for highest $(BH)_{max}$ between room temperature and 300 °C. The range of chemical compositions of these high energy magnets is given in Table 4.3-46.

Tables 4.3-48 and 4.3-50 list the magnetic properties of typical $Co_{17}Sm_2$ based magnets and their physical properties, respectively. Figures 4.3-69a–c show characteristic demagnetization curves.

Two modifications of 17/2 magnets can be obtained from identical chemical compositions according to the annealing treatment applied (see Table 4.3-47). 17/2 magnets with reduced TC of B_r are obtained by substituting part of Sm by Gd. Table 4.3-45 shows some typical values of the magnetic properties of high energy $TM_{17}Sm_2$ magnets.

Magnets for High Temperature Applications. Special applications in aircraft and spacecraft require magnets with linear demagnetization curves up to 550 °C. Suitable high temperature magnets are obtained by increasing Sm and Cu and decreasing Fe. Data are given in Table 4.3-49 and Fig. 4.3-69a.

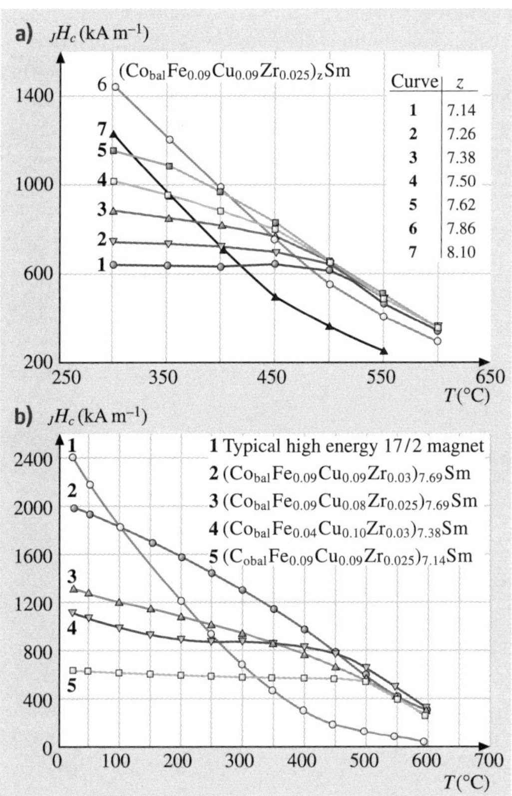

Fig. 4.3-68a,b Temperature dependence of $_JH_c$ for $(Co_{bal}Fe_vCu_yZr_x)_z$ as varied by the concentration of the substitutional elements and of Sm [3.10] **(a)** Dependence on the TM/Sm ratio; **(b)** Dependence on other variations in chemical composition

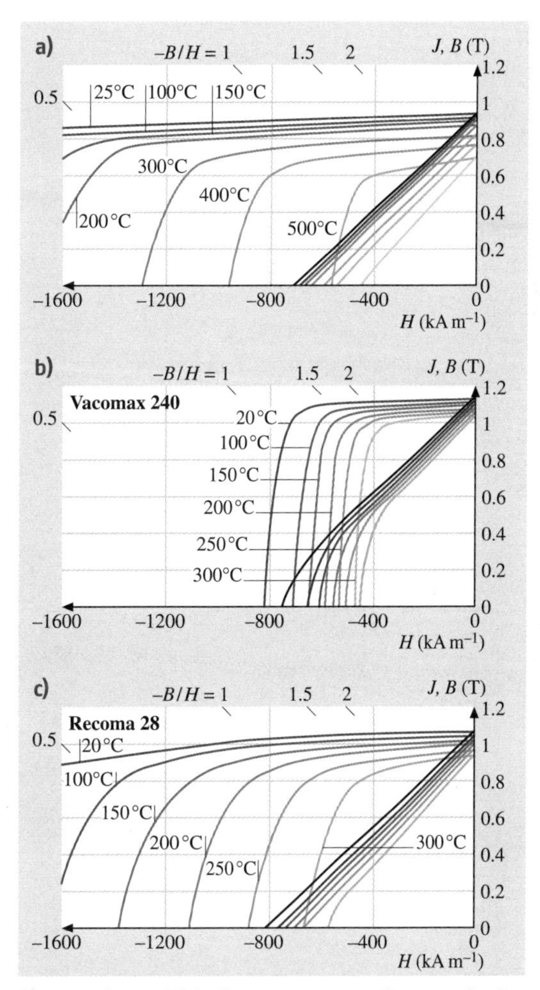

Fig. 4.3-69a–c $TM_{17}Sm_2$ magnets. Demagnetization curves [3.10] **(a)** Magnet may be used up to 500 °C. Coating is needed for protection against oxidation > 300 °C. **(b)** Low coercivity, easy-to-magnetize magnet. **(c)** High energy, high coercivity magnet

4.3.3.6 Mn–Al–C

In the binary Mn–Al system the ferromagnetic, metastable τ-Mn–Al phase containing 55 wt% Mn is obained by annealing at 923 K, or by controlled cooling of the stable ε phase from ≥ 1073 K. The τ phase has a high magnetocrystalline anisotropy $K_1 \cong 10^6\,\mathrm{J\,m^{-3}}$ and $I_s = 0.6$ T. The magnetic easy axis is the c axis. The τ phase can be stabilized by adding 0.5 wt% C. Hot deformation at about 973 K permits formation of anisotropic magnets. Figure 4.3-70 shows schematically the 873 K isotherm of the Mn–Al–C phase diagram.

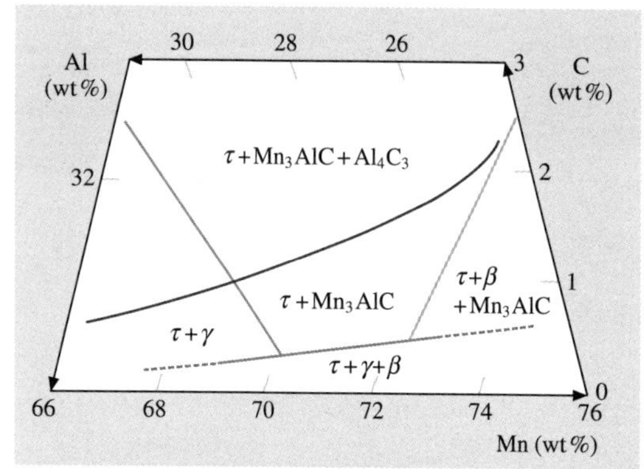

Fig. 4.3-70 Schematic representation of the phase relations of Mn–Al–C at 873 K [3.10]

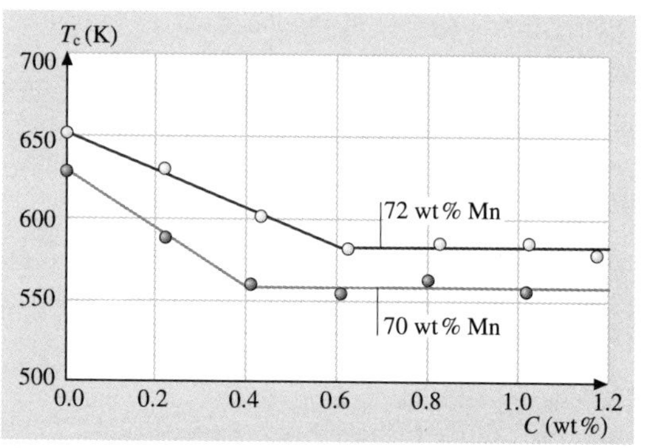

Fig. 4.3-71 Dependence of Curie temperature of τ-Mn–Al–C on the carbon concentration [3.10]

Carbides coexist with the metallic ferromagnetic phases. This is unfavorable for the mechanical properties. The Curie temperature of the τ phase decreases with increasing C content to a constant value (see Fig. 4.3-71) according to the phase diagram.

An alternative process to form τ-Mn–Al–C magnets is to use gas-atomized powder which is canned

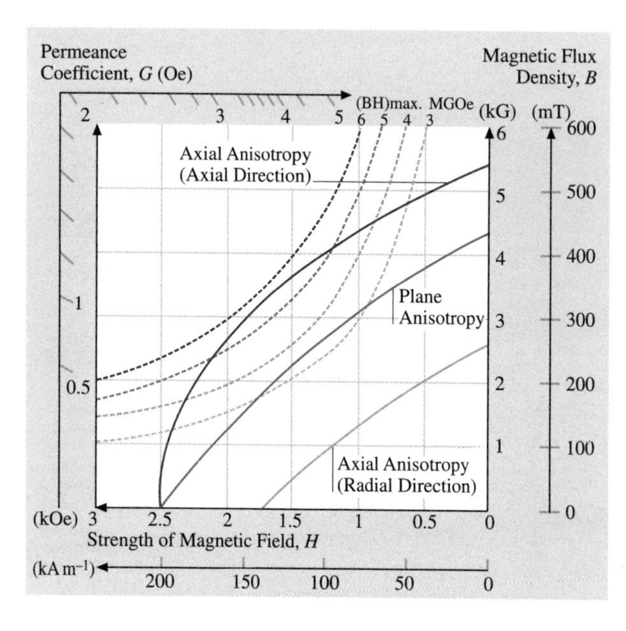

Fig. 4.3-72 Magnetic properties of an extruded Mn–Al–C magnet [3.10]

Table 4.3-51 Properties of a Mn–Al–C magnet [3.10]

		Axial Anisotropy		Plane anisotropy
		Axial direction	Radial direction	
Maximum energy product $(BH)_{max}$	(kJ m^{-3})	44	10	28
Residual magnetic flux density B_r	(mT)	550	270	440
Coercivity H_c	(kA m^{-1})	200	144	200
Optimum permeance coefficient	(μH m^{-1})	1.9	1.9	1.9
Average reversal permeability	(μH m^{-1})	1.4 ~ 1.6	1.4 ~ 1.6	1.4 ~ 1.6
Temperature coefficient of B_r	(% K^{-1})	−0.11	−0.11	−0.11
Curie point T_c	(K)	573	573	573
Maximum operating temperature	(K)	773	773	773
Density	(kg m^{-3})	5100	5100	5100
Hardness	(HR$_c$)	49 ~ 56	49 ~ 56	49 ~ 56
Tensile strength R_m	(N m^{-2})	$> 290 \times 10^6$	$> 290 \times 10^6$	$> 290 \times 10^6$
Compression strength	(N m^{-2})	$> 2000 \times 10^6$	$> 2000 \times 10^6$	$> 2000 \times 10^6$

and extruded at 973 K. By atomizing, the powder is rapidly quenched and consists of the high temperature ε phase. During hot extrusion it transforms to the axially anisotropic τ phase. Repeated die pressing results in plane anisotropy. The magnetic properties are shown in Fig. 4.3-72 and listed in Table 4.3-51.

4.3.4 Magnetic Oxides

The magnetic properties of oxides and related compounds have been tabulated in comprehensive data collections [3.4]. A review of the basic magnetic properties of garnets $\{A_3\}[B_2](Si_3)O_{12}$ and spinel ferrites $MeOFe_2O_3$ or $Me^{II}Fe_2^{III}O_4$ is given by *Guillot* in [3.3].

4.3.4.1 Soft Magnetic Ferrites

Both MnZn and NiZn ferrites are the common designations of the two main groups of soft magnetic oxide materials. A more extensive account is given in [3.10]. The chemical formula of MnZn and NiZn ferrites is $M^{2+}Fe_2O_4$ and they have spinel structures. The divalent ions (M^{2+}) are elements such as Mn, Fe, Co, Ni, Cu, Mg, Zn, and Cd. They are located at tetrahedral or octahedral sites of the spinel structures. Ceramic processing methods are applied to produce the magnetic parts such as ring-shaped cores for inductive components.

The use of different types of MnZn and NiZn ferrite materials varies with the operating frequency in the application concerned. Figure 4.3-73 and Table 4.3-52 indicate the typical ranges of use and the magnetic properties of characteristic materials. Typically, MnZn ferrites are used in the range of several MHz. Table 4.3-53 lists the major applications with the pertinent operating frequencies. Since a phase shift occurs in an ac-excited magnetic field, the permeability μ is expressed as a complex number, $\mu' - i\mu''$. The imaginary

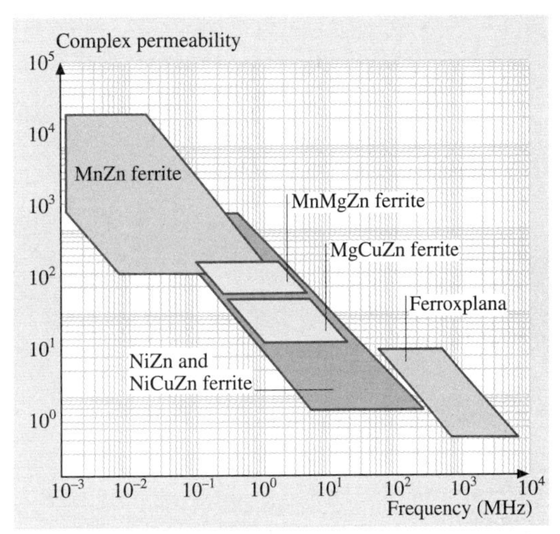

Fig. 4.3-73 Relationships between frequency and complex permeability of groups of characteristic soft magnetic ferrites [3.10]

Table 4.3-52 Typical magnetic properties of characteristic materials [3.10]

Type	MnZn ferrite					NiCuZn ferrite		
Designation	PC40	PC44	PC45	PC50	H5C4	L7H	HF70	L6
μ_i	2300	2400	2500	1400	12 000	800	1300	1500
B_s (mT) at 25 °C	510	510	530	470	380	390	270	280
H_c (A m^{-1}) at 25 °C	14	13	12	37	4	16	16	16
T_c (°C)	>215	>215	>230	>240	>110	>180	>110	>110
ρ_v (m)	6.5					10^6	10^6	10^6
Pcv (kW m^{-3})	410	300	250	80[a]				
At 100 kHz, 200 mT	(100 °C)	(100 °C)	(75 °C)	(100 °C)				

[a] At 500 kHz, 50 mT

Table 4.3-53 Ferrite applications [3.10]

Application	Frequency	Ferrite material
Communication coils	1 kHz ~ 1 MHz	MnZn
	0.5 ~ 80 MHz	NiZn
Pulse transformers		MnZn, NiZn
Transformers	~ 300 kHz	MnZn
Flyback transformers	15.75 kHz	MnZn
Deflection yoke cores	15.75 kHz	MnZn, MnMgZn, NiZn
Antennas	0.4 ~ 50 MHz	NiZn
Intermediate frequency transformers	0.4 ~ 200 MHz	NiZn
Magnetic heads	1 kHz ~ 10 MHz	MnZn
Isolators		MnMgAl
Circulators	30 MHz ~ 30 GHz	YIG
Splitters		YIG
Temperature responsive switches		MnCuZn

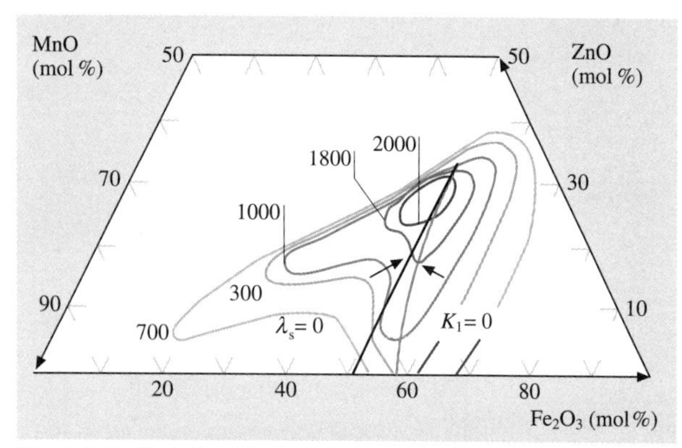

Fig. 4.3-74 Relationship between composition and complex permeability in MnZn ferrites [3.10]

part, μ'', is related to magnetic losses and is high at the resonance frequency. For a specific composition range, the values of magnetic anisotropy and magnetostriction are reduced to near zero leading to a maximum in complex permeability, as shown in Fig. 4.3-74. By suitable control of the microstructure it is possible to favor the formation and mobility of fast-moving domain walls. Consequently, the permeability of MnZn ferrite is the highest among the ferrites with spinel structure. However, the increase of the loss factor is considerably enhanced with increasing frequency because of the lower resistivity of MnZn ferrites compared to other ferrite materials.

The properties of NiZn and NiCuZn ferrites are designed for applications in the radio wave band. Figure 4.3-75 shows the effect of the Zn content on

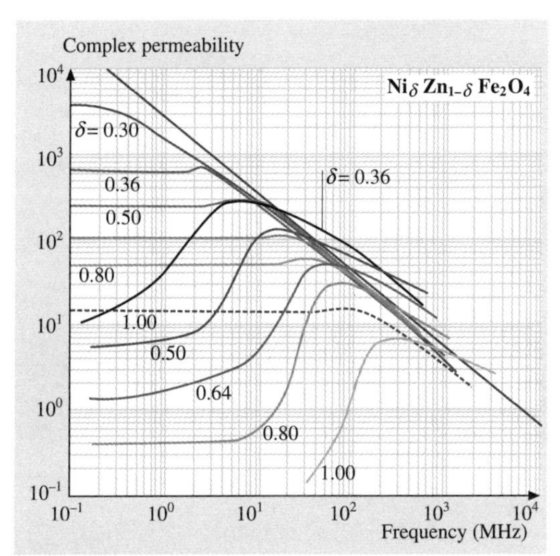

Fig. 4.3-75 Frequency dependence of the complex permeability in NiZn ferrites. *Solid* and *dashed lines* refer to the real and the imaginary part, respectively [3.10]

the complex permeability spectra of NiZn ferrites. In general, the high-permeability materials cause lower resonance frequencies.

4.3.4.2 Hard Magnetic Ferrites

A hard magnetic ferrite, also called ferrite magnet, is a magnetic material based on iron oxide. The composition of the typical hard magnetic ferrite compounds is shown in Table 4.3-54. M type material is used most widely and BaO can be replaced by SrO.

Figure 4.3-76 shows the quasi-binary phase diagram $BaO–Fe_2O_3$. The composition of the actual industrial hard magnetic ferrite is selected to deviate slightly from the stoichiometric composition in order to provide easy wetting and permit liquid-phase sintering. The ferromagnetic phase $SrO·6Fe_2O_3$ has a higher magnetocrystalline anisotropy constant K_1 and, thus, a higher intrinsic coercive force $_JH_c$ than the BaO-based compound.

Table 4.3-55 shows the basic magnetic properties of hard magnetic ferrites. The properties of actual products are shown in Fig. 4.3-77. The demagnetization curves of type YBM-9B, which have the best magnetic properties, are shown in Fig. 4.3-78.

Comparing hard ferrites and rare earth magnets, the ratio of remanence B_r is about 1 : 3, that of the coercivity $_JH_c$ is also about 1 : 3, such that the ratio of the energy product $(BH)_{max}$ is about 1 : 10. From the cost/performance point of view, the rare earth magnets are used where weight and size are essential.

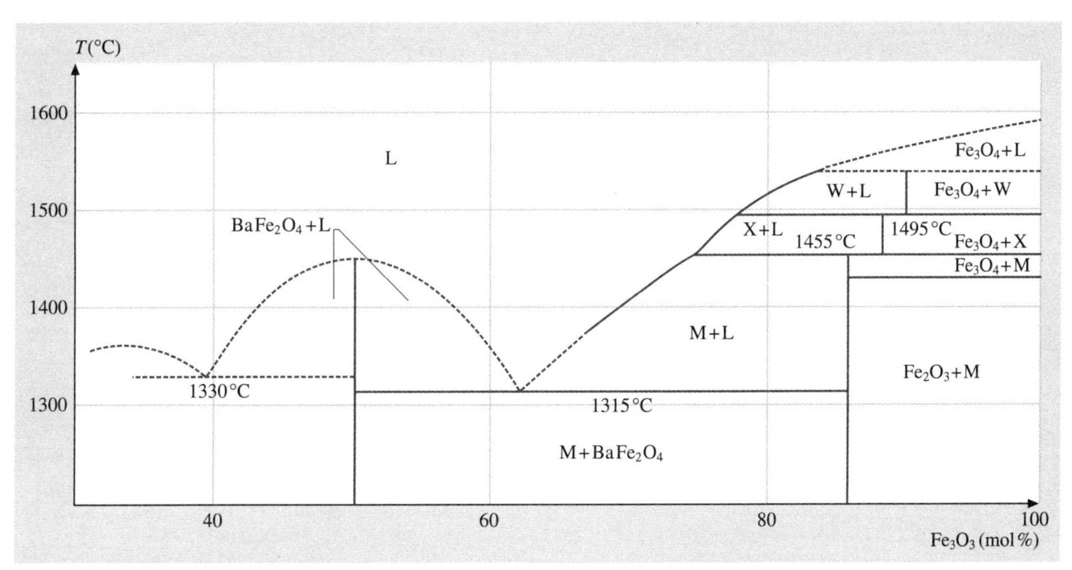

Fig. 4.3-76 Quasi-binary phase diagram $BaO–Fe_2O_3$. $p(O_2) = 1$ atm; X = $BaO·FeO·7Fe_2O_3$; W = $BaO·2FeO·8Fe_2O_3$. *van Hook* 1964 [3.10, 59]

Table 4.3-54 $BaO-MeO-Fe_2O_3$ hexagonal magnetic compounds and their technical designations as hard ferrite materials 3.10

Designation symbol	Molecular formula	Chemical composition (mol%)		
		MeO	BaO	Fe_2O_3
M	$BaO \cdot 6Fe_2O_3$	–	14.29	85.71
W	$2MeO \cdot BaO \cdot 8Fe_2O_3$	18.18	9.09	72.71
Y	$2MeO \cdot 2BaO \cdot 6Fe_2O_3$	20	20	60
Z	$2MeO \cdot 3BaO \cdot 12Fe_2O_3$	11.76	17.65	70.59

Table 4.3-55 Magnetic properties of hard ferrites 3.10

Composition	σ_s (10^{-4} Wb m kg^{-1})	(emu g^{-1})	σ_0 (10^{-4} Wb m kg^{-1})	(emu g^{-1})	J_s (wb m^{-3})	T_c (K)	K_1 (10^3 J m^{-2})	H_a (kA m^{-1})	D_c (μm)
$BaFe_{17}O_{19}$	0.89	71	1.257	100	47.8	723	3.2	1.350	0.90
$SrFe_{17}O_{19}$	0.905	72	1.357	108	47.8	735	3.5	1.590	0.94

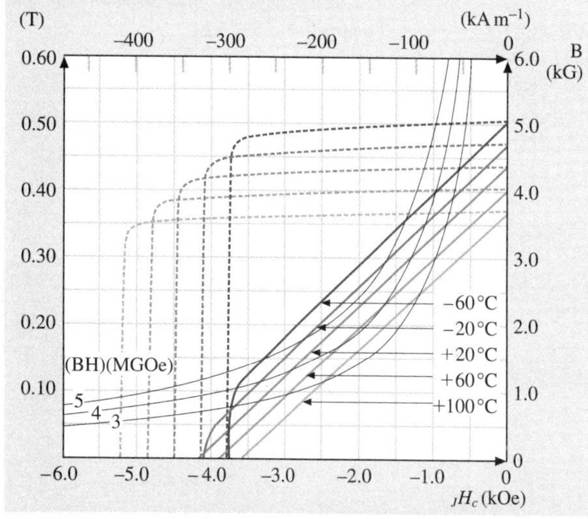

Fig. 4.3-78 Demagnetization behavior and its temperature dependence of a hard ferrite YBM9B (Hitachi Metals) 3.10

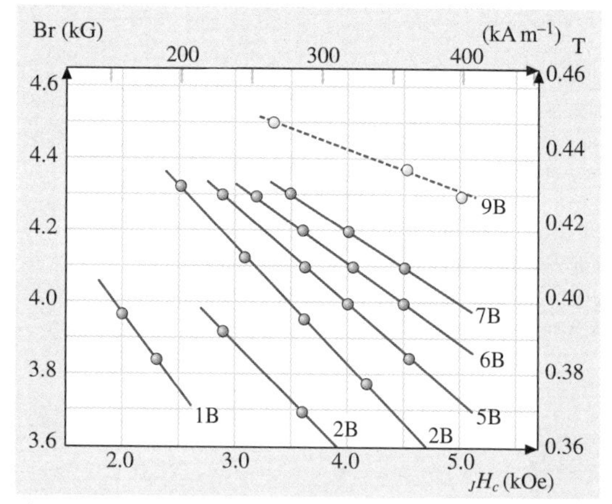

Fig. 4.3-77 Magnetic properties of ferrite magnets (Hitachi Metals series YBM 1B-9B) 3.10

References

3.1 K. H. J. Buschow: Electronic and Magnetic Properties of Metals and Ceramics, Part I. In: *Materials Science and Technology*, Vol. 3, ed. by R. W. Cahn, P. Haasen, E. J. Kramer (VCH, Weinheim 1991)

3.2 H. P. J. Wijn (Ed.): *Magnetic Properties of Metals*, Landolt–Börnstein, New Series III/19, 32 (Springer, Berlin, Heidelberg 1986–2001)

3.3 K. H. J. Buschow: Electronic and Magnetic Properties of Metals and Ceramics, Part II. In: *Materials Science and Technology*, Vol. 3, ed. by R. W. Cahn, P. Haasen, E. J. Kramer (VCH, Weinheim 1994)

3.4 K.-H. Hellwege, A. M. Hellwege (Eds.): *Magnetic and Other Properties of Oxides and Related Compounds*, Landolt–Börnstein, New Series III/4, 12 (Springer, Berlin, Heidelberg 1970–1982)

3.5 H. P. J. Wijn (Ed.): *Magnetic Properties of Non-Metallic Inorganic Compounds Based on Transition Metals*, Landolt–Börnstein, New Series III/27 (Springer, Berlin, Heidelberg 1988–2001)

3.6 K. Adachi, D. Bonnenberg, J. J. M. Franse, R. Gersdorf, K. A. Hempel, K. Kanematsu, S. Misawa, M. Shiga, M. B. Stearns, H. P. J. Wijn: *Magnetic*

Properties of Metals: 3d, 4d, 5d Elements, Alloys, Compounds, Landolt–Börnstein, New Series III/19 (Springer, Berlin, Heidelberg 1986)
3.7 G. Bertotti, A. R. Ferchmin, E. Fiorello, K. Fukamichi, S. Kobe, S. Roth: *Magnetic Alloys for Technical Applications. Soft Magnetic Alloys, Invar, Elinvar Alloys*, Landolt–Börnstein, New Series III/19, ed. by H. P. J. Wijn (Springer, Berlin, Heidelberg 1994)
3.8 R. Boll: *Soft Magnetic Materials* (Heyden & Son, London 1979)
3.9 J. Evetts: Concise Encyclopedia of Magnetic & Superconducting Materials. In: *Advances in Materials Science, Engineering* (Pergamon Press, Oxford 1992)
3.10 P. Beiss, R. Ruthardt, H. Warlimont (Eds.): *Advanced Materials, Technologies: Materials: Powder Metallurgy Data: Metals, Magnets*, Landolt–Börnstein, New Series VIII/2 (Springer, Berlin, Heidelberg 2003)
3.11 K. Günther: SMM Conference 2003
3.12 R. Boll: Soft Magnetic Metals and Alloys. In: *Materials Science and Technology*, Vol. 3B, ed. by K. H. J. Buschow (VCH, Weinheim 1994) pp. 399–450
3.13 Bölling and Hastenrath
3.14 G. Rassmann, U. Hofmann: Magnetismus, Struktur und Eigenschaften magnetischer Festkörper. In: (VEB Deutscher Verlag für Grundstoffindustrie, Leipzig 1967) pp. 176–198
3.15 G. Rassmann, U. Hofmann: J. Appl. Phys. **39**, 603 (1968)
3.16 R. Boll: *Weichmagnetische Werkstoffe*, 4 edn. (Vacuumschmelze, 1990)
3.17 H. Saito (Ed.): *Physics and Application of Invar Alloys* (Maruzen Company, Ltd., Tokyo 1978)
3.18 O'Handley 1977
3.19 Hilzinger 1980
3.20 Miyazaki 1972
3.21 Fujimori 1976
3.22 Allied Signal: Technical Bulletin 9 (1998)
3.23 H. Herzer: Nanocrystalline Soft Magnetic Alloys. In: *Handbook of Magnetic Materials*, ed. by K. H. J. Buschow (Elsevier, Amsterdam 1997) pp. 416–462
3.24 Hono and Sakurai 1995
3.25 M. Müller, N. Mattern, L. Illgen, H. R. Hilzinger, G. Herzer: Key Eng. Mater. **81–83**, 221–228 (1991)
3.26 M. Müller, N. Mattern, L. Illgen: Z. Metallkunde **82**, 895–901 (1991)
3.27 G. Herzer: IEEE Trans. Magn. Mag. **26**, 1397 (1990)
3.28 *Herzer* 1990
3.29 *Herzer* 1990–1995
3.30 *Herzer* and *Warlimont* 1992
3.31 *Sawa* and *Takahashi* 1990
3.32 *Suzuki* et al. 1991
3.33 *Guo* et al. 1991
3.34 *Pfeifer* and *Radeloff* 1980
3.35 *Arai* et al. 1984
3.36 Y. Yoshizawa, S. Oguma, K. Yamauchi: J. Appl. Phys. **64**, 6044 (1988)
3.37 K. Suzuki, A. Makino, N. Kataoka, A. Inoue, T. Masumoto: Mater. Trans. Japn. Inst. Met. (JIM) **32**, 93 (1991)
3.38 K. Suzuki, A. Makino, N. Kataoka, A. Inoue, T. Masumoto: J. Appl. Phys. **74**, 3316 (1993)
3.39 *Masumoto*
3.40 *Kussmann* and *Jessen*
3.41 K. Fukamichi: *Magnetic Properties of Metal*, Landolt–Börnstein, New Series III/19, ed. by H. P. J. Wijn (Springer, Berlin, Heidelberg 1994) pp. 193–238
3.42 M. Shiga: Invar Alloys. In: *Materials Science and Technology*, Vol. 3B, ed. by K. H. J. Buschow (VCH, Weinheim 1994) pp. 161–207
3.43 *Patrick*
3.44 *Fujimori* and *Kaneko*
3.45 C. E. Guillaume: C.R. hebd. Séances Acad. Sci. **1244**, 176–179 (1897)
3.46 P. Chevenard, X. Waché, A. Villechon: Ann. France Chronométrie , 259–294 (1937)
3.47 M. Müller: . Ph.D. Thesis (Akademie der Wissenschaften der DDR, Berlin 1977)
3.48 M. Kersten: Z. Phys. **85**, 708 (1933)
3.49 R. Becker, W. Döring: *Ferromagnetismus* (Springer, Berlin 1939) p. 340
3.50 G. Hausch: Phys. Stat. Sol. (a) **15**, 501 (1973) **16 (1973), 371**
3.51 M. Müller: . Ph.D. Thesis (Technische Universität Dresden, Dresden 1969)
3.52 H. Masumoto, S. Sawaya, M. Kikuchi: J. Japn. Inst. Met. (JIM) **35**, 1143, 1150 (1971) **36 (1972), 57, 176, 492, 498, 881, 886, 1116**
3.53 M. Müller: J. Magn. Magn. Mater. **78**, 337 (1989)
3.54 P. E. Armstrong, J. M. Dickinson, A. L. Brown: Trans. AIME **236**, 1404 (1966)
3.55 E. P. Wohlfahrt: Hard Magnetic Materials. Advances in Physics, Phil. Mag. **A8** (1959)
3.56 D. Bonnenberg, K. Burzo, K. Fukamichi, H. P. Kirchmayr, T. Nakamichi, H. P. J. Wijn: *Magnetic Properties of Metals: Magnetic Alloys for Technical Applications. Hard Magnetic Alloys*, Landolt–Börnstein, New Series III/19 (Springer, Berlin, Heidelberg 1992)
3.57 WIDIA Magnettechnik, Gesinterte Dauermagnete Koerzit und Koerox, d.2000.D.5
3.58 C. Abache, H. Oesterreicher: J. Appl. Phys. **57**, 4112 (1985)
3.59 *van Hook* 1964

4.4. Dielectrics and Electrooptics

Dielectrics an

The present section describes the physical properties of dielectrics and includes the following data:

1. *low-frequency properties*, i.e. density and Mohs hardness, thermal conductivity, static dielectric constant, dissipation factor (loss tangent), elastic stiffness and elastic compliance, and piezoelectric strain;
2. *high-frequency (optical) properties*, i.e. elastooptic and electrooptic coefficients, optical transparency range, two-photon absorption coefficient, refractive indices and their temperature variation, dispersion relations (Sellmeier equations), and second and/or third-order nonlinear dielectric susceptibilities.

4.4.1 **Dielectric Materials: Low-Frequency Properties** ... 822
4.4.1.1 General Dielectric Properties ... 822
4.4.1.2 Static Dielectric Constant (Low-Frequency) ... 823
4.4.1.3 Dissipation Factor ... 823
4.4.1.4 Elasticity ... 823
4.4.1.5 Piezoelectricity ... 824

4.4.2 **Optical Materials: High-Frequency Properties** ... 824
4.4.2.1 Crystal Optics: General ... 824
4.4.2.2 Photoelastic Effect ... 824
4.4.2.3 Electrooptic Effect ... 825
4.4.2.4 Nonlinear Optical Effects ... 825

4.4.3 **Guidelines for Use of Tables** ... 826

4.4.4 **Tables of Numerical Data for Dielectrics and Electrooptics** ... 828
Isotropic Materials ... 828
Cubic, Point Group $m3m$ (O_h) Materials ... 828
Cubic, Point Group $\bar{4}3m$ (T_d) Materials ... 834
Cubic, Point Group 23 (T) Materials ... 838
Hexagonal, Point Group $\bar{6}m2$ (D_{3h}) Materials ... 838
Hexagonal, Point Group $6mm$ (C_{6v}) Materials ... 840
Hexagonal, Point Group 6 (C_6) Materials ... 842
Trigonal, Point Group $\bar{3}m$ (D_{3d}) Materials ... 844
Trigonal, Point Group 32 (D_3) Materials ... 844
Trigonal, Point Group $3m$ (C_{3v}) Materials ... 848
Tetragonal, Point Group $4mmm$ (D_{4h}) Materials ... 852
Tetragonal, Point Group $4/m$ (C_{4h}) Materials ... 854
Tetragonal, Point Group 422 (D_4) Materials ... 854
Tetragonal, Point Group $\bar{4}2m$ (D_{2d}) Materials ... 856
Tetragonal, Point Group $4mm$ (C_{4v}) Materials ... 864
Orthorhombic, Point Group mmm (D_{2h}) Materials ... 866
Orthorhombic, Point Group 222 (D_2) Materials ... 866
Orthorhombic, Point Group $mm2$ (C_{2v}) Materials ... 872
Monoclinic, Point Group 2 (C_2) Materials ... 884

References ... 890

A large amount of the information in this section is taken from the compilations of low- and high-frequency properties of dielectric crystals in Landolt–Börnstein, Group III, Vols. 29 and 30, especially Vol. 30b. Since 1992–1993, the date of publication of the first of these volumes, a large amount of new data on the physical properties of dielectrics has appeared in the literature. In particular, various linear and nonlinear optical properties of new crystals in the borate family (BBO, LBO, CBO, and CLBO) and of new organic crystals (DLAP, MNMA, etc.) have been (re)measured in recent years. This situation has encouraged us to "refresh" the knowledge about these crystals by adding new data from publications from the last decade. We have also included the most commonly used isotropic materials. The criteria for the selection of 124 dielectrics out of several hundred were their wide range of application and the availability of most of the above-mentioned data.

One of our aims was to produce a "reader-friendly" compilation. For this purpose, all data on dielectrics are presented in a unified form in tables that are similar for both isotropic materials and crystals of various symmetry classes. Moreover, *all* data for each particular material are collected together in one place. Sections 4.4.1–4.4.2 serve as a brief introduction to the various physical phenomena and to the definitions, symbols, and abbreviations used. All numerical data are presented in Sect. 4.4.4. The tables in this section are arranged according to piezoelectric classes in order of decreasing symmetry. Guidelines for searching for (and finding!) the required parameter in these tables can be found in Sect. 4.4.3.

The following table presents a list of the 124 different substances which have been selected to be described in Sect. 4.4.4.

Alphabetical List of Described Crystals and Isotropic Dielectrics

Name of material	Formula	Symbol	Page
α-Aluminium oxide (sapphire)	Al_2O_3		844
Aluminium phosphate (berlinite)	$AlPO_4$		844
Ammonium dideuterium phosphate	$ND_4D_2PO_4$	AD*P or DADP	856
Ammonium dihydrogen arsenate	$NH_4H_2AsO_4$	ADA	856
Ammonium dihydrogen phosphate	$NH_4H_2PO_4$	ADP	856
Ammonium sulfate	$(NH_4)_2SO_4$		872
β-Barium borate	BaB_2O_4	BBO	848
Barium fluoride	BaF_2		828
Barium formate	$Ba(COOH)_2$		866
Barium magnesium fluoride	$BaMgF_4$	BMF	872
Barium nitrite monohydrate	$Ba(NO_2)_2 \cdot H_2O$		842
Barium sodium niobate ("banana")	$Ba_2NaNb_5O_{15}$		872
Barium titanate	$BaTiO_3$		864
Beryllium oxide (bromellite)	BeO		840
Bismuth germanium oxide	$Bi_{12}GeO_{20}$	BGO	838
Bismuth silicon oxide (sillenite)	$Bi_{12}SiO_{20}$	BSO	838
Bismuth triborate	BiB_3O_6	BIBO	884
BK7 Schott glass		BK7	828
Cadmium germanium arsenide	$CdGeAs_2$		856
Cadmium germanium phosphide	$CdGeP_2$		856
Cadmium selenide	$CdSe$		840
Cadmium sulfide (greenockite)	CdS		840
Cadmium telluride (Irtran-6)	$CdTe$		834
Calcite (calcspar, Iceland spar)	$CaCO_3$		844
Calcium fluoride (fluorite, fluorspar, Irtran-3)	CaF_2		828
Calcium tartrate tetrahydrate	$Ca(C_4H_4O_6) \cdot 4H_2O$	L-CTT	874
Cesium dideuterium arsenate	CsD_2AsO_4	CD*A or DCDA	856
Cesium dihydrogen arsenate	CsH_2AsO_4	CsDA or CDA	858

Name of material	Formula	Symbol	Page
Cesium lithium borate	$CsLiB_6O_{10}$	CLBO	858
Cesium triborate	CsB_3O_5	CBO	868
m-Chloronitrobenzene	$ClC_6H_4NO_2$	CNB	875
Copper bromide	CuBr		834
Copper chloride (nantokite)	CuCl		834
Copper gallium selenide	$CuGaSe_2$		858
Copper gallium sulfide	$CuGaS_2$		858
Copper iodide	CuI		834
2-Cyclooctylamino-5-nitropyridine	$C_{13}H_{19}N_3O_2$	COANP	874
Deuterated L-arginine phosphate	$(ND_xH_{2-x})_2^+(CND)(CH_2)_3CH$ $(ND_yH_{3-y})^+COO^- \cdot D_2PO_4^- \cdot D_2O$	DLAP	884
Diamond	C		830
4-(*N*1*N*-Dimethylamino)-3-acetamido-nitrobenzene (*N*-[2-(dimethylamino)-5-nitrophenyl]-acetamide)	$C_{10}H_{13}N_3O_3$	DAN	884
N, 2-Dimethyl-4-nitrobenzenamine	$C_8H_{10}N_2O_2$	MNMA	874
Dipotassium tartrate hemihydrate	$K_2C_4H_4O_6 \cdot 0.5H_2O$	DKT	886
Gadolinium molybdate	$Gd_2(MoO_4)_3$	GMO	876
Gallium antimonide	GaSb		834
Gallium arsenide	GaAs		834
Gallium nitride	GaN		840
Gallium phosphide	GaP		834
Gallium selenide	GaSe		838
Gallium sulfide	GaS		838
Germanium	Ge		830
Indium antimonide	InSb		836
Indium arsenide	InAs		836
Indium phosphide	InP		836
α-Iodic acid	HIO_3		868
Lead molybdate	$PbMoO_4$		854
Lead titanate	$PbTiO_3$		864
Lithium fluoride	LiF		830
Lithium formate monohydrate	$LiCOOH \cdot H_2O$	LFM	876
Lithium gallium oxide (lithium metagallate)	$LiGaO_2$		876
α-Lithium iodate	$LiIO_3$		842
Lithium niobate	$LiNbO_3$		848

Name of material	Formula	Symbol	Page
Lithium niobate (MgO-doped)	$MgO:LiNbO_3$		848
Lithium sulfate monohydrate	$Li_2SO_4 \cdot H_2O$		886
Lithium tantalate	$LiTaO_3$		848
Lithium tetraborate	$Li_2B_4O_7$		864
Lithium triborate	LiB_3O_5	LBO	878
Magnesium fluoride	MgF_2		852
Magnesium oxide	MgO		830
Magnesium silicate (forsterite)	Mg_2SiO_4		866
α-Mercurie sulfide (cinnabar)	HgS		846
3-Methyl 4-nitropyridine 1-oxide	$C_6N_2O_3H_6$	POM	868
m-Nitroaniline (3-nitrobenzenamine, meta-nitroaniline)	$C_6H_4(NO_2)NH_2$	mNA	878
Poly(methyl methacrylate) (Plexiglas®)	$(C_5H_8O_2)_n$	PMMA	828
Potassium acid phthalate	$KH(C_8H_4O_4)$		878
Potassium bromide	KBr		830
Potassium chloride (sylvine, sylvite)	KCl		830
Potassium dideuterium arsenate	KD_2AsO_4	KD*A or DKDA	858
Potassium dideuterium phosphate	KD_2PO_4	KD*P or DKDP	858
Potassium dihydrogen arsenate	KH_2AsO_4	KDA	860
Potassium dihydrogen phosphate	KH_2PO_4	KDP	860
Potassium fluoroboratoberyllate	$KBe_2BO_3F_2$	KBBF	846
Potassium iodide	KI		830
Potassium lithium niobate	$K_3Li_2Nb_5O_{15}$	KLINBO	864
Potassium niobate	$KNbO_3$		880
Potassium pentaborate tetrahydrate	$KB_5O_8 \cdot 4H_2O$	KB5	880
Potassium sodium tartrate tetrahydrate (Rochelle salt)	$KNa(C_4H_4O_6) \cdot 4H_2O$		870
Potassium titanate (titanyl) phosphate	$KTiOPO_4$	KTP	880
Potassium titanyl arsenate	$KTiOAsO_4$	KTA	880
Rubidium dideuterium arsenate	RbD_2AsO_4	RbD*A, RD*A, or DRDA	860
Rubidium dideuterium phosphate	RbD_2PO_4	RbD*P, RD*P, or DRDP	860
Rubidium dihydrogen arsenate	RbH_2AsO_4	RbDA or RDA	860
Rubidium dihydrogen phosphate	RbH_2PO_4	RbDP or RDP	862
Rubidium titanate (titanyl) phosphate	$RbTiOPO_4$	RTP	882

Name of material	Formula	Symbol	Page
D(+)-Saccharose (sucrose)	$C_{12}H_{22}O_{11}$		888
Silicon	Si		830
α-Silicon carbide	SiC		840
α-Silicon dioxide (quartz)	SiO_2		846
Silicon dioxide (fused silica)	SiO_2		828
Silver antimony sulfide (pyrargyrite)	Ag_3SbS_3		850
Silver arsenic sulfide (proustite)	Ag_3AsS_3		850
Silver gallium selenide	$AgGaSe_2$		862
Silver gallium sulfide (silver thiogallate)	$AgGaS_2$		862
Sodium ammonium tartrate tetrahydrate (ammonium Rochelle salt)	$Na(NH_4)C_4H_4O_6 \cdot 4H_2O$		870
Sodium chlorate	$NaClO_3$		838
Sodium chloride (rock salt, halite)	NaCl		832
Sodium fluoride	NaF		832
Sodium nitrite	$NaNO_2$		882
Strontium fluoride	SrF_2		832
Strontium titanate	$SrTiO_3$		832
Tellurium	Te		846
Tellurium dioxide (paratellurite)	TeO_2		854
Thallium arsenic selenide	Tl_3AsSe_3	TAS	850
Titanium dioxide (rutile)	TiO_2		852
Tourmaline	$(Na, Ca)(Mg, Fe)_3B_3Al_6Si_6(O, OH, F)_{31}$		850
Triglycine sulfate	$(CH_2NH_2COOH)_3 \cdot H_2SO_4$	TGS	888
Urea	$(NH_2)_2CO$		862
Yttrium aluminate	$YAlO_3$	YAP or YALO	866
Yttrium aluminium garnet	$Y_3Al_5O_{12}$	YAG	832
Yttrium lithium fluoride	$YLiF_4$	YLF	854
Yttrium vanadate	YVO_4	YVO	852
Zinc germanium diphosphide	$ZnGeP_2$		862
Zinc oxide (zincite)	ZnO		840
Zinc selenide	ZnSe		836
α-Zinc sulfide (wurtzite)	ZnS		842
Zinc telluride	ZnTe		836

The physical quantities used to describe the properties of the dielectric substances are drawn up as follows.

Used Physical Qunatities, their Symbols and their Units		
B_{ij} (B_m)	relative dielectric impermeability	dimensionless
C	capacitance C/V	
c_{ijkl} (c_{mn})	elastic stiffness tensor	10^9 Pa = GPa
D	electric displacement field (or electric flux density)	C/m^2
d_{ijk} (d_{in})	piezoelectric strain tensor	10^{-12} C/N
E	electric field	V/m
n	refractive index	dimensionless
P	polarization (dipole moment per unit volume of matter)	C/m^2
p_{ijkl} (p_{mn})	elastooptic tensor	dimensionless
q_{ijkl} (q_{mn})	piezooptic tensor	10^{-9} Pa^{-1} = $(GPa)^{-1}$
r_{ijk} (r_{mk})	electrooptic coefficient	10^{-12} m/V = pm/V
r^S_{mk}, r^T_{mk}	electrooptic coefficient at constant strain or stress, respectively	10^{-12} m/V = pm/V
S_{ij} (S_m)	strain tensor	dimensionless
s_{ijkl} (s_{mn})	elastic compliance tensor	10^{-12} Pa^{-1} = $(TPa)^{-1}$
T_{ij} (T_m)	stress tensor	10^9 Pa = GPa
$\tan\delta$	dissipation factor (loss tangent)	dimensionless
ϱ	density	g/cm^3
ε_0	dielectric constant (or permittivity) of free space (vacuum),	8.854×10^{-12} C/V m
ε	relative dielectric constant (permittivity)	dimensionless
κ_{ij}	thermal conductivity	W/m K
v^s_{mn}	sound velocity in the direction *mn*	m/s
χ_{ij} $\left(\chi^{(1)}_{ij}\right)$	linear dielectric susceptibility	dimensionless
$\chi^{(2)}_{ijk}$, d_{ijk} (d_{im})	second-order nonlinear dielectric susceptibility	10^{-12} m/V = pm/V
$\chi^{(3)}_{ijkl}$	third-order nonlinear dielectric susceptibility	10^{-22} m^2/V^2

4.4.1 Dielectric Materials: Low-Frequency Properties

4.4.1.1 General Dielectric Properties

Density

The density of a substance is defined as the mass per unit volume of the substance:

$$\varrho = \frac{m}{V}\,, \qquad (4.1)$$

where V is the volume occupied by a mass m. The density is thus a measure of the volume concentration of mass.

Mohs Hardness Scale

In 1832, Mohs introduced a hardness scale ranging from 1 to 10, based on ten minerals:

1. talc, $Mg_3H_2SiO_{12}$;
2. gypsum, $CaSO_4 \cdot 2H_2O$;
3. iceland spar, $CaCO_3$;
4. fluorite, CaF_2;
5. apatite, $Ca_5F(PO_4)_3$;
6. orthoclase, $KAlSi_3O_8$;

7. quartz, SiO_2;
8. topaz, $Al_2F_2SiO_4$;
9. corundum, Al_2O_3;
10. diamond, C.

Thermal Conductivity

A temperature gradient between different parts of a solid causes a flow of heat. In an isotropic medium, the heat flux of thermal energy h (i. e. the heat transfer rate per unit area normal to the direction of heat flow) is given by

$$h = -\kappa \operatorname{grad} T \,, \quad (4.2)$$

where κ is the thermal conductivity. In a crystal, this expression is replaced by

$$h_i = -\kappa_{ij} \frac{\partial T}{\partial x_j} \,. \quad (4.3)$$

Here κ_{ij} is the thermal-conductivity tensor. Note that other notations are also used in the literature, e.g. $\kappa \equiv \lambda$ or $\kappa \equiv k$.

4.4.1.2 Static Dielectric Constant (Low-Frequency)

In isotropic and cubic dielectric materials, the electric displacement field D, the electric field E, and the polarization P are connected by the relation

$$D = \varepsilon_0 E + P = \varepsilon_0 (1+\chi) E \,, \quad (4.4)$$

where $\varepsilon_0 = 8.854 \times 10^{-12}$ C/V m is the dielectric constant (or permittivity) of free space (vacuum), and χ is the *dielectric susceptibility*. The relative dielectric constant of the material is defined as

$$\varepsilon = 1 + \chi \,, \quad (4.5)$$

and therefore (4.4) becomes

$$D = \varepsilon_0 \varepsilon E \,. \quad (4.6)$$

In anisotropic crystals, these equations should be written in tensor form:

$$D_i = \varepsilon_0 \varepsilon_{ij} E_j \,, \quad \varepsilon_{ij} = 1 + \chi_{ij} \,. \quad (4.7)$$

The following relations are valid:

$$\varepsilon_{ij} = \varepsilon_{ji} \,, \quad \chi_{ij} = \chi_{ji} \,. \quad (4.8)$$

Note that other notations are also used in the literature, e.g. $\varepsilon \equiv \varepsilon_r$, or $\varepsilon \equiv \kappa$ and $\varepsilon_0 \equiv \kappa_0$.

4.4.1.3 Dissipation Factor

The capacitance C of a capacitor filled with a dielectric is

$$C = \frac{\varepsilon_0 \varepsilon A}{d} \,, \quad (4.9)$$

where A is the area of the two parallel plates and d is the spacing between them. For a lossy dielectric, the relative dielectric constant ε can be represented in a complex form,

$$\varepsilon = \varepsilon' - \mathrm{i}\varepsilon'' \,. \quad (4.10)$$

The imaginary part is the frequency-dependent conductivity

$$\sigma(\omega) = \omega \varepsilon_0 \varepsilon'' \,, \quad (4.11)$$

where ω is the frequency. The dissipation factor (or loss tangent) is defined as

$$\tan\delta = \varepsilon''/\varepsilon' \,. \quad (4.12)$$

and in anisotropic crystals

$$\tan\delta_1 = \frac{\varepsilon''_{11}}{\varepsilon'_{11}} \,; \quad \tan\delta_2 = \frac{\varepsilon''_{22}}{\varepsilon'_{22}} \,; \quad \tan\delta_3 = \frac{\varepsilon''_{33}}{\varepsilon'_3} \,; \quad (4.13)$$

The quality factor Q of the dielectric is the reciprocal of the dissipation factor:

$$Q = 1/\tan\delta \,. \quad (4.14)$$

4.4.1.4 Elasticity

Hooke's law states that for sufficiently small deformations the strain is directly proportional to the stress. Thus the strain tensor **S** and the stress tensor **T** obey the relation

$$S_{ij} = s_{ijkl} T_{kl} \,, \quad (4.15)$$

where s_{ijkl} is called the *elastic compliance constant* (or compliance, or elastic constant). The *elastic stiffness constant* (or stiffness, or Young's modulus) is the reciprocal tensor

$$c_{ijkl} = s^{-1}_{ijkl} \,, \quad (4.16)$$

Table 4.4-1 The relations between ij (tensor notation) and m (matrix notation), jk and n, and kl and n

Tensor notation	11	22	33	23 or 32	31 or 13	12 or 21
Matrix notation	1	2	3	4	5	6

and for the stress tensor we have

$$T_{ij} = c_{ijkl} S_{kl} \,. \tag{4.17}$$

In the matrix notation for the elastic compliance and stiffness, we have

$$S_m = s_{mn} T_n \text{ and } T_m = c_{mn} S_n \,, \tag{4.18}$$

where

$$\begin{aligned} s_{ijkl} &= s_{mn} && \text{when both } m \text{ and } n \text{ are 1, 2, or 3} \,, \\ 2s_{ijkl} &= s_{mn} && \text{when either } m \text{ or } n \text{ is 4, 5, or 6} \\ 4s_{ijkl} &= s_{mn} && \text{when both } m \text{ and } n \text{ are 4, 5, or 6} \,, \\ c_{ijkl} &= c_{mn} && \text{for all } m \text{ and } n \,; \end{aligned} \tag{4.19}$$

and

$$\begin{aligned} S_{ij} &= S_m && \text{when } m \text{ is 1, 2, or 3} \,, \\ S_{ij} &= \frac{1}{2} S_m && \text{when } m \text{ is 4, 5, or 6} \,, \\ T_{ij} &= T_m && \text{for all } m \,. \end{aligned} \tag{4.20}$$

For relations between tensor and matrix notation, see Table 4.4-1. The sound velocity v^{s}_{mn} in the direction mn in a crystal is given by

$$v^{s}_{mn} = \sqrt{c_{mn}/\varrho} \,. \tag{4.21}$$

4.4.1.5 Piezoelectricity

The phenomenon of the development of an electric moment P_i if a stress T_{jk} is applied to a crystal is called the direct piezoelectric effect:

$$P_i = d_{ijk} T_{jk} \,, \tag{4.22}$$

where d_{ijk} is the piezoelectric strain tensor (or the piezoelectric moduli). The relation $d_{ijk} = d_{ikj}$ reduces the number of independent tensor components to 18. The matrix notation is introduced for the piezoelectric strain as follows:

$$\begin{aligned} d_{ijk} &= d_{in} && \text{when } n = 1, 2, \text{or } 3 \,, \\ 2d_{ijk} &= d_{in} && \text{when } n = 4, 5, \text{or } 6 \,, \end{aligned} \tag{4.23}$$

and thus

$$P_i = d_{in} T_n \,. \tag{4.24}$$

The relations between jk and n are presented in Table 4.4-1.

The converse piezoelectric effect is described by

$$S_{jk} = d_{ijk} E_i \tag{4.25}$$

and, correspondingly,

$$S_n = d_{in} E_i \,. \tag{4.26}$$

4.4.2 Optical Materials: High-Frequency Properties

4.4.2.1 Crystal Optics: General

The dielectric properties of a medium at optical frequencies are given by

$$D = \varepsilon_0 \varepsilon E \,, \tag{4.27}$$

where ε_0 is the dielectric constant of free space and ε is the relative dielectric constant of the material. From Maxwell's equations, the velocity of propagation of electromagnetic waves through the medium is given by

$$v = c/\sqrt{\varepsilon} \,, \tag{4.28}$$

where c is the velocity in vacuum (the relative magnetic permeability is taken as 1). The refractive index $n = c/v$ is therefore $n = \sqrt{\varepsilon}$.

In an anisotropic medium,

$$D_i = \varepsilon_0 \varepsilon_{ij} E_j \,. \tag{4.29}$$

In this general case, two waves of different velocity may propagate through the crystal. The relative dielectric impermeabilities are defined as the reciprocals of the principal dielectric constants:

$$B_i = 1/\varepsilon_i = 1/n_i^2 \,. \tag{4.30}$$

4.4.2.2 Photoelastic Effect

The photoelastic effect is the effect in which a change of the refractive index is caused by stress. The changes in the relative dielectric impermeabilities are

$$\Delta B_{ij} = q_{ijkl} T_{kl} , \quad (4.31)$$

where the q_{ijkl} are the piezooptic coefficients. The photoelastic effect can also be expressed in terms of the stress:

$$\Delta B_{ij} = p_{ijrs} S_{rs} , \quad (4.32)$$

where the $p_{ijrs} = q_{ijkl} c_{klrs}$ are the (dimensionless) elastooptic coefficients. In matrix notation,

$$\Delta B_m = q_{mn} T_n \quad \text{and } \Delta B_m = p_{mn} S_n . \quad (4.33)$$

Note that $q_{mn} = q_{ijkl}$ when $n = 1, 2$, or 3 and $q_{mn} = 2q_{ijkl}$ when $n = 4, 5$, or 6; $p_{mn} = p_{ijrs}$ (see Table 4.4-1).

4.4.2.3 Electrooptic Effect

The electrooptic effect is the effect in which a change in the refractive index of a crystal is produced by an electric field:

$$n = n_0 + aE_0 + bE_0^2 + \dots , \quad (4.34)$$

where a and b are constants and n_0 is the refractive index at $E_0 = 0$. The linear electrooptic effect (Pockels effect) is due to the first-order term aE_0. In isotropic dielectrics and in crystals with a center of symmetry, $a = 0$, and only the second-order term bE_0^2 and higher even-order terms exist (Kerr effect).

The changes in the relative dielectric impermeabilities are

$$\Delta B_{ij} = r_{ijk} E_k , \quad (4.35)$$

where the r_{ijk} are the electrooptic coefficients. Since $r_{ijk} = r_{jik}$, the number of independent tensor components is 18, and the above formula can be written in matrix notation (see Table 4.4-1):

$$\Delta B_m = r_{mk} E_k \quad (m = 1, 2, \dots , 6,\ k = 1, 2, 3) . \quad (4.36)$$

4.4.2.4 Nonlinear Optical Effects

The dielectric polarization $\boldsymbol{P}$ is related to the electromagnetic field $\boldsymbol{E}$ at optical frequencies by the material equation of the medium:

$$\boldsymbol{P}(\boldsymbol{E}) = \varepsilon_0 \left(\chi^{(1)} \boldsymbol{E} + \chi^{(2)} \boldsymbol{E}^2 + \chi^{(3)} \boldsymbol{E}^3 + \dots \right) , \quad (4.37)$$

where $\chi^{(1)} = n^2 - 1$ is the linear dielectric susceptibility, and $\chi^{(2)}$, $\chi^{(3)}$, etc. are the nonlinear dielectric susceptibilities.

The Miller delta formulation is

$$\varepsilon_0 E_i(\omega_3) = \delta_{ijk} P_j(\omega_1) P_k(\omega_2) , \quad (4.38)$$

where the Miller coefficient,

$$\delta_{ijk} = \frac{1}{2\varepsilon_0} \frac{\chi_{ijk}^{(2)}(\omega_3)}{\chi_{ii}^{(1)}(\omega_1)\chi_{jj}^{(1)}(\omega_2)\chi_{kk}^{(1)}(\omega_3)} , \quad (4.39)$$

has a small dispersion and is almost constant for a wide range of crystals.

For anisotropic media, the coefficients $\chi^{(1)}$ and $\chi^{(2)}$ are, in general, second- and third-rank tensors, respectively. In practice, the tensor

$$d_{ijk} = (1/2)\chi_{ijk} \quad (4.40)$$

is used instead of χ_{ijk}. Usually the "plane" representation of d_{ijk} in the form d_{il} is used; the relations between jk and l are presented in Table 4.4-1.

The Kleinman symmetry conditions

$$d_{21} = d_{16},\quad d_{23} = d_{34},\quad d_{14} = d_{25} = d_{36},$$
$$d_{26} = d_{12},\quad d_{31} = d_{15},\quad d_{32} = d_{24},\quad d_{35} = d_{13} \quad (4.41)$$

are valid in the case of no dispersion of the electronic nonlinear polarizability.

The following three-wave interactions in crystals with a square nonlinearity ($\chi^{(2)} \neq 0$) are possible:

- second-harmonic generation (SHG), $\omega + \omega = 2\omega$;
- sum frequency generation (SFG) or up-conversion, $\omega_1 + \omega_2 = \omega_3$;
- difference frequency generation (DFG) or down-conversion, $\omega_3 - \omega_2 = \omega_1$;
- optical parametric oscillation (OPO), $\omega_3 = \omega_2 + \omega_1$.

For efficient frequency conversion, the phase-matching condition $\boldsymbol{k}_1 + \boldsymbol{k}_2 = \boldsymbol{k}_3$, where the $\boldsymbol{k}_i$ are the wave vectors for ω_1, ω_2, and ω_3, respectively, must be satisfied. Two types of phase matching can be defined:

type I: o + o → e or e + e → o ;

and

type II: o + e → e or o + e → o .

These can be represented with a shortened notation as follows:

ooe: o + o → e or e → o + o ;

eeo: e + e → o or o → e + e ;

eoe: e + o → e or e → e + o ;

oeo: o + e → o or o → e + o .

In the shortened notation (ooe, eoe, ...), the frequencies satisfy the condition $\omega_1 < \omega_2 < \omega_3$, i.e. the first symbol refers to the longest-wavelength radiation, and the last symbol refers to the shortest-wavelength radiation. Here the ordinary beam, or o-beam is the beam with its polarization normal to the principal plane of the crystal, i.e. the plane containing the wave vector $\boldsymbol{k}$ and the crystallophysical axis Z (or the optical axis, for uniaxial crystals). The extraordinary beam, or e-beam is the beam with its polarization in the principal plane. The third-order term $\chi^{(3)}$ is responsible for the optical Kerr effect.

Uniaxial Crystals

For uniaxial crystals, the difference between the refractive indices of the ordinary and extraordinary beams, the birefringence Δn, is zero along the optical axis (the crystallophysical axis Z) and maximum in a direction normal to this axis. The refractive index for the ordinary beam does not depend on the direction of propagation. However, the refractive index for the extraordinary beam $n^{\mathrm{e}}(\theta)$, is a function of the polar angle θ between the Z axis and the vector $\boldsymbol{k}$:

$$n^{\mathrm{e}}(\theta) = n_{\mathrm{o}} \left(\frac{1+\tan^2\theta}{1+(n_{\mathrm{o}}/n_{\mathrm{e}})^2 \tan^2\theta} \right)^{1/2} , \quad (4.42)$$

where n_{o} and n_{e} are the refractive indices of the ordinary and extraordinary beams, respectively in the plane normal to the Z axis, and are termed the *principal values*.

If $n_{\mathrm{o}} > n_{\mathrm{e}}$ the crystal is called *negative*, and if $n_{\mathrm{o}} < n_{\mathrm{e}}$ it is called *positive*. For the o-beam, the indicatrix of the refractive indices is a sphere with radius n_{o}, and for the e-beam it is an ellipsoid of rotation with semiaxes n_{o} and n_{e}. In the crystal, in general, the beam is divided into two beams with orthogonal polarizations; the angle between these beams ρ is the *birefringence* (or *walk-off*) angle.

Equations for calculating phase-matching angles in uniaxial crystals are given in [4.1–4].

Biaxial Crystals

For biaxial crystals, the optical indicatrix is a bilayer surface with four points of interlayer contact, which correspond to the directions of the two optical axes. In the simple case of light propagation in the principal planes XY, YZ, and XZ, the dependences of the refractive indices on the direction of light propagation are represented by a combination of an ellipse and a circle. Thus, in the principal planes, a biaxial crystal can be considered as a uniaxial crystal; for example, a biaxial crystal with $n_Z > n_Y > n_X$ in the XY plane is similar to a negative uniaxial crystal with $n_{\mathrm{o}} = n_Z$

$$n^{\mathrm{e}}(\varphi) = n_Y \left(\frac{1+\tan^2\varphi}{1+(n_Y/n_X)^2 \tan^2\varphi} \right)^{1/2} , \quad (4.43)$$

where φ is the azimutal angle. Equations for calculating phase-matching angles for propagation in the principal planes of biaxial crystals are given in [4.3–6].

4.4.3 Guidelines for Use of Tables

Tables 4.4-3 – 4.4-21 are arranged according to piezoelectric classes in order of decreasing symmetry (see Table 4.4-2), and alphabetically within each class. They contain a number of columns placed on two pages, even and odd. The following properties are presented for each dielectric material: density ϱ, Mohs hardness, thermal conductivity κ, static dielectric constant ε_{ij}, dissipation factor $\tan\delta$ at various temperatures and frequencies, elastic stiffness c_{mn}, elastic compliance s_{mn} (for isotropic and cubic materials only), piezoelectric strain tensor d_{in}, elastooptic tensor p_{mn}, electrooptic coefficients r_{mk} (the latter two at 633 nm unless otherwise stated), optical transparency range, temperature variation of the refractive indices dn/dT, refractive indices n (the latter two at 1.064 μm unless otherwise stated), dispersion relations (Sellmeier equations), second-order nonlinear dielectric susceptibility d_{ij}, and third-order nonlinear dielectric susceptibility $\chi^{(3)}_{ijk}$ (for isotropic and cubic materials only) (the latter two at 1.064 μm unless otherwise stated). For isotropic materials, the two-photon absorption coefficient β is also included

The numerical values of the elastic and elastooptic constants are often averages of three or more measurements, as presented in [4.7–11]. In such cases, the corresponding Landolt–Börnstein volume is cited together with the most reliable (latest) reference. The standard deviation of the averaged value is given in parentheses. Vertical bars || mean the modulus of the corresponding quantity. The absolute scale for the second-order nonlinear susceptibilities of crystals is based on [4.12–14]. The second-order susceptibilities for all crystals measured relative to a standard crystal have been recalculated accordingly. In particular, all previous measurements relative to KDP and quartz have been normalized to $d_{36}(\mathrm{KDP}) = 0.39\,\mathrm{pm/V}$ and $d_{11}(\mathrm{SiO_2}) = 0.30\,\mathrm{pm/V}$. These data lead to an accurate, self-consistent set of absolute second-order nonlinear coefficients [4.14]. All numerical data are for room temperature (300 K) and in SI units.

Table 4.4-2 Number of independent components of the various property tensors

Symmetry point group	Dielectric tensor[a] ε_{ij}	Elastic tensor c_{mn} or s_{mn}	Piezoelectric tensor d_{imn}	Elasto-optic tensor p_{mn} or q_{mn}	Electro-optic tensor r_{mk}	Nonlinear susceptibility tensors[b] $\chi^{(2)}$	$\chi^{(3)}$	Table number
Isotropic	1(1)	2	0	2	0	0(0)	1(1)	4.4-3
Cubic:								
432 (O)	1(1)	3	0	3	0	0(0)	2(2)	
$m3m$ (O_h)	1(1)	3	0	3	0	0(0)	2(2)	4.4-4
$\bar{4}3m$ (T_d)	1(1)	3	1	3	1	1(1)	2(2)	4.4-5
23 (T)	1(1)	3	1	4	1	1(1)	3(2)	4.4-6
Hexagonal:								
$6/mmm$ (D_{6h})	2(2)	5	0	6	0	0(0)	4(3)	
$6/m$ (C_{6h})	2(2)	5	0	8	0	0(0)	6(3)	
622 (D_6)	2(2)	5	1	6	1	1(1)	4(3)	
$\bar{6}m2$ (D_{3h})	2(2)	5	1	6	1	1(1)	4(3)	4.4-7
$\bar{6}$(C_6)	2(2)	5	2	8	2	2(2)	6(3)	
$6mm$ (C_{6v})	2(2)	5	3	6	3	3(2)	4(3)	4.4-8
6 (C_6)	2(2)	5	4	8	4	4(3)	6(3)	4.4-9
Trigonal:								
$\bar{3}m$ (D_{3d})	2(2)	6	0	8	0	0(0)	5(4)	4.4-10
$\bar{3}$ (C_{3i})	2(2)	7	0	12	0	0(0)	10(5)	
32 (D_3)	2(2)	6	2	8	2	2(2)	5(4)	4.4-11
3m (C_{3v})	2(2)	6	4	8	4	4(3)	5(4)	4.4-12
3 (C_3)	2(2)	7	6	12	6	6(5)	10(5)	
Tetragonal:								
$4/mmm$ (D_{4h})	2(2)	6	0	7	0	0(0)	5(4)	4.4-13
$4/m$ (C_{4h})	2(2)	7	0	10	0	0(0)	8(6)	4.4-14
422 (D_4)	2(2)	6	1	7	1	1(1)	5(4)	4.4-15
$\bar{4}2m$ (D_{2d})	2(2)	6	2	7	2	2(1)	5(4)	4.4-16
$4mm$ (C_{4v})	2(2)	6	3	7	3	3(2)	5(4)	4.4-17
$\bar{4}$(S_4)	2(2)	7	4	10	4	4(2)	8(6)	
4 (C_4)	2(2)	7	4	10	4	4(3)	8(6)	
Orthorhombic:								
mmm (D_{2h})	3(3)	9	0	12	0	0(0)	9(6)	4.4-18
222 (D_2)	3(3)	9	3	12	3	3(1)	9(6)	4.4-19
$mm2$ (C_{2v})	3(3)	9	5	12	5	5(3)	9(6)	4.4-20
Monoclinic:								
$2/m$ (C_{2h})	4(3)	13	0	20	0	0(0)	16(9)	
2 (C_2)	4(3)	13	8	20	8	8(4)	16(9)	4.4-21
m (C_s)	4(3)	13	10	20	10	10(6)	16(9)	
Triclinic:								
$\bar{1}$ (C_i)	9(3)	21	0	36	0	0(0)	30(15)	
1 (C_1)	9(3)	21	18	36	18	18(10)	30(15)	

[a] The number of principal refractive indices n_i is given in parentheses.

[b] The number of independent components for the case of Kleinman symmetry conditions is given in parentheses.

4.4.4 Tables of Numerical Data for Dielectrics and Electrooptics

Table 4.4-3 Isotropic materials

	General	Static dielectric constant	Dissipation factor	Elastic stiffness tensor	Elastic compliance tensor	Elastooptic tensor
Material	ϱ (g/cm^3) Mohs hardness κ (W/m K)	ε_{11}^{S} ε_{11}^{T}	$\tan\delta_1$ (f (Hz))	c_{11} c_{12} (10^9 Pa)	s_{11} s_{12} (10^{-12} Pa^{-1})	p_{11} p_{12}
BK7 Schott glass	2.510 – 1.114 [4.17]			92.325 (145) – [4.15]	12[b] −2[b] [4.16]	– 0.198 (22) [4.15]
Poly(methylmethacrylate), $(C_5H_8O_2)_n$ (PMMA, Plexiglas)	1.190 2–3 0.2 [4.19]	3.65 (19)[a] [4.18]	0.06 (50 Hz) [4.18]	9.282 (30) – [4.15]	300[b] 8.9[b] [4.18]	
Silicon dioxide, SiO_2 (Fused silica, fused quartz, vitreous quartz)	2.202 5–6 1.38 [4.21]	3.5[a] [4.20]	14 (5) × 10^{-4} (1 MHz) [4.20]	77.806 (185) – [4.15]	14[b] −2.1[b] [4.20]	– 0.243 (17) [4.15]

[a] This value for ε_{11} is neither at constant stress nor at constant strain.
[b] The elastic compliances are at constant electric field (s^E) and have been calculated from the Young's modulus via $Y_0 = 1/s^E$ and from the shear modulus via $G = \left[2\left(s_{11}^E - s_{12}^E\right)\right]^{-1}$.

Table 4.4-4 Cubic, point group $m3m$ (O_h) materials

	General	Static dielectric constant	Dissipation factor	Elastic stiffness tensor	Elastic compliance tensor	Elastooptic tensor
Material	ϱ (g/cm^3) Mohs hardness κ (W/m K)	ε_{11}	$\tan\delta_1$ (f (Hz))	c_{11} c_{44} c_{12} (10^9 Pa)	s_{11} s_{44} s_{12} (10^{-12} Pa^{-1})	p_{11} p_{12} p_{44} $p_{11} - p_{12}$
Barium fluoride, BaF_2	4.89 3 11 [4.21, 22]	7.33 [4.21]		91.1(10) 25.3(4) 41.2(15) [4.7]	15.2(1) 39.6(7) −4.7(1) [4.7]	0.11 0.26 0.02 −0.14 (589–633 nm) [4.23–25]
Calcium fluoride, CaF_2 (fluorite, fluorspar, Irtran-3)	3.179 4.0 9.71 [4.21, 22]	7.4 [4.26]		165(2) 33.9(3) 47(3) [4.7]	6.93(14) 29.5(3) −1.52(11) [4.7]	0.027 0.198 0.02 −0.17 (550–650 nm) [4.23, 24, 27]

Table 4.4-3 Isotropic materials, cont.

General optical properties Transparency range (μm) dn/dT (10^{-5} K^{-1})	Refractive index n	Two-photon absorption coefficient β (10^{-9} cm/W)	Nonlinear dielectric susceptibility $\chi^{(3)}_{1111}$ (10^{-22} m^2/V^2)
0.35–2.8 [4.16, 28] 0.28 (0.546 μm) [4.16]	**1.5067** [4.16] For dispersion relation, see[c] [4.28]	0.006 (351 nm) [4.29] 0.0029 (532 nm) [4.31]	2.7 ± 0.2 [4.30]
0.27–1.1 –	**1.503** (0.436 μm) **1.493** (0.546 μm) **1.489** (0.633 μm) **1.481** (1.052 μm) [4.32]		
0.17–4.0 [4.22] 1.5 (0.21 μm) 1.0 (0.4 μm) 1.2 (2.0 μm) 1.0 (3.7 μm) [4.39]	**1.5343** (0.2139 μm) **1.4872** (0.3022 μm) **1.4601** (0.5461 μm) **1.4494** (1.083 μm) **1.4372** (2.0581 μm) **1.3994** (3.7067 μm) For dispersion relation, see[d] [4.39]	0.75 (212.8 nm) [4.33] 0.5 (216 nm) [4.35] 0.08 (248 nm) [4.37] 0.045 (266 nm) [4.38]	1.8 [4.34] 1.7 ± 0.3 [4.36]

[cd] Dispersion relations (λ (μm), $T = 20\,°$C):

[c] $$n^2 = 1 + \frac{1+1.03961212\lambda^2}{\lambda^2 - 0.00600069867} + \frac{0.231792344\lambda^2}{\lambda^2 - 0.0200179144} + \frac{1.01046945\lambda^2}{\lambda^2 - 103.560653};$$

[d] $$n^2 = 1 + \frac{1+0.6961663\lambda^2}{\lambda^2 - (0.0684043)^2} + \frac{0.4079426\lambda^2}{\lambda^2 - (0.1162414)^2} + \frac{0.8974794\lambda^2}{\lambda^2 - (9.896161)^2}.$$

Table 4.4-4 Cubic, point group $m3m$ (O_h) materials, cont.

General optical properties	Refractive index[a]		Nonlinear dielectric susceptibility
Transparency range (μm) dn/dT (10^{-5} K^{-1})	n A B C	D E F G	$\chi^{(3)}_{1111}$ $\chi^{(3)}_{1122}$ $\chi^{(3)}_{1133}$ (10^{-22} m^2/V^2)
0.14–12.2 [4.40] −1.60 (0.633 μm) [4.41] −0.6 (0.150 μm) −1.0 (0.590 μm) −1.1 (15.0 μm) [4.43]	**1.678** (0.150 μm) **1.557** (0.200 μm) **1.5118** (0.266 μm) **1.4744** (0.590 μm) **1.4683** (1.05 μm) **1.4441** (6.00 μm)	**1.4027** (9.00 μm) **1.3865** (11.0 μm) **1.305** (15.0 μm) dispersion relations, see [4.43][a]	1.548 ± 0.17 – 0.636 ± 0.06 (0.575; 0.613 μm) [4.42]
0.135–9.4 [4.40] −1.04 (0.633 μm) [4.44] −0.1 (0.150 μm) −1.5 (0.590 μm) −1.0 (12.0 μm)	**1.577** (0.150 μm) **1.495** (0.200 μm) **1.4621** (0.266 μm) **1.4338** (0.590 μm) **1.4286** (1.05 μm) **1.3856** (6.00 μm)	**1.3268** (9.00 μm) **1.268** (11.0 μm) **1.230** (12.0 μm) dispersion relations, see [4.43][b]	0.8 ± 0.24 0.36 ± 0.1 (0.575; 0.613 μm) [4.42]

Table 4.4-4 Cubic, point group $m3m$ (O_h) materials, cont.

	General	Static dielectric constant	Dissipation factor	Elastic stiffness tensor	Elastic compliance tensor	Elastooptic tensor
Material	ϱ (g/cm^3) Mohs hardness κ (W/m K)	ε_{11}	$\tan\delta_1$ (f (Hz))	c_{11} c_{44} c_{12} (10^9 Pa)	s_{11} s_{44} s_{12} (10^{-12} Pa^{-1})	p_{11} p_{12} p_{44} $p_{11}-p_{12}$
Diamond, C	3.52 10 660 (0 °C) [4.22, 45]	5.5–10.0 [4.26]		1077(2) 577(1) 124.7(6) [4.7]	0.951(2) 1.732(3) −0.0987(6) [4.7]	−0.25 0.04 0.17 −0.30(1) (514–633 nm) [4.46]
Germanium, Ge	5.33 6 58.61 [4.21]	16.6 [4.21]		129(3) 67.1(5) 48(3) [4.7]	9.73(5) 14.90(12) −2.64(11) [4.7]	
Lithium fluoride, LiF	2.639 4 4.01 [4.21]	9.1 [4.21]		112(2) 63.5(6) 46(3) [4.7]	11.6(1) 15.8(2) −3.35(13) [4.7]	0.02 0.13 −0.04(1) −0.10 (589–633 nm) [4.27, 47]
Magnesium oxide, MgO	3.58 – 42 [4.21]	9.7 [4.26]	0.009 (10 GHz) [4.48]	294(6) 155(3) 93(5) [4.7]	4.01(4) 6.46(12) −0.96(2) [4.7]	−0.25 −0.01 −0.10 −0.24 (589 nm) [4.27, 49]
Potassium bromide, KBr	2.753 1.5 4.816 [4.21]	4.9 [4.21]		34.5(3) 5.1(2) 5.5(4) [4.7]	30.3(6) 196(6) −4.2(3) [4.7]	0.22 0.17 (546–600 nm) [4.50] 0.02 0.04 (488–600 nm) [4.51]
Potassium chloride, KCl (sylvine, sylvite)	1.99 2 6.53 [4.21]	4.6 [4.26]		40.5(4) 6.27(6) 6.9(3) [4.7]	25.9(1) 159(1) −3.8(3) [4.7]	0.23 0.17 0.02 0.05 (546–600 nm) [4.50–53]
Potassium iodide, KI	3.12 – 2.1 [4.21]	5.6 [4.26]		27.4(5) 3.70(4) 4.3(2) [4.7]	38.2(8) 270(3) −5.2(3) [4.7]	1.21 0.15 −0.031 [4.41] –
Silicon, Si	2.329 7 163.3 [4.21]	11.0–12.0 [4.26]		165(2) 79.1(6) 63(1) [4.7]	7.73(8) 12.70(9) −2.15(4) [4.7]	−0.094 0.017 −0.051 −0.111 (3390 nm) [4.54–57]

Table 4.4-4 Cubic, point group $m3m$ (O_h) materials, cont.

General optical properties	**Refractive index** [a]		**Nonlinear dielectric susceptibility**
Transparency range (μm) dn/dT $(10^{-5}\,K^{-1})$	n A B C	D E F G	$\chi^{(3)}_{1111}$ $\chi^{(3)}_{1122}$ $\chi^{(3)}_{1133}$ $(10^{-22}\,m^2/V^2)$
0.24–27 1.01 (0.546 μm) 0.96 (30 μm) [4.59, 60]	**2.7151** (0.2265 μm) **2.4190** (0.578 μm) **2.3914** (1.064 μm)	For dispersion relation, see[c] [4.58]	$\chi^{(3)}_{1122} = 4.87(12)$ $\chi^{(3)}_{1111} + 3\chi^{(3)}_{1122} = 30 \pm 0.8$ (both at 0.407 μm) [4.42] $\chi^{(3)}_{1111} + 3\chi^{(3)}_{1133} = 65 \pm 20$ (0.53 μm) [4.61]
1.8–15 [4.62] –	**4.0038** (10.6 μm) 9.28156 6.72880 0.44105	0.21307 3870.1 0 0 [4.63][d]	56 000 ± 28 000 34 000 ± 1200 – (10.6 and 9.5 μm) [4.64]
0.120–6.60 –	**1.387** 1 0.9259 0.005441	6.96747 1075 0 0 [4.65][d]	1.12(24) 0.50 – (0.6943 and 0.7456 μm) [4.66]
0.35–6.8 [4.67] 1.95 (0.365 μm) 1.65 (0.546 μm) 1.36 (0.768 μm) [4.68]	**1.7217** 1 1.111033 0.00507606	0.8460085 0.01891186 7.808527 723.2345 [4.69][d]	+1.4(2) – +0.77(15) (0.6943 and 0.7456 μm) [4.66]
0.200–30 [4.40] −3.93 (0.458 μm) −4.19 (1.15 μm) −4.11 (10.6 μm) [4.41]	**1.5435** For dispersion relation, see[e] [4.65]		15.7 [4.70] 5.8 [4.71] 6.2 (0.6943 and 0.7456 μm) [4.66]
0.18–23.3 [4.40] −3.49 (0.458 μm) −3.62 (1.15 μm) −3.48 (10.6 μm) [4.41]	**1.4792** For dispersion relation, see[f] [4.65]		6.7 [4.70] 1.9 [4.71] 0.8 (0.6943 and 0.7456 μm) [4.66]
0.32–42 [4.21] −4.15 (0.458 μm) −4.47 (1.15 μm) −3.08 (30 μm) [4.41]	**1.6393** For dispersion relation, see[g] [4.65]		1.4 ± 0.3 (0.6943 and 0.7456 μm) [4.66] – –
1.1–6.5 [4.62] –	**3.4176** (10.6 μm) 1 10.6684293 0.09091219	0.003043475 1.2876602 1.54133408 1 218 820 [4.73][d] [4.74]	24 300 [4.72] – 33 000 [4.71] at 10.6 and 11.8 μm $\chi^{(3)}_{1111} = 2800$ $\chi^{(3)}_{1122} = 1340$ [4.75]

Table 4.4-4 Cubic, point group $m3m$ (O_h) materials, cont.

	General	Static dielectric constant	Dissipation factor	Elastic stiffness tensor	Elastic compliance tensor	Elastooptic tensor
Material	ϱ (g/cm^3) Mohs hardness κ (W/m K)	ε_{11}	$\tan\delta_1$ (f (Hz))	c_{11} c_{44} c_{12} (10^9 Pa)	s_{11} s_{44} s_{12} (10^{-12} Pa^{-1})	p_{11} p_{12} p_{44} $p_{11}-p_{12}$
Sodium chloride, NaCl (rock salt, halite)	2.17 2 1.15 [4.21]	5.9 [4.26]		49.1(5) 12.8(1) 12.8(1) [4.7]	22.9(5) 78.3(8) −4.8(1) [4.7]	0.128 0.171 −0.01 −0.04 (589–633 nm) [4.52, 53, 76]
Sodium fluoride, NaF	2.558 – 3.746 [4.21]	6 [4.21]		97.0(4) 28.1(3) 24.2(10) [4.7]	11.50(6) 35.6(4) −2.30(7) [4.7]	0.03 0.14 – −0.10 (589–633 nm) [4.76–78]
Strontium fluoride, SrF_2	4.24 – 1.42 [4.21]	7.69 [4.21]		124(1) 31.8(3) 44(1) [4.7]	9.86(4) 31.5(3) −2.57(5) [4.7]	0.080 0.269 0.018 −0.189 [4.23, 24]
Strontium titanate, $SrTiO_3$	5.12 6–6.5 12 [4.79]	300 [4.48]	0.02 (10 GHz) [4.48]	316(2) 123(1) 102(1) [4.7]	3.75(3) 8.15(8) −0.92(1) [4.7]	|0.15| |0.095| |0.072| – [4.80]
Yttrium aluminium garnet, $Y_3Al_5O_{12}$ (YAG)	4.56 8–8.5 13.4 [4.81]			333(3) 114(1) 111(3) [4.7]	3.61(4) 8.74(4) −0.90(2) [4.7]	−0.029 0.0091 −0.0615 −0.038 [4.82]

[a] The refractive index is given in **bold** type to distinguish it from the constants A, B, etc. in the dispersion relation.

[abcdefghij] Dispersion relations (λ (μm), $T = 20\,°C$):

[a] $n^2 = 1.33973 + \frac{0.81070\lambda^2}{\lambda^2 - 0.10065^2} + \frac{0.19652\lambda^2}{\lambda^2 - 29.87^2} + \frac{4.52469\lambda^2}{\lambda^2 - 53.82^2}$;

[b] $n^2 = 1.33973 + \frac{0.69913\lambda^2}{\lambda^2 - 0.09374^2} + \frac{0.11994\lambda^2}{\lambda^2 - 21.18^2} + \frac{4.35181\lambda^2}{\lambda^2 - 38.46^2}$;

[c] $n^2 = 2.37553 + \frac{0.0336440\lambda^2}{\lambda^2 - 0.028} - \frac{0.0887524}{(\lambda^2 - 0.028)^2} - 2.40455 \times 10^{-6}\lambda^2 + 2.21390 \times 10^{-9}\lambda^4$;

[d] $n^2 = A + \frac{B\lambda^2}{\lambda^2 - C} + \frac{D\lambda^2}{\lambda^2 - E} + \frac{F\lambda^2}{\lambda^2 - G}$;

[e] $n^2 = 1.39408 + \frac{0.79221\lambda^2}{\lambda^2 - 0.0213} + \frac{0.01981\lambda^2}{\lambda^2 - 0.0299} + \frac{0.15587\lambda^2}{\lambda^2 - 0.0350} + \frac{0.17673\lambda^2}{\lambda^2 - 3674} + \frac{2.0621\lambda^2}{\lambda^2 - 7695}$.

Table 4.4-4 Cubic, point group $m3m$ (O_h) materials, cont.

General optical properties	**Refractive index**		**Nonlinear dielectric susceptibility**
Transparency range (μm) dn/dT $(10^{-5}\,K^{-1})$	n A B C	D E F G	$\chi^{(3)}_{1111}$ $\chi^{(3)}_{1122}$ $\chi^{(3)}_{1133}$ $(10^{-22}\,m^2/V^2)$
0.2–16 [4.40] −3.42 (0.458 μm) −3.54 (0.633 μm) −3.63 (3.39 μm) [4.41]	**1.5313** For dispersion relation, see [h] [4.65]		6.7 [4.70] 2.78 [4.71] at 0.6943 and 0.7456 μm – $\chi^{(3)}_{1111} = 9.5 \pm 2$, $\chi^{(3)}_{1133} = 4.0 \pm 1.2$ [4.66]
0.135–11.2 [4.83] −1.19 (0.458 μm) −1.32 (0.633 μm) −1.25 (3.39 μm) [4.41]	**1.32** 1.41572 0.32785 0.0137	3.18248 1646 0 0 [4.65] [d]	1.4 [4.70] – –
0.13–11 −0.9 (0.150 μm) −1.2 (0.590 μm) −0.4 (14.0 μm) [4.43]	**1.59** (0.150 μm) **1.50** (0.200 μm) **1.47** (0.266 μm) **1.4380** (0.590 μm) **1.433** (1.05 μm) **1.404** (6.00 μm)	**1.37** (9.00 μm) **1.33** (11.0 μm) **1.25** (14.0 μm) dispersion relations, see [4.43] [i]	0.82 ± 0.2 – 0.56 ± 0.8 (0.575; 0.613 μm) [4.42]
0.41–5.1 [4.84, 85] –	**2.3104** 1 3.042143 0.02178287	1.170065 0.08720717 30.83326 1101.3146 [4.86] [d]	≈ 2000 – ≈ 1000 (0.6943 and 0.7456 μm) [4.66]
0.21–5.3 [4.87] 0.905 [4.87]	**1.8422** (0.5 μm) **1.8258** (0.7 μm) **1.8186** (0.9 μm) **1.8152** (1.05 μm) **1.8149** (1.064 μm)	For dispersion relation, see [j] [4.88]	

[f] $n^2 = 1.26486 + \frac{0.30523\lambda^2}{\lambda^2 - 0.0100} + \frac{0.41620\lambda^2}{\lambda^2 - 0.0172} + \frac{0.18870\lambda^2}{\lambda^2 - 0.0262} + \frac{2.6200\lambda^2}{\lambda^2 - 4959}$;

[g] $n^2 = 1.47285 + \frac{0.16512\lambda^2}{\lambda^2 - 0.0166} + \frac{0.41222\lambda^2}{\lambda^2 - 0.0306} + \frac{0.44163\lambda^2}{\lambda^2 - 0.0350} + \frac{0.16076\lambda^2}{\lambda^2 - 0.0480} + \frac{0.33571\lambda^2}{\lambda^2 - 4822} + \frac{1.92474\lambda^2}{\lambda^2 - 9612}$;

[h] $n^2 = 1.00055 + \frac{0.19800\lambda^2}{\lambda^2 - 0.00250} + \frac{0.48398\lambda^2}{\lambda^2 - 0.0100} + \frac{0.38696\lambda^2}{\lambda^2 - 0.0164} + \frac{0.25998\lambda^2}{\lambda^2 - 0.0250} + \frac{0.08796\lambda^2}{\lambda^2 - 1640} + \frac{3.17064\lambda^2}{\lambda^2 - 3719} + \frac{0.30038\lambda^2}{\lambda^2 - 14482}$;

[i] $n^2 = 1.33973 + \frac{0.7097\lambda^2}{\lambda^2 - 0.09597^2} + \frac{0.1788\lambda^2}{\lambda^2 - 26.03^2} + \frac{3.8796\lambda^2}{\lambda^2 - 45.60^2}$;

[j] $n^2 = 3.2968230 - 0.0166197\lambda^2 + 0.0126503\lambda^{-2} + 0.0069986\lambda^{-4} - 0.0013968\lambda^{-6} + 0.0001088\lambda^{-8}$.

Table 4.4-5 Cubic, point group $\bar{4}3m$ (T_d) materials

	General	Static dielectric constant	Dissipation factor	Elastic stiffness tensor	Piezoelectric strain tensor	Elastooptic tensor
Material	ϱ (g/cm^3) Mohs hardness κ (W/m K)	ε_{11}^{S} ε_{11}^{T}	$\tan\delta_1$ (f (Hz))	c_{11} c_{44} c_{12} (10^9 Pa)	d_{14} (10^{-12} C/N)	p_{11} p_{12} p_{44} $p_{11}-p_{12}$
Cadmium telluride, CdTe (Irtran-6)	5.855 3 6.28 [4.79]	9.65 (−196 °C) [4.89] 11.00 (23 °C) [4.90]		53.5(2) 20.2(2) 36.9(3) [4.7]	1.68(8) (−196 °C) [4.89]	−0.152 −0.017 −0.057 −0.135 (10 600 nm) [4.91]
Copper bromide, CuBr		7.9(8) (22 °C) [4.92] 6.6 (−193 °C) [4.93]		43.5 14.7 34.9 [4.95]	16(1) [4.94]	0.072 0.195 −0.083 −0.123 [4.96]
Copper chloride, CuCl (nantokite)	4.136 – –	8.3(5) 9.2(5) [4.97, 98]		45.4 13.6[a] 36.3 [4.99]	27.2(5) [4.94]	0.120 0.250 −0.082 −0.130 [4.96]
Copper iodide, CuI	5.60 – –	– 6.5 (−269 °C) [4.93]		45.1 18.2[a] 30.7 [4.93]	7(1) [4.94]	3.032 0.151 −0.068 −0.119 [4.96]
Gallium antimonide, GaSb	5.614 4.5 32 [4.101]	15 [4.100] 15.7[b] [4.101]		88.4(9) 43.4(9) 40.3(8) [4.7]	−2.9 [4.100]	
Gallium arsenide, GaAs	5.3169 4.5 46.05 [4.16]	12.95(10) [4.102] 13.08 [4.90]	0.001 (2.5–10 GHz) [4.103]	118(1) 59.4(2) 53.5(5) [4.7]	−2.7 [4.100]	−0.16 −0.13 −0.05 −0.03 (1150 nm) [4.96, 104]
Gallium phosphide, GaP	4.138 5 110 [4.101]	11.1[b] [4.105]		141(3) 71.2(21) 62.4(12) [4.7]		−0.151 −0.082 −0.074 −0.069 [4.82]

Table 4.4-5 Cubic, point group $\bar{4}3m$ (T_d) materials, cont.

Electrooptic tensor	General optical properties	Refractive index	Nonlinear dielectric susceptibility Second order	Third order
r^{T}_{41} (10^{-12} m/V)	Transparency range (μm) dn/dT (10^{-5} K^{-1})	n A B C D E	d_{14} (10^{-12} m/V)	$\chi^{(3)}_{1111}$ $\chi^{(3)}_{1122}$ $\chi^{(3)}_{1133}$ (10^{-22} m^2/V^2)
6.8 (10.6 μm) [4.106]	0.85–29.9 [4.85] For temperature-dependent dispersion relations, see [4.110]	**2.693** (14 μm) 1 6.1977889 0.100533 3.2243821 5193.55 [4.111][c]	170 ± 60 (10.6 μm) 60 ± 24 (28 μm) [4.107–109]	
−2.5(5) [4.112][f]	0.50–20 [4.113] –	**2.3365** (0.4358 μm) **2.152** (0.532 μm) **2.045** (1.064 μm) **2.025** (10.6 μm) [4.92, 113, 114]	−6.5 ± 1.3 (1.064 μm) [4.114] −5.0 ± 1.5 (10.6 μm) [4.113]	
−4.97(50) [4.115]	0.45–15 [4.113] –	**1.9216** 3.580 0.03162 0.1642 0.09288 0 [4.116][c]	−5.7 ± 1.1 (1.064 μm) [4.114] −4.15 ± 1.2 (10.6 μm) [4.113]	
18 [4.117, 118][g]	0.50–20 [4.113] –	**2.5621** (0.4358 μm) **2.378** (0.532 μm) **2.245** (1.064 μm) [4.113, 114]	−4.7 ± 1.0 (1.064 μm) [4.114] −5.0 ± 1.5 (10.6 μm) [4.113]	
	1.8–20 [4.22, 85] –	**3.820** (1.8 μm) **3.898** (3.0 μm) **3.824** (5.0 μm) **3.843** (10 μm) [4.121]	+628 ± 63 (10.6 μm) [4.119, 120]	
1.43(7) (1.15 μm) 1.24(4) (3.39 μm) 1.51(5) (10.6 μm) [4.126]	0.900–17.3 [4.122] 25 (1.15 μm) 20 (3.39 μm) 20 (10.6 μm) [4.127]	**3.5072** 3.5 7.497 0.167 1.935 1382 [4.129][c]	$d_{36} = 170$ (1.064 μm) [4.13, 14, 123, 124] $d_{36} = 83$ (10.6 μm) [4.12, 14, 128]	6700 [4.64] 1400 [4.125] 1700 [4.64] (all at 10.6 μm)
−0.76(3) [4.130]	0.54–10.5 20 (0.546 μm) 16 (0.633 μm) [4.131]	**3.1057** For dispersion relation, see [4.132][d]	100 [4.105, 131] 58 (10.6 μm) [4.131]	

Table 4.4-5 Cubic, point group $\bar{4}3m$ (T_d) materials, cont.

	General	Static dielectric constant	Dissipation factor	Elastic stiffness tensor	Piezoelectric strain tensor	Elastooptic tensor
Material	ϱ (g/cm^3) Mohs hardness κ (W/m K)	ε_{11}^S ε_{11}^T	tan δ_1 (f (Hz))	c_{11} c_{44} c_{12} (10^9 Pa)	d_{14} (10^{-12} C/N)	p_{11} p_{12} p_{44} $p_{11}-p_{12}$
Indium antimonide, InSb	5.78 – 18 [4.101]	16.8[b] [4.101] 17 [4.100]		66.2(7) 30.2(3) 35.9(22) [4.7]	−2.35 [4.100]	0.46 0.58 0.064 (10.6 μm) – [4.133, 134]
Indium arsenide, InAs	5.70 3.8 27 [4.101]	15.5[b] [4.101] 14.5 [4.100]		84.4(17) 39.6(1) 46.4(19) [4.7]	−1.14 [4.100]	
Indium phosphide, InP	4.78 – 68 [4.101]	12.56(20)[b]	See [4.135]	101.1 45.6 56.1 [4.136]		
Zinc selenide, ZnSe	5.26 3–4 19 [4.137]	9.12 9.12 [4.89]		86.4(39) 40.2(18) 51.5(34) [4.7]	1.1(1) [4.89]	|0.100| |0.065| |0.065| (633 nm) [4.138] −0.10 (10 600 nm) [4.139]
Zinc telluride, ZnTe	6.34 6 18 [4.140]	10.10 10.10 [4.89]		71.5(6) 31.1(3) 40.8(1) [4.7]	0.91(5) [4.89]	−0.144 −0.094 −0.046 [4.141] −0.040 [4.142]

[a] The elastic stiffness was measured at constant electric field (c^E).

[b] This value for ε_{11} is neither at constant stress nor at constant strain.

[cde] Dispersion relations (λ (μm), $T = 20\,°\mathrm{C}$):

[c] $n^2 = A + \frac{B\lambda^2}{\lambda^2 - C} + \frac{D\lambda^2}{\lambda^2 - E}$.

[d] $n^2 = 1 + \frac{1.390\lambda^2}{\lambda^2 - 0.0296} + \frac{4.131\lambda^2}{\lambda^2 - 0.05476} + \frac{2.570\lambda^2}{\lambda^2 - 0.1190} + \frac{2.056\lambda^2}{\lambda^2 - 757.35}$.

Table 4.4-5 Cubic, point group $\bar{4}3m$ (T_d) materials, cont.

Electrooptic tensor	**General optical properties**	**Refractive index**	**Nonlinear dielectric susceptibility**	
			Second order	**Third order**
r^T_{41} (10^{-12} m/V)	Transparency range (μm) dn/dT (10^{-5} K^{-1}) –	n A B C D E	d_{14} (10^{-12} m/V)	$\chi^{(3)}_{1111}$ $\chi^{(3)}_{1122}$ $\chi^{(3)}_{1133}$ (10^{-22} m^2/V^2)
	8–30 [4.85] –	**3.904** (14 μm) **3.745** (28 μm) [4.107]	660 [4.123] 2280 ± 270 (10.6 μm) [4.145] 580 (28 μm) [4.107]	See [4.143, 144]
	3.9–20 [4.85] 50 (4 μm) 40 (6 μm) 30 (10 μm) [4.127]	**3.49** (10.6 μm) 11.1 0.71 6.5076 2.75 2084.8 [4.146][c]	346 (1.058 μm) [4.123] 249 (10.6 μm) [4.119, 120]	≈ 25 000 (10.6 and 9.5 μm) [4.64]
1.45 (1.06 μm) [4.149]	0.98–20 [4.147] 8.3 (5 μm) 8.2 (10.6 μm) 7.7 (20 μm) [4.150]	**3.44** 7.255 2.316 0.39225 2.765 1084.7 [4.151, 152][c]	136 (1.058 μm) [4.148] 105 ± 11 (10.6 μm) [4.145]	
1.8 [4.153, 154] 2.2 (10.6 μm) [4.155]	0.47–19 [4.122] –	**2.48** For dispersion relation, see [4.41][e]	+103 [4.122] +80 (10.6 μm) [4.109, 156]	1.05×10^7 (0.4606 μm) 2650 (0.532 μm) 1680 (1.064 μm) (all for $\|\chi^{(3)}\|$) [4.157]
4.27 (0.616 μm) [4.153, 154]	0.59–25 [4.14] –	**2.69** (10.6 μm) 4.27 3.01 0.142 0 0 [4.158][c]	3.47 [4.122] +90 (10.6 μm) [4.108, 109]	

[e] $n^2 = 1 + \frac{4.2980149\lambda^2}{\lambda^2 - 0.03688810} + \frac{0.62776557\lambda^2}{\lambda^2 - 0.14347626} + \frac{2.8955633\lambda^2}{\lambda^2 - 2208.4920}$.

[f] r^S_{41}.

[g] $n^3 r_{41}$.

Table 4.4-6 Cubic, point group 23 (T) materials

	General	Static dielectric constant	Dissipation factor	Elastic stiffness tensor	Piezoelectric strain tensor	Elastooptic tensor
Material	ϱ (g/cm^3) Mohs hardness κ (W/m K)	ε_{11}^S ε_{11}^T	$\tan\delta_1$ (f (Hz))	c_{11} c_{44} c_{12} (10^9 Pa)	d_{14} (10^{-12} C/N)	p_{11} p_{12} p_{44} $p_{11}-p_{12}$
Bismuth germanium oxide, $Bi_{12}GeO_{20}$ (BGO)	9.239 5 [4.159] –	38.0(4) [4.160] 40 [4.161]	0.0035 [4.161]	126.0 26.9 34.2 [4.163]	37.58(4) [4.162]	0.12 0.10 0.09(1) 0.01 [4.164–169]
Bismuth silicon oxide, $Bi_{12}SiO_{20}$ (BSO, sillenite)	9.2 5 [4.159] –	42(1) [4.170] 64(2) [4.171]	0.0004 (> 1 MHz) [4.171]	129(2) 24.7(2) 29.4(12) [4.7]		0.16 0.13 0.12 0.015 [4.172, 173]
Sodium chlorate, $NaClO_3$	2.488 – –	4.8 5.85 [4.175]		49.6(5) 11.6(2) 14.7(6) [4.7]	−1.74 [4.174]	0.162 0.24 −0.0198 −0.078 [4.176]

Table 4.4-7 Hexagonal, point group $\bar{6}m2$ (D_{3h}) materials

	General	Static dielectric tensor	Elastic stiffness tensor	Elastooptic tensor
Material	ϱ (g/cm^3) Mohs hardness $\kappa \perp c$ $\kappa \parallel c$ (W/m K)	ε_{11}^S ε_{33}^S ε_{11}^T ε_{33}^T	c_{11} c_{33} c_{44} c_{12} c_{13} (10^9 Pa)	p_{11} p_{12} p_{13} p_{31} p_{33} p_{44}
Gallium selenide, GaSe	5.0 2 9.0 ($\perp c$) 8.25 ($\parallel c$) [4.178]	7.45[a] 7.1[a] 9.80[a] 8.0(3) [4.179]	106.4(37) 35.8(15) 10.2(5) 30.0(25) 12.1(4) [4.7]	$\lvert p_{12}/p_{13}\rvert < 0.05$ [4.177]
Gallium sulfide, GaS	3.86 2 – –		127(19) 42(11) 12.0(73) 35.7(45) 14.3(81) [4.7]	

[a] From IR measurements. "ε^S" is ε at optical frequencies.

Table 4.4-6 Cubic, point group 23 (T) materials, cont.

Electrooptic tensor	General optical properties	Refractive index	Nonlinear dielectric susceptibility
r^T_{41} $(10^{-12}\,\mathrm{m/V})$	Transparency range (μm) $dn/dT\ (10^{-5}\,\mathrm{K}^{-1})$	n A B C D E	d_{14} $(10^{-12}\,\mathrm{m/V})$
4.1(1)	0.385–7 [4.159, 180] –	**2.55** (0.633 μm) [4.159]	
4.25–5.0 [4.181–183]	0.390–6 [4.159, 180] –	**2.54** (0.633 μm) [4.159]	
0.36 (0.4 μm) 0.39 (0.59 μm) [4.185]		**1.512** (0.6943 μm) For dispersion relation, see [4.186][a]	0.43 (0.6943 μm) [4.184]

[a] Dispersion relation (λ (μm), $T = 20\,°\mathrm{C}$):

$$n^2 = A + \frac{B\lambda^2}{\lambda^2 - C} + \frac{D\lambda^2}{\lambda^2 - E} + \frac{F\lambda^2}{\lambda^2 - G}.$$

Table 4.4-7 Hexagonal, point group $\bar{6}m2$ (D_{3h}) materials, cont.

Electrooptic tensor	General optical properties	Refractive index		Nonlinear dielectric susceptibility
r^T_{13} r^T_{33} r^T_{41} r^T_{51} $(10^{-12}\,\mathrm{m/V})$ at $\lambda = 633\,\mathrm{nm}$	Transparency range (μm)	n_o A B C D E F	n_e A B C D E F	d_{31} d_{33} $(10^{-12}\,\mathrm{m/V})$
22 ($n_o^3 r_{11}$) [4.188, 189]	0.62–20 [4.187]	**2.9082** 7.443 0.4050 0.0186 0.0061 3.1485 2194 [4.193][b]	**2.5676** 5.76 0.3879 0.2288 0.1223 1.8550 1780 [4.193][b]	$d_{22} = 54\,\mathrm{pm/V}$ (10.6 μm) [4.190–192]
	0.420–			$d_{16} = 84\,\mathrm{pm/V}$ (0.6943 μm) [4.194]

[b] Dispersion relations (λ (mm), $T = 20\,°\mathrm{C}$):

$$n^2 = A + \frac{B}{\lambda^2} + \frac{C}{\lambda^4} + \frac{D}{\lambda^6} + \frac{E\lambda^2}{\lambda^2 - F}.$$

Table 4.4-8 Hexagonal, point group $6mm$ (C_{6v}) materials

	General	Static dielectric tensor	Elastic stiffness tensor	Piezoelectric tensor	Elastooptic tensor
Material	ϱ (g/cm^3) Mohs hardness $\kappa \perp c$ $\kappa \parallel c$ (W/m K)	ε_{11}^{S} ε_{33}^{S} ε_{11}^{T} ε_{33}^{T}	c_{11} c_{33} c_{44} c_{12} c_{13} (10^9 Pa)	d_{31} d_{33} d_{15} d_{h} (10^{-12} C/N)	p_{11} p_{12} p_{13} p_{31} p_{33} p_{44}
Beryllium oxide, BeO (bromellite)	3.010 – 370 – [4.45]	6.82[a] 7.62[a] [4.195]	460.6 491.6 147.7 126.5 88.4 [4.196]	−0.12(3) +0.24(6) – – [4.197]	
Cadmium selenide, CdSe	5.67 3.25 6.2 ($\perp c$) 6.9 ($\parallel c$) [4.200]	9.53 10.20 9.70 10.65 [4.89]	74.1 83.6 13.17 45.2 39.3 [4.89][b]	−3.80(11) 7.81(23) −10.1(4) – [4.201, 202]	See [4.198, 199]
Cadmium sulfide, CdS (greenockite)	4.82 3–3.5 14 ($\perp c$) 16 ($\parallel c$) [4.200]	8.67(7) 9.53(7) 8.92(7) 10.20(7) [4.203, 204]	88.4(47) 95.2(22) 15.0(3) 55.4(39) 48.0(22) [4.7, 205]	−5.09(9) [4.203, 204] +9.71(29) [4.203, 204] −11.91(39) [4.203, 204] <0.05 [4.8]	−0.142 −0.066 −0.057 −0.041 −0.20 ≈|0.054| [4.82, 206]
Gallium nitride, GaN	6.15 – 1300 – [4.208]	9.5(3) 10.4(3) [4.207][a]	296 267 24.1 130 158 [4.209]		
α-Silicon carbide, SiC	3.217 – 490 [4.45] –	$\varepsilon_{33} = 6.65$ [4.213]	502 565 169 95 56 [4.214]		See [4.210–212]
Zinc oxide, ZnO (zincite)	5.605 – 29 [4.45] –	8.33(8) 8.81(10) 8.67(9) 11.26(12) [4.215, 216]	207.0 209.5 44.8 117.7 106.1 [4.215, 216][b]	−5.12(6) [4.215, 216] 12.3(2) [4.215, 216] −8.3(3) [4.215, 216] <0.2 [4.8]	|0.221| |0.099| −0.090 |0.089| −0.263 −0.061 [4.90]

Table 4.4-8 Hexagonal, point group $6mm$ (C_{6v}) materials, cont.

Electrooptic tensor	**General optical properties**	**Refractive index**		**Nonlinear dielectric susceptibility**
r^{T}_{13} r^{T}_{33} r^{T}_{51} (10^{-12} m/V) at $\lambda = 633$ nm	Transparency range (μm) dn_o/dT (10^{-5} K^{-1}) dn_e/dT (10^{-5} K^{-1})	n_o A B C D E	n_e A B C D E	d_{31} d_{33} d_{15} (10^{-12} m/V)
1.33 (r^{T}_{51}) [4.218, 219]	0.21–7, 15–25 [4.217] 0.818 [4.220] 1.34 [4.220]	**1.7055** 1 1.92274 0.0062536 1.24209 94.344 [4.222][c]	**1.7204** 1 1.96939 0.0073788 1.24209 109.82 [4.222][c]	0.23 0.32 – [4.156, 221]
1.8 (r^{S}_{13}) 4.3 (r^{S}_{33}) [4.225]	0.75–20 [4.223, 224] – –	**2.5375** 4.2243 1.768 0.227 3.12 3380 [4.227][c]	**2.5572** 4.2009 1.8875 0.2171 3.6461 3629 [4.227][c]	−29 (10.6 μm) [4.109] +55 (10.6 μm) [4.109] 23 ± 3 (1.054 mm) [4.226]
2.45(8) 2.75(8) 1.7(3) (all at 10.6 μm) [4.126]	0.53–16 [4.122] 5.86 (10.6 μm) 6.24 (10.6 μm) [4.229]	**2.212** (10.6 μm) 5.235 −0.1819 0.1651 0 [4.230][d] –	**2.225** (10.6 μm) 5.239 0.2076 0.1651 0 [4.230][d] –	16 [4.228] 32 [4.228] 18 [4.228] −16 (d_{31}, 10.6 μm) [4.12, 14] −16 (d_{32}, 10.6 μm) [4.12, 14] 32 (d_{33}, 10.6 μm) [4.12, 14]
0.57 ± 0.11 (r_{31}) 1.91 ± 0.35 (r_{33}) [4.231]	0.37– – –	**2.33** 3.60 1.75 0.0655 4.1 319 [4.207]	**2.35** 5.35 5.08 315.4 0 [4.207] –	21 44 25.5 [4.232]
	0.51–4 – –	**2.5830** 1 5.5515 0.026406 0 [4.213][c] –	**2.6225** 1 5.7382 0.028551 0 [4.213][c] –	5 50 [4.233] –
−1.4 (r^{S}_{13}) +2.6 (r^{S}_{33}) [4.225, 234, 235] −3.1 (r^{T}_{51}) at 396 nm [4.236]	0.38–6.0 – –	**1.939** 2.81418 0.87968 0.09254 0.00711 [4.222][e] –	**1.955** 2.80333 0.94470 0.09024 0.00714 [4.222][e] –	+1.7 ± 0.15 −5.6 ± 0.15 1.8 ± 0.15 [4.156, 228]

Table 4.4-8 Hexagonal, point group $6mm$ (C_{6v}) materials, cont.

	General	Static dielectric tensor	Elastic stiffness tensor	Piezoelectric tensor	Elastooptic tensor
Material	ϱ (g/cm^3) Mohs hardness $\kappa \perp c$ $\kappa \parallel c$ (W/m K)	ε_{11}^{S} ε_{33}^{S} ε_{11}^{T} ε_{33}^{T}	c_{11} c_{33} c_{44} c_{12} c_{13} (10^9 Pa)	d_{31} d_{33} d_{15} d_h (10^{-12} C/N)	p_{11} p_{12} p_{13} p_{31} p_{33} p_{44}
α-Zinc sulfide, α-ZnS (wurtzite, zinc blende)	4.09 – 27 [4.45] –	8.58 8.52 8.60(5) 8.57(7) [4.239, 240]	122.0 140.2 28.5 58.0 46.8 [4.239, 240][b]	−1.1(0) +3.2(1) −2.8(1) +1.0 [4.237, 238]	−0.115 0.017 0.025 0.0271 −0.13 −0.0627 [4.241]

[a] These values for ε_{ij} are neither at constant stress nor at constant strain.

[b] The elastic stiffness constant were measured at constant electric field (c^E).

Table 4.4-9 Hexagonal, point group 6 (C_6) materials

	General	Static dielectric tensor	Elastic stiffness tensor	Piezoelectric strain tensor	Elastooptic tensor
Material	ϱ (g/cm^3) Mohs hardness $\kappa \perp c$ $\kappa \parallel c$ (W/m K)	ε_{11}^{S} ε_{33}^{S} ε_{11}^{T} ε_{33}^{T}	c_{11} c_{33} c_{44} c_{12} c_{13} (10^9 Pa)	d_{31} d_{33} d_{14} d_{15} (10^{-12} C/N)	p_{11} p_{12} p_{13} p_{16} p_{31} p_{33} p_{44} p_{45}
Barium nitrite monohydrate, $Ba(NO_2)_2 \cdot H_2O$	3.179 – – – –	– – 7.56(8) 6.78(7) [4.242]	54.2 29.9 11.2 27.5 17.8 [4.242][b]	−1.73(17) +3.27(16) +0.47(9) −1.03(10) [4.242]	
α-Lithium iodate, $LiIO_3$	4.490 4 1.27 ($\perp c$) 0.65 ($\parallel c$) [4.247, 248]	7.9(4) (−20 °C) [4.245] 5.9(3) (−20 °C) [4.245] 8.25[a] (0 °C) [4.246] 6.53[a] (0 °C) [4.246]	82.5(8) 55.9(23) 18.0(3) 31.9(8) 20.8(87) [4.7, 249][b]	3.5 [4.8, 243, 244] 48.5 [4.8, 243, 244] 73(7) [4.246] 55.5 [4.8, 243, 244]	|0.32| – |0.24| |0.03| |0.41| |0.23| [4.250] – –

[a] The values for ε_{ij} are neither at constant stress nor at constant strain.

[b] The elastic stiffness constants were measured at constant electric field (c^E).

Table 4.4-8 Hexagonal, point group $6mm$ (C_{6v}) materials, cont.

Electrooptic tensor	General optical properties	Refractive index		Nonlinear dielectric susceptibility
r^T_{13} r^T_{33} r^T_{51} (10^{-12} m/V) at $\lambda = 633$ nm	Transparency range (μm) dn_o/dT (10^{-5} K^{-1}) dn_e/dT (10^{-5} K^{-1})	n_o A B C D E	n_e A B C D E	d_{31} d_{33} d_{15} (10^{-12} m/V)
0.92 1.85 [4.234, 235] –	0.35–23 [4.122] – –	**2.213** (10.6 μm) 3.4175 1.7396 0.07166 0 [4.222][e] –	**2.219** (10.6 μm) 3.4264 1.7491 0.07150 0 [4.222][e] –	−18.9±6.3 +37.3±12.6 21.4±8.4 (all at 10.6 μm) [4.109, 251]

[cde] Dispersion relations (λ (mm), $T = 20\,°$C):

[c] $n^2 = A + \frac{B\lambda^2}{\lambda^2 - C} + \frac{D\lambda^2}{\lambda^2 - E}$.

[d] $n^2 = A + \frac{B}{\lambda^2 - C} - D\lambda^2$.

[e] $n^2 = A + \frac{B\lambda^2}{\lambda^2 - C} - D\lambda^2$.

Table 4.4-9 Hexagonal, point group 6 (C_6) materials, cont.

Electrooptic tensor	General optical properties	Refractive index		Nonlinear dielectric susceptibility
r^T_{13} r^T_{33} r^T_{41} r^T_{51} (10^{-12} m/V) at $\lambda = 633$ nm	Transparency range (μm) dn_o/dT (10^{-5} K^{-1}) dn_e/dT (10^{-5} K^{-1})	n_o A B C D	n_e A B C D	d_{31} d_{33} (10^{-12} m/V)
+3.47(10) +3.31(13) −0.85(21) −0.41(10) [4.242]	0.4–2 [4.252] – –	**1.6266** (0.5 μm) 1.99885 0.542910 0.04128 0.01012 [4.252][c]	**1.5238** (0.5 μm) 1.48610 0.775625 0.01830 0.00090 [4.252][c]	1.14 [4.252] –
6.4 4.2 3.1 7.9 (r^S_{51}) [4.256, 257]	0.3–6.0 [4.253, 254] −9.38 [4.255] −8.25 [4.255]	**1.8571** 3.415716 0.047031 0.035306 0.008801 [4.258][c]	**1.7165** 2.918692 0.035145 0.028224 0.003641 [4.258][c]	4.4 4.5 [4.12, 14]

[c] Dispersion relation (λ (mm), $T = 20\,°$C):

$n^2 = A + \frac{B}{\lambda^2 - C} - D\lambda^2$.

Table 4.4-10 Trigonal, point group $\overline{3}m$ (D_{3d}) materials

	General	Static dielectric tensor	Dissipation factor	Elastic stiffness tensor	Elastic compliance tensor	Elastooptic tensor
Material	ϱ (g/cm^3) Mohs hardness $\kappa \perp c$ $\kappa \parallel c$ (W/m K)	ε_{11}^{S} ε_{33}^{S} ε_{11}^{T} ε_{33}^{T}	$\tan\delta_1$ $\tan\delta_3$	c_{11} c_{33} c_{44} c_{12} c_{13} c_{14} (10^9 Pa)	s_{11} s_{33} s_{44} s_{12} s_{13} s_{14} (10^{-12} Pa^{-1})	p_{11} p_{12} p_{13} p_{14} p_{31} p_{33} p_{41} p_{44}
α-Aluminium oxide, Al_2O_3 (sapphire)	3.98 9 33 ($\perp c$) 35 ($\parallel c$) [4.259]	9.4 11.5 [4.48][a]	0.0001 [4.48] –	496(3) 499(4) 146(3) 159(10) 114(3) −23(1) [4.7,260]	2.35 2.17 6.95 −0.70 −0.38 0.46 [4.260]	−0.23 −0.03 0.02 0.00 0.04 −0.20 0.01 −0.10 (644 nm) [4.261]
Calcite, $CaCO_3$ (calcspar, Iceland spar)	2.71 3 3.00 ($\perp c$) 3.40 ($\parallel c$) (50 °C) [4.262]	8.0[ab] [4.26]		144(3) 84.3(26) 33.5(7) 54.2(47) 51.2(34) −20.5(2) [4.7,263]	11.4(2) 17.3(5) 41.4(12) −4.0(4) 4.5(4) 9.5(5) [4.7]	0.062 0.147 0.186 −0.011 0.241 0.139 −0.036 −0.058 (514 nm) [4.264]

[a] These values for ε_{11} are neither at constant stress nor at constant strain.
[b] It is not clear which relative dielectric constant ε_{ij} is specified.

Table 4.4-11 Trigonal, point group 32 (D_3) materials

	General	Static dielectric tensor	Elastic stiffness tensor	Piezoelectric tensor	Elastooptic tensor
Material	ϱ (g/cm^3) Mohs hardness $\kappa \perp c$ $\kappa \parallel c$ (W/m K)	ε_{11}^{S} ε_{33}^{S} ε_{11}^{T} ε_{33}^{T}	c_{11} c_{33} c_{44} c_{12} c_{13} c_{14} (10^9 Pa)	d_{11} d_{14} (10^{-12} C/N)	p_{11} p_{12} p_{13} p_{14} p_{31} p_{33} p_{41} p_{44}
Aluminium phosphate, $AlPO_4$ (berlinite)	2.620 6.5 ≈6 [4.45] –	5.88 [4.8,265] – 4.73 4.62 [4.268,269]	67.0(26) 87.2(10) 42.9(4) 9.3(14) 13.1(21) −12.7(5) [4.7,270][a]	−3.0 1.3 [4.266,267]	

Table 4.4-10 Trigonal, point group $\overline{3}m$ (D_{3d}) materials, cont.

General optical properties	Refractive index	
Transparency range (μm) dn_o/dT (10^{-5} K^{-1}) dn_e/dT (10^{-5} K^{-1})	n_o A B C D E F	n_e A B C D E F
0.147–5.2 [4.87] – –	**1.7655** (0.7 μm) 1.077 0.0033 1.025 0.0114 5.04 151.2 [4.271][cd]	**1.7573** (0.7 μm) 1.041 0.0004 1.030 0.0141 3.55 123.8 [4.271][cd]
0.2–2.3 0.21 (0.633 μm) 1.19 (0.633 μm) [4.22]	**1.6428** (1.042 μm) [4.272]	**1.4799** (1.042 μm) [4.272]

[c] Dispersion relation (λ (μm), $T = 20\,°\mathrm{C}$):

$$n^2 = 1 + \frac{A\lambda^2}{\lambda^2 - B} + \frac{C\lambda^2}{\lambda^2 - D} + \frac{E\lambda^2}{\lambda^2 - F}.$$

[d] For crystals grown by heat exchanger method (HEM).

Table 4.4-11 Trigonal, point group 32 (D_3) materials, cont.

Electrooptic tensor	General optical properties	Refractive index		Nonlinear dielectric susceptibility
r^T_{11} r^T_{41} (10^{-12} m/V) at $\lambda = 633$ nm	Transparency range (μm) dn_o/dT (10^{-5} K^{-1}) dn_e/dT (10^{-5} K^{-1})	n_o A B C D E	n_e A B C D E	d_{11} (10^{-12} m/V)
		1.5161 (1 μm) [4.273]	**1.5285** (1 μm) [4.273]	0.53 [4.228] $d_{14} = 0.013$ [4.228]

Table 4.4-11 Trigonal, point group 32 (D_3) materials, cont.

	General	Static dielectric tensor	Elastic stiffness tensor	Piezoelectric tensor	Elastooptic tensor
Material	ϱ (g/cm^3) Mohs hardness $\kappa \perp c$ $\kappa \parallel c$ (W/m K)	ε^S_{11} ε^S_{33} ε^T_{11} ε^T_{33}	c_{11} c_{33} c_{44} c_{12} c_{13} c_{14} (10^9 Pa)	d_{11} d_{14} (10^{-12} C/N)	p_{11} p_{12} p_{13} p_{14} p_{31} p_{33} p_{41} p_{44}
α-Mercuric sulfide, HgS (cinnabar)	8.05 2–2.5 – –	14.0(3) 25.5(5) 15.0(3) 25.5(5) [4.275]	35.36 50.92 21.40 7.02 8.66 11.3 [4.277][b]	19.1 ≈|1.7| [4.8, 274]	– – |0.445| – – |0.115| [4.276] – –
Potassium fluoroboratoberyllate, $KBe_2BO_3F_2$ (KBBF)					
α-Silicon dioxide, SiO_2 (quartz)	2.6485 7 6.5 (⊥ c) 11.7 (‖ c) [4.200]	4.435 4.640 4.520 4.640 [4.280]	86.6(3) 106.4(12) 58.0(7) 6.7(9) 12.4(16) 17.8(3) [4.7][a]	+2.3 −0.67 [4.278, 279]	0.16 0.27 0.27 −0.030 0.29 0.10 −0.047 −0.079 (589 nm) [4.281]
Tellurium, Te	6.25 2–2.5 2.0 (⊥ c) 3.4 (‖ c) [4.285]	33 [4.282] 53 [4.282] – 37 [4.284]	33(1) 71(3) 31.8(25) 9.1(7) 24.5(15) (−)12.4(15) [4.7]	+55 +50 [4.283]	0.164 0.138 0.146 −0.04 −0.086 0.038 0.28 0.14 (10.6 μm) [4.286]

[a] These values are approximately the constant-field values c^E.

[b] c^E at constant electric field.

[cde] Dispersion relations (λ (μm), $T = 20\,°C$):

[c] $n^2 = A + \frac{B\lambda^2}{\lambda^2 - C} + \frac{D\lambda^2}{\lambda^2 - E}$.

Table 4.4-11 Trigonal, point group 32 (D_3) materials, cont.

Electrooptic tensor	General optical properties	Refractive index		Nonlinear dielectric susceptibility
r^T_{11} r^T_{41} (10^{-12} m/V) at $\lambda = 633$ nm	Transparency range (μm) dn_o/dT (10^{-5} K^{-1}) dn_e/dT (10^{-5} K^{-1})	n_o A B C D E	n_e A B C D E	d_{11} (10^{-12} m/V)
3.1 1.55 [4.289]	0.63–13.5 [4.287] – –	**2.7041** 4.1506 2.7896 0.1328 1.1378 705 [4.227][c]	**2.9909** 4.0101 4.3736 0.1284 1.5604 705 [4.227][c]	50 (10.6 μm) [4.288]
	0.155–3.5 [4.290] – –	**1.487** (0.4 μm) 1 1.169725 0.00624 0.009904 [4.290][d] –	**1.410** (0.4 μm) 1 0.956611 0.0061926 0.027849 [4.290][d] –	0.76 [4.290]
−0.445(10) +0.1904(50) [4.293]	0.15–4.5 [4.262] −0.62 (546 nm) −0.7 (546 nm) [4.68]	**1.5350** (1 μm) For dispersion relation, see [4.68][e]	**1.5438** (1 μm) For dispersion relation, see [4.68][e]	0.30 [4.12–14, 128, 291, 292]
	3.8–32 [4.85] – –	**4.7979** (10.6 μm) 18.5346 4.3289 3.9810 3.78 11.813 [4.227][c]	**6.2483** (10.6 μm) 29.5222 9.3068 2.5766 9.235 13.521 [4.227][c]	598 (10.6 μm) [4.294]

[d] $n^2 = A + \dfrac{B\lambda^2}{\lambda^2 - C} - D\lambda^2$.

[e] $n_o^2 = 1 + \dfrac{0.663044\lambda^2}{\lambda^2 - 0.0036} + \dfrac{0.517852\lambda^2}{\lambda^2 - 0.0112} + \dfrac{0.175912\lambda^2}{\lambda^2 - 0.0142} + \dfrac{0.565380\lambda^2}{\lambda^2 - 78.216} + \dfrac{1.675299\lambda^2}{\lambda^2 - 430.23}$;

$n_e^2 = 1 + \dfrac{0.665721\lambda^2}{\lambda^2 - 0.0036} + \dfrac{0.503511\lambda^2}{\lambda^2 - 0.0112} + \dfrac{0.214792\lambda^2}{\lambda^2 - 0.0142} + \dfrac{0.539173\lambda^2}{\lambda^2 - 77.30} + \dfrac{1.807613\lambda^2}{\lambda^2 - 390.85}$.

Part 4 | 4.4

Table 4.4-12 Trigonal, point group $3m$ (C_{3v}) materials

	General	Static dielectric tensor	Dissipation factor	Elastic stiffness tensor	Piezoelectric tensor	Elastooptic tensor
Material	ϱ (g/cm^3) Mohs hardness $\kappa \perp c$ $\kappa \parallel c$ (W/m K)	ε_{11}^{S} ε_{33}^{S} ε_{11}^{T} ε_{33}^{T}	tan δ_1 tan δ_3 (f (kHz))	c_{11} c_{33} c_{44} c_{12} c_{13} c_{14} (10^9 Pa)	d_{22} d_{31} d_{33} d_{15} d_{h} (10^{-12} C/N)	p_{11} p_{12} p_{13} p_{14} p_{31} p_{33} p_{41} p_{44}
β-Barium borate, BaB_2O_4 (BBO)	3.849 4 1.2 ($\perp c$) 1.6 ($\parallel c$) [4.200]	– – 6.7 8.1 [4.295]	$< 10^{-3}$ $< 10^{-3}$ (10–50 °C, 1–100 kHz) [4.295]	123.8 53.3 7.8 60.3 49.4 12.3 [4.295]		
Lithium niobate, $LiNbO_3$	4.628 5–5.5 4.6 [4.137]	43.9(22) [4.296] 23.7(12) [4.296] 85.2 [4.298] 28.7 [4.298]	0.0015(4) 0.0011(3) (1 GHz) [4.296]	202(2) 244(5) 60.2(6) 55(2) 72(5) 8.5(7) [4.7][a]	20.7(1) [4.297] −0.86(2) [4.297] +16.2(1) [4.297] 74.0(3) [4.297] 6.310(14) [4.299]	−0.031(9) 0.082(9) 0.135(7) −0.076(15) 0.170(14) 0.069(5) −0.150(10) 0.147(100) [4.11][b]
Lithium niobate (5% MgO-doped) $MgO:LiNbO_3$						
Lithium tantalate, $LiTaO_3$	7.45 5.5	42.6 42.8 53.6 43.4 [4.298]	0.0013(4) 0.0007(3) (1 GHz) [4.296]	230(2) 276(6) 96(1) 42(6) 79(4) −11(1) [4.7,301,302]	8.5 [4.300] −3.0 [4.300] 9.2 [4.300] 26 [4.300] 2.000(12) [4.299]	−0.081 0.081 0.093 −0.026 0.089 −0.044 −0.085 0.028 [4.303,304]

Table 4.4-12 Trigonal, point group $3m$ (C_{3v}) materials, cont.

	Electrooptic tensor	General optical properties	Refractive index		Nonlinear dielectric susceptibility
	r^T_{13} r^T_{22} r^T_{33} r^T_{51} (10^{-12} m/V) at $\lambda = 633$ nm	Transparency range (μm) dn_o/dT (10^{-5} K^{-1}) dn_e/dT (10^{-5} K^{-1})	n_o A B C D E F	n_e A B C D E F	d_{31} d_{33} d_{22} d_{15} (10^{-12} m/V)
β-Barium borate, BaB_2O_4 (BBO)	+0.27(2) −2.41(3) +0.29(3) +1.7(1) [4.310]	0.189–3.5 [4.295, 305–308] −1.66 [4.295] −0.93 [4.295]	**1.6551** 2.7359 0.01878 0.01822 0.01471 0.0006081 0.0000674 [4.311][h]	**1.5426** 2.3753 0.01224 0.01667 0.01627 0.0005716 0.00006305 [4.311][h]	2.16 ± 0.08 (d_{22}) [4.12, 14, 309]
Lithium niobate, $LiNbO_3$	+9.6 +6.8 +30.9 +32.6 [4.315, 316]	0.33–5.5 [4.312] $dn_o/dT = 0.141$ [4.313][f] $dn_o/dT = 2.0$ [4.314][g] $dn_e/dT = 3.85$ [4.313][f] $dn_e/dT = 7.6$ [4.314][g]	**2.2340** 4.91296 0.116275 0.048398 0.0273 [4.313][i] – –	**2.1554** 4.54528 0.091649 0.046079 0.0303 [4.313][i] – –	−4.6 −25 [4.13, 14] – –
Lithium niobate (5% MgO-doped) $MgO:LiNbO_3$	11.2(5) – 36.0(5) – (for 10.7% MgO) [4.320]	0.4–5 [4.317–319] – –	**2.2272** 4.9017 0.112280 0.049656 0.039636 [4.317, 318][i] – –	**2.1463** 4.5583 0.091806 0.048086 0.032068 [4.317, 318][i] – –	−4.69 [4.309] |25| [4.13] – –
Lithium tantalate, $LiTaO_3$	4.5 0.3 27 15 (all for r^S at 3.39 μm) [4.323]	0.4–5.5 – –	**2.1366** (1.058 μm) [4.321]	**2.1406** (1.058 μm) [4.321]	−1.0 ± 0.2 −16 ± 2 +1.7 ± 0.2 [4.156, 322] –

Table 4.4-12 Trigonal, point group $3m$ (C_{3v}) materials, cont.

	General	Static dielectric tensor	Dissipation factor	Elastic stiffness tensor	Piezoelectric tensor	Elastooptic tensor
Material	ϱ (g/cm^3) Mohs hardness $\kappa \perp c$ $\kappa \parallel c$ (W/m K)	ε^S_{11} ε^S_{33} ε^T_{11} ε^T_{33}	$\tan\delta_1$ $\tan\delta_3$ (f (kHz))	c_{11} c_{33} c_{44} c_{12} c_{13} c_{14} (10^9 Pa)	d_{22} d_{31} d_{33} d_{15} d_h (10^{-12} C/N)	p_{11} p_{12} p_{13} p_{14} p_{31} p_{33} p_{41} p_{44}
Silver antimony sulfide, Ag_3SbS_3 (pyrargyrite)	5.83 2–2.5 – –	– – 21.7(7) 24.7(7) [4.324]		52.7 35.6 11.6 26.3 28.0 −0.4 [4.324, 327][c]	16.2 −19.5 62.2 31.6 – [4.325, 326]	
Silver arsenic sulfide, Ag_3AsS_3 (proustite)	5.63 2–2.5 – –	20.2(2) 20.2(2) 21.5 22.0 [4.324]		56.76 37.0 9.12 30.4 30.4 −0.16 [4.324]	12.1 −13.8 30 −20 [4.324] –	|0.056| |0.082| |0.068| – |0.103| |0.100| | ≈ 0.01| (1150 nm) – [4.328, 329]
Thallium arsenic selenide, Tl_3AsSe_3 (TAS)	7.83 2–3 – –			See [4.330]		See [4.330]
Tourmaline, (Na, Ca)(Mg, Fe)$_3$B$_3$Al$_6$ Si$_6$(O, OH, F)$_{31}$	3.0–3.2 7–7.5 3.03 (⊥ c) 2.92 (∥ c) (125 °C) [4.332][d]	6.3[e] [4.278] 7.1[e] [4.278] 8.2 [4.331] 7.5 [4.331]		277(19) 163(11) 64(3) 65(24) 32(20) −6.9(25) [4.7]	−0.3 [4.331] −0.34 [4.331] −1.8 [4.331] −3.6 [4.331] 4.00(6) [4.333, 334] [4.333, 334]	

[a] These values are approximately the constant-field values c^E.
[b] p^E at constant electric field.
[c] c^E at constant electric field.
[d] The thermal conductivities have been interpolated from $\kappa(\perp c) = 0.108 \times T^{0.556}$ and $\kappa(\parallel c) = 0.492 \times T^{0.297}$, valid for temperature ranges of 398.2–723.2 K and 393.2–729.2 K, respectively [4.332].
[e] These values for ε_{ij} are neither at constant stress nor at constant strain.
[f] Fabricated by vapor transport equilibration (VTE).
[g] Stoichiometric melt.

Table 4.4-12 Trigonal, point group $3m$ (C_{3v}) materials, cont.

Electrooptic tensor	General optical properties	Refractive index		Nonlinear dielectric susceptibility
r^T_{13}	Transparency range (μm)	n_o	n_e	d_{31}
r^T_{22}	dn_o/dT (10^{-5} K^{-1})	A	A	d_{33}
r^T_{33}	dn_e/dT (10^{-5} K^{-1})	B	B	d_{22}
r^T_{51}		C	C	d_{15}
(10^{-12} m/V)		D	D	(10^{-12} m/V)
at $\lambda = 633$ nm		E	E	
		F	F	
	0.7–14 [4.335]	**2.9458**	**2.7956**	7.8 (10.6 μm)
	–	1	1	–
	–	6.585	5.845	8.2 (10.6 μm)
		0.16	0.16	–
		0.1133	0.0202	[4.3,4,294,336]
		225 [4.335][j]	225 [4.335][j]	
		–	–	
2.0 [4.337,338]	0.6–13 [4.339]	**2.8163**	**2.5822**	10.4 (10.6 μm)
1.05(4) [4.340]	–	9.220	7.007	–
0.22 [4.337,338]	–	0.4454	0.3230	16.6 (10.6 μm)
3.4(5) [4.340]		0.1264	0.1192	
		1733	660	10.8 (10.6 μm)
		1000 [4.341][k]	1000 [4.341][k]	[4.339,342]
		–	–	
	1.28–17 [4.343]	**3.331** (11 μm)	**3.152** (11 μm)	d_{eff} SHG 10.6 μm:
	−4.52 [4.344]	1	1	67.5 [4.342,343]
	+3.55 [4.344]	10.210	8.993	36.5 [4.294,343]
		0.197136	0.197136	29 [4.345]
		0.522	0.308	20 [4.346]
		625 [4.344][j]	625 [4.344][j]	
		–	–	
$r^S_{13} = 1.7$ [4.347]		**1.6274**	**1.6088**	0.13(3)
$r^T_{22} = 0.3$ (589 nm)		1	1	0.47(6)
[4.348]		1.6346	1.57256	0.07(1)
$r^S_{33} = 1.7$ [4.347]		0.010734	0.011346	0.23(4) [4.321]
		0 [4.349][j]	0 [4.349][j]	
		–	–	
		–	–	

[hijk] Dispersion relations (λ (μm), $T = 20\,°$C):

[h] $n^2 = A + \frac{B}{\lambda^2 - C} - D\lambda^2 + E\lambda^4 - F\lambda^6$.

[i] $n^2 = A + \frac{B}{\lambda^2 - C} - D\lambda^2$.

[j] $n^2 = A + \frac{B\lambda^2}{\lambda^2 - C} + \frac{D\lambda^2}{\lambda^2 - E}$.

[k] $n^2 = A + \frac{B}{\lambda^2 - C} + \frac{D}{\lambda^2 - E}$.

Part 4 | 4.4

Table 4.4-13 Tetragonal, point group $4/mmm$ (D_{4h}) materials

	General	Static dielectric tensor	Dissipation factor	Elastic stiffness tensor	Elastic compliance tensor	Elastooptic tensor
Material	ϱ (g/cm^3) Mohs hardness $\kappa \perp c$ $\kappa \parallel c$ (W/m K)	ε_{11}^{S} ε_{33}^{S} ε_{11}^{T} ε_{33}^{T}	$\tan \delta_1$	c_{11} c_{33} c_{44} c_{66} c_{12} c_{13} (10^9 Pa)	s_{11} s_{33} s_{44} s_{66} s_{12} s_{13} (10^{-12} Pa^{-1})	p_{11} p_{12} p_{13} p_{31} p_{33} p_{44} p_{66}
Magnesium fluoride, MgF_2	3.177 6 21 ($\perp c$) 30 ($\parallel c$) [4.200]	5.85 4.87 [4.21][a]		138(7) 201(10) 56.5(5) 96(2) 88(7) 62(4) [4.7]	12.6(1) 6.0(3) 17.7(2) 10.5(2) −7.2(2) −1.7(1) [4.7]	0.041 0.119 0.078 −0.099 0.014 – −0.0475 [4.350]
Titanium dioxide, TiO_2 (rutile)	4.26 6–6.5 8.8 ($\perp c$) (40 °C) 12.6 ($\parallel c$) [4.262]	85 190 [4.79][a]	0.017 [4.79] –	269(3) 480(5) 124(1) 192(2) 177(4) 146(6) [4.7]	6.8(3) 2.60(1) 8.06(5) 5.21(5) −4.0(3) −0.85(3) [4.7]	0.012(4) 0.162(16) −0.157(12) −0.092(9) −0.059(3) 0.020 −0.060 [4.351]
Yttrium vanadate, YVO_4 (YVO)	4.22 5 5.32 ($\perp c$) 5.10 ($\parallel c$) [4.48]					

[a] These values for ε_{ij} are neither at constant stress nor at constant strain.

Table 4.4-13 Tetragonal, point group $4mmm$ (D_{4h}) materials, cont.

General optical properties	Refractive index		Nonlinear dielectric susceptibility
Transparency range (μm) dn_o/dT (10^{-5} K^{-1}) dn_e/dT (10^{-5} K^{-1})	n_o A B C D E	n_e A B C D E	$\chi^{(3)}$
0.12–7.5 0.17 (0.4 μm) 0.23 (0.4 μm) [4.21]	**1.48** (0.150 μm) **1.423** (0.200 μm) **1.3776** (0.590 μm) **1.373** (1.05 μm) **1.32** (6.00 μm) **1.29** (7.50 μm) dispersion relations, see [4.43] [b]	**1.49** (0.150 μm) **1.437** (0.200 μm) **1.3894** (0.590 μm) **1.385** (1.05 μm) **1.33** (6.00 μm) **1.30** (7.50 μm) dispersion relations, see [4.43] [b]	[4.21]
0.42–4.0 [4.84] 0.4 (0.4 μm) [4.353] −0.9 (0.4 μm) [4.353]	**2.4851** 5.913 0.2441 0.0803 0 [4.353] [c] –	**2.7488** 7.197 0.3322 0.0843 0 [4.353] [c] –	See [4.352]
0.4–5 0.85 0.3 [4.354]	**1.9573** 3.77834 0.069736 0.04724 0.0108133 [4.354] [d] –	**2.1652** 4.59905 0.110534 0.04813 0.0122676 [4.354] [d] –	

[bcd] Dispersion relations (λ (μm), $T = 20\,°$C):

[b] $$n_o^2 = 1.27620 + \frac{0.60967\lambda^2}{\lambda^2 - 0.08636^2} + \frac{0.0080\lambda^2}{\lambda^2 - 18.0^2} + \frac{2.14973\lambda^2}{\lambda^2 - 25.0^2}\,;$$

$$n_e^2 = 1.25385 + \frac{0.66405\lambda^2}{\lambda^2 - 0.088504^2} + \frac{1.0899\lambda^2}{\lambda^2 - 22.2^2} + \frac{0.1816\lambda^2}{\lambda^2 - 24.4^2} + \frac{2.1227\lambda^2}{\lambda^2 - 40.6}\,;$$

[c] $$n^2 = A + \frac{B\lambda^2}{\lambda^2 - C} + \frac{D\lambda^2}{\lambda^2 - E}\,;$$

[d] $$n^2 = A + \frac{B}{\lambda^2 - C} - D\lambda^2\,.$$

Table 4.4-14 Tetragonal, point group $4/m$ (C_{4h}) materials

	General	Elastic stiffness tensor	Elastic compliance tensor	Elastooptic tensor
Material	ϱ (g/cm^3) Mohs hardness $\kappa \perp c$ $\kappa \parallel c$ (W/m K)	c_{11} c_{33} c_{44} c_{66} c_{12} c_{13} c_{16} (10^9 Pa)	s_{11} s_{33} s_{44} s_{66} s_{12} s_{13} s_{16} (10^{-12} Pa^{-1})	p_{11} p_{12} p_{13} p_{16} p_{31} p_{33} p_{44} p_{45} p_{61} p_{66}
Lead molybdate, $PbMoO_4$	6.95 2.5–3 150 ($\perp c$) 150 ($\parallel c$) [4.355]	107.2(19) 93.2(13) 26.4(3) 34.8(8) 61.9(55) 52.0(8) −15.8(18) [4.7]	20.8(2) 16.3(8) 37.9(4) 42.3(20) −11.8(6) −5.0(5) 14.8(12) [4.356]	0.253(23) 0.253(23) 0.273(43) 0.015(4) 0.163(20) 0.298(12) 0.04(0) −0.01(0) 0.025(21) 0.046(5) (477–633 nm) [4.357]
Yttrium lithium fluoride, $YLiF_4$ (YLF)	3.995 4–5 7.2 ($\perp c$) 5.8 ($\parallel c$) [4.358]	121 156 40.9 17.7 60.9 52.6 −7.7 [4.359]	12.8 7.96 24.4 63.6 −6.0 −2.3 8.16 [4.359]	

[a] Dispersion relation (λ (µm), $\lambda > 0.5$ µm, $T = 20$ °C):

$$n^2 = A + \frac{B\lambda^2}{\lambda^2 - C} .$$

Table 4.4-15 Tetragonal, point group 422 (D_4) materials

	General	Static dielectric tensor	Dissipation factor	Elastic stiffness tensor	Piezoelectric tensor	Elastooptic tensor
Material	ϱ (g/cm^3) Mohs hardness $\kappa \perp c$ $\kappa \parallel c$ (W/m K)	ε^S_{11} ε^S_{33} ε^T_{11} ε^T_{33}	$\tan\delta_1$ $\tan\delta_3$ (f (kHz))	c_{11} c_{33} c_{44} c_{66} c_{12} c_{13} (10^9 Pa)	d_{14} (10^{-12} C/N)	p_{11} p_{12} p_{13} p_{31} p_{33} p_{44} p_{66}
Tellurium dioxide, TeO_2 (paratellurite)	5.99 4 3.0(3) [4.362] –	22.7 [4.360] 24.9[a] [4.361] 22.9(10) [4.360] 24.7(15) [4.360]	0.0011 0.012 (100 kHz) [4.360]	55.9(2) 105.5(4) 26.7(2) 66.3(4) 51.6(3) 23.9(22) [4.7]	8.13(70) [4.360]	0.007 0.187 0.340 0.091 0.240 −0.17 −0.046 [4.363]

[a] This value for ε_{33} is neither at constant stress nor at constant strain.

Table 4.4-14 Tetragonal, point group $4/m$ (C_{4h}) materials, cont.

General optical properties	Refractive index	
Transparency range (μm) dn_o/dT (10^{-5} K^{-1}) dn_e/dT (10^{-5} K^{-1})	n_o A B C	n_e A B C
0.42–5.5 [4.364] $-(3.0\pm0.2)n_o$ $-(1.8\pm0.2)n_e$ [4.355]	**2.38** (633 nm) [4.364] 1 4.0650407 0.0536585 [4.355][a]	**2.25** (633 nm) [4.364] 1 3.7037037 0.0400000 [4.355][a]
0.18–6.7 −0.2 −0.43 [4.354]	**1.448** 1.38757 0.70757 0.00931 0.18849 50.99741 [4.354][c]	**1.470** 1.31021 0.84903 0.00876 0.53607 134.9566 [4.354][c]

[c] $n^2 = A + \frac{B\lambda^2}{\lambda^2 - C} + \frac{D\lambda^2}{\lambda^2 - E}$.

Table 4.4-15 Tetragonal, point group 422 (D_4) materials, cont.

Electrooptic tensor	General optical properties	Refractive index		Nonlinear dielectric susceptibility
r^T_{41} (10^{-12} m/V) at $\lambda = 633$ nm	Transparency range (μm) dn_o/dT (10^{-5} K^{-1}) dn_e/dT (10^{-5} K^{-1})	n_o A B C D E	n_e A B C D E	d_{14} (10^{-12} m/V)
0.62 [4.365]	0.35–6 [4.366] 0.9 (0.644 μm) 0.8 (0.644 μm) [4.368]	**2.2005** 1 2.584 0.0180 1.157 0.0696 [4.368][b]	**2.3431** 1 2.823 0.0180 1.542 0.0692 [4.368][b]	0.39 [4.367]

[b] Dispersion relation (λ (μm), $T = 20\,°$C):
$n^2 = A + \frac{B\lambda^2}{\lambda^2 - C} + \frac{D\lambda^2}{\lambda^2 - E}$

Table 4.4-16 Tetragonal, point group $\overline{4}2m$ (D_{2d}) materials

	General	**Static dielectric tensor**	**Dissipation factor**	**Elastic stiffness tensor**	**Piezoelectric tensor**	**Elastooptic tensor**
Material	ϱ (g/cm^3) Mohs hardness $\kappa \perp c$ $\kappa \parallel c$ (W/m K)	ε_{11}^{S} ε_{33}^{S} ε_{11}^{T} ε_{33}^{T}	$\tan\delta_1$ $\tan\delta_3$ (f (kHz))	c_{11} c_{33} c_{44} c_{66} c_{12} c_{13} (10^9 Pa)	d_{14} d_{36} (10^{-12} C/N)	p_{11} p_{12} p_{13} p_{31} p_{33} p_{44} p_{66}
Ammonium dideuterium phosphate, $ND_4D_2PO_4$ (AD*P, DADP)	1.885 [4.370, 371] – – –	70[a] [4.369] 26[a] [4.369] 72 [4.372] 22 [4.372]	0.003 0.015 (1 kHz) [4.369]	62.1 29.9 9.1 6.1 −5 14 [4.372][b]	10 75 [4.372]	– – – – – – 0.04 [4.373, 374]
Ammonium dihydrogen arsenate, $NH_4H_2AsO_4$ (ADA)	2.310 – – –	74 [4.375] 13 [4.375] 75 [4.376] 14 [4.376]		67.5 30.2 6.85 6.39 −10.6 16.5 [4.379]	36.5 27.5 [4.377, 378]	
Ammonium dihydrogen phosphate, $NH_4H_2PO_4$ (ADP)	1.80 2.0 1.26 ($\perp c$) 0.71 ($\parallel c$) (315 K) [4.262]	55.5(15) 15.0(5) 56.0(15) 15.5(5) [4.97, 98]	0.001 0.004 (1 kHz) [4.369]	67.3(27) 33.7(7) 8.6(2) 6.02(6) 5.0(13) 19.8(6) [4.7]	1.76 48.31 [4.380]	0.302(12) 0.252(17) 0.204(33) 0.191(6) 0.219(49) −0.058 −0.088(17) (589–633 nm) [4.381]
Cadmium germanium arsenide, $CdGeAs_2$	5.60 3.5–4 4.18 [4.140] –			98.0 86.6 43.2 42.3 60.5 59.6 [4.382]		
Cadmium germanium phosphide, $CdGeP_2$						
Cesium dideuterium arsenate, CsD_2AsO_4 (CD*A, DCDA)		– – 74 61 [4.369][cd]	0.57 0.38 (1 kHz) [4.369][cd]		10.5 125.4 [4.383][c]	0.223 0.271 0.160 0.206 0.133 −0.061 [4.384, 385] –

Table 4.4-16 Tetragonal, point group $\overline{4}2m$ (D_{2d}) materials, cont.

Electrooptic tensor	General optical properties	Refractive index		Nonlinear dielectric susceptibility
r^T_{41} r^T_{63} (10^{-12} m/V)	Transparency range (μm) dn_o/dT (10^{-5} K^{-1}) dn_e/dT (10^{-5} K^{-1})	n_o *A* *B* *C* *D* *E*	n_e *A* *B* *C* *D* *E*	d_{36} (10^{-12} m/V)
	0.22–1.7 [4.386–389] – –	**1.5049** 2.279481 1.215879 57.97555433 0.010761 0.013262977 [4.386][e]	**1.4659** 2.151161 1.199009 126.6005279 0.009652 0.009712103 [4.386][e]	0.43 (0.6943 μm) [4.390, 391]
33.5 (550 nm) 9.2 (550 nm) [4.284, 369]	0.22–1.2 [4.386] −4.56 [4.392] 1.25 [4.392]	**1.5550** 2.443449 2.017752 57.83514282 0.016757 0.018272821 [4.386][e]	**1.5081** 2.275962 1.59826 126.8851303 0.014296 0.016560859 [4.386][e]	0.43 [4.386]
26.0 (633 nm) 8.7 (633 nm)	0.18–1.53 [4.386, 393] −4.93 [4.392] 0 [4.392]	**1.5065** 2.302842 15.102464 400 0.011125165 0.013253659 [4.394][e]	**1.4681** 2.163510 5.919896 400 0.009616676 0.012989120 [4.394][e]	0.47 [4.12, 14]
	2.5–15 [4.395] – –	**3.5046** (10.6 μm) 10.1064 2.2988 1.0872 1.6247 1370 [4.227][f]	**3.5911** (10.6 μm) 11.8018 1.2152 2.6971 1.6922 1370 [4.227][f]	282 (10.6 μm) [4.395, 396]
	0.9–12 [4.396] – –	**3.1422** (10.6 μm) 5.9677 4.2286 0.2021 1.6351 671.33 [4.397][f]	**3.1563** (10.6 μm) 6.1573 4.0970 0.2330 1.4925 671.33 [4.397][f]	100 (10.6 μm) [4.396]
– 38.8 (700 nm) [4.399]	0.27–1.66 [4.398] −2.33 [4.392] −1.67 [4.392]	**1.5499** 2.40817 2.2112173 126.871163 0.015598 0.019101728 [4.386][e]	**1.5341** 2.345809 0.651843 127.3304614 0.015141 0.016836101 [4.386][e]	0.402 [4.398]

Table 4.4-16 Tetragonal, point group $\bar{4}2m$ (D_{2d}) materials, cont.

	General	Static dielectric tensor	Dissipation factor	Elastic stiffness tensor	Piezoelectric tensor	Elastooptic tensor
Material	ϱ (g/cm^3) Mohs hardness $\kappa \perp c$ $\kappa \parallel c$ (W/m K)	ε^S_{11} ε^S_{33} ε^T_{11} ε^T_{33}	$\tan\delta_1$ $\tan\delta_3$ (f (kHz))	c_{11} c_{33} c_{44} c_{66} c_{12} c_{13} (10^9 Pa)	d_{14} d_{36} (10^{-12} C/N)	p_{11} p_{12} p_{13} p_{31} p_{33} p_{44} p_{66}
Cesium dihydrogen arsenate, CsH_2AsO_4 (CsDA, CDA)	3.53 – – –	58.0[a] [4.400][d] 34.0[a] [4.400][d] 61 [4.369][d] 29 [4.369][d]	15 13 (1 kHz) [4.369]	51.6 39.9 6.66 1.7 0.56 1.33 [4.403, 404]	5.6(10) 123(8) [4.401, 402]	0.238(58) 0.225(0) 0.211(19) 0.200(4) 0.220(12) −0.042 |0.065| (633 nm) [4.384, 385]
Cesium lithium borate, $CsLiB_6O_{10}$ (CLBO)	– 4 – –					
Copper gallium selenide, $CuGaSe_2$						
Copper gallium sulfide, $CuGaS_2$	4.45 – – –	9.3[a] 10[a] [4.112]				
Potassium dideuterium arsenate, KD_2AsO_4 (KD*A, DKDA)	2.890 [4.405][c] – – –	71[a] 33[ad] [4.369] (≈ 90% deuterated)	0.02 0.20 (1 kHz) [4.369][d] (≈ 90% deuterated)	74.6 69.3 9.88 6.17 12.4 33.9 [4.406]	22.5 24.8 [4.406] (≈ 90% deuterated)	
Potassium dideuterium phosphate, KD_2PO_4 (KD*P, DKDP)	2.355 2.5 2.09 ($\perp c$) 1.86 ($\parallel c$) [4.386]	47.1(24)[a] [4.177][c] 48 [4.375][c] – 50(2) [4.407][c]	0.0068(17) 0.0072(18) (1 GHz) [4.296][c]	67.4 54.5 12.6 5.94 −5.8 12.2 [4.408]	– 58 [4.407][c]	0.241 0.247 0.245 0.236 0.245 −0.035 −0.072 [4.409, 410]

Table 4.4-16 Tetragonal, point group $\bar{4}2m$ (D_{2d}) materials, cont.

Electrooptic tensor	General optical properties	Refractive index		Nonlinear dielectric susceptibility
r^T_{41} r^T_{63} (10^{-12} m/V)	Transparency range (μm) dn_o/dT (10^{-5} K^{-1}) dn_e/dT (10^{-5} K^{-1})	n_o A B C D E	n_e A B C D E	d_{36} (10^{-12} m/V)
– 19.1 (700 nm) [4.284]	0.26–1.43 [4.398] −2.87 [4.392] −2.21 [4.392]	**1.5514** 2.420405 1.403336 57.82416181 0.016272 0.018005614 [4.386][e]	**1.5356** 2.350262 0.685328 127.2688578 0.015645 0.014820871 [4.386][e]	0.402 [4.398]
	0.18–2.75 [4.411] $-0.104\lambda^2+0.035\lambda-1.291$ [4.412] $0.331\lambda^2-0.243\lambda-0.84$ [4.412]	**1.4854** 2.2145 0.00890 0.02051 0.01413 [4.412][g] –	**1.4352** 2.0588 0.00866 0.01202 0.00607 [4.412][g] –	0.86 [4.411]
	0.73–17 [4.413] – –	**2.8358** (1.0 μm) **2.7430** (2.0 μm) **2.7133** (6.0 μm) [4.413]	**2.8513** (1.0 μm) **2.7510** (2.0 μm) **2.7192** (6.0 μm) [4.413]	27 (10.6 μm) [4.413]
1.76 (633 nm) – [4.112]	0.52–12 [4.414] 5.9 [4.415] 6.0 [4.415]	**2.4360** (10.6 μm) 4.0984 2.1419 0.1225 1.5755 738.43 [4.414][f]	**2.4201** (10.6 μm) 4.4834 1.7316 0.1453 1.7785 738.43 [4.414][f]	9.0 (10.6 μm) [4.414]
– 18.2 (550 nm) [4.284]	0.22–2.3 [4.416] – –			0.39 [4.386]
8.8(4) (546 nm) [4.418] 25.8 (633 nm) [4.419]	0.2–2.1 [4.387, 417] −3.1 [4.392] −2.1 [4.392]	**1.4928** 1.661145 0.586015 0.016017 0.691194 30 [4.397][f]	**1.4555** 1.687499 0.44751 0.017039 0.596212 30 [4.397][f]	0.37 [4.12, 14]

Table 4.4-16 Tetragonal, point group $\bar{4}2m$ (D_{2d}) materials, cont.

	General	**Static dielectric tensor**	**Dissipation factor**	**Elastic stiffness tensor**	**Piezoelectric tensor**	**Elastooptic tensor**
Material	ϱ (g/cm^3) Mohs hardness $\kappa \perp c$ $\kappa \parallel c$ (W/m K)	ε_{11}^{S} ε_{33}^{S} ε_{11}^{T} ε_{33}^{T}	$\tan\delta_1$ $\tan\delta_3$ (f (kHz))	c_{11} c_{33} c_{44} c_{66} c_{12} c_{13} (10^9 Pa)	d_{14} d_{36} (10^{-12} C/N)	p_{11} p_{12} p_{13} p_{31} p_{33} p_{44} p_{66}
Potassium dihydrogen arsenate, KH_2AsO_4 (KDA)	2.87 – – –	53 [4.420] 18 [4.420] 53.7 [4.421] 21.0 [4.421]	0.007 0.008 (10 GHz) [4.420]	64.8 48.2 10.75 6.63 0.77 13.6 [4.422]	17.1 19.1 [4.406]	– – – – – – 0.020 [4.373, 374]
Potassium dihydrogen phosphate, KH_2PO_4 (KDP)	2.3383 2.5 1.34 ($\perp c$) (42 °C) 1.21 ($\parallel c$) [4.262]	42.5(15) 20.0(5) 43.2(15) 20.8(5) [4.97, 98]	0.004 (9.2 GHz) [4.423] 0.0005 (1 GHz) [4.97, 98]	71.2(14) 57(1) 12.6(4) 6.2(1) −5.0(11) 14.1(14) [4.7]	3.5(15) 22.4(10) [4.401, 402]	0.256(18) 0.254(19) 0.224(34) 0.227(8) 0.200(54) −0.265(10) −0.063(6) (589–633 nm) [4.409, 410]
Rubidium dideuterium arsenate, RbD_2AsO_4 (RbD*A, RD*A, DRDA)	3.333 – – –	72[a] 41[ac] [4.369]	1.6 2.3 (1 kHz) [4.369][c]	49.3 38.6 9.48 4.08 −19.3 4.9 [4.424]	14.6 45.8[c] [4.424]	
Rubidium dideuterium phosphate, RbD_2PO_4 (RbD*P, RD*P, DRDP)		– 72[a] [4.425, 426]		See [4.427, 428][c]	3.3(3) 52(1)	
Rubidium dihydrogen arsenate, RbH_2AsO_4 (RbDA, RDA)	3.28 – – –	54.5[a] 28.5[a] [4.429]	2.1 5.2 [4.429]	51.0 39.2 10.4 4.31 −18.9 2.3 [4.424]	9.9 28.5 [4.400, 424]	0.227 0.239 0.200 0.205 0.182 [4.430, 431] – 0.023 [4.373, 374]

Table 4.4-16 Tetragonal, point group $\overline{4}2m$ (D_{2d}) materials, cont.

Electrooptic tensor	General optical properties	Refractive index		Nonlinear dielectric susceptibility
r^{T}_{41} r^{T}_{63} $(10^{-12}\,\mathrm{m/V})$	Transparency range (μm) dn_o/dT $(10^{-5}\,\mathrm{K}^{-1})$ dn_e/dT $(10^{-5}\,\mathrm{K}^{-1})$	n_o A B C D E	n_e A B C D E	d_{36} $(10^{-12}\,\mathrm{m/V})$
12.5 (550 nm) 10.9 (550 nm) [4.369, 399]	0.216–1.67 [4.386, 387, 390, 391] −3.95 [4.392] −2.27 [4.392]	**1.5509** 2.424647 3.742954 126.9036045 0.015841 0.018624061 [4.386][e]	**1.5059** 2.262579 0.769288 127.0537007 0.013461 0.016165851 [4.386][e]	0.41 [4.228]
−8.277(7) (589 nm) +10.22 (589 nm) [4.433]	0.174–1.57 [4.393, 417, 432] −3.4 [4.392] −2.87 [4.392]	**1.4938** 2.259276 0.01008956 0.012942625 13.00522 400 [4.434, 435][f]	**1.4599** 2.132668 0.008637494 0.012281043 3.2279924 400 [4.434, 435][f]	0.39 [4.12, 14]
– 21.4 (550 nm) [4.399]	0.26–1.7 [4.386] – –	**1.5392** 2.373255 1.979528 126.9867549 0.01543 0.015836964 [4.386][e]	**1.5091** 2.270806 0.275372 58.08499107 0.013592 0.01596609 [4.386][e]	0.31 [4.386]
	0.22–1.5 [4.386, 436] – –	**1.4913** 2.235596 0.010929 0.001414783 2.355322 126.8547185 [4.386][e]	**1.4681** 2.152727 0.010022 0.001379157 0.691253 127.0144778 [4.386][e]	0.38 [4.386]
13.5 (550 nm) [4.369] 14.8 (550 nm) [4.399]	0.26–1.46 [4.437] −3.37 [4.392] −2.21 [4.392]	**1.5405** 2.390661 3.487176 126.7648558 0.015513 0.018112315 [4.386][e]	**1.5105** 2.27557 0.720099 126.6309092 0.013915 0.01459264 [4.386][e]	0.4 (0.694 μm) [4.438, 439]

Table 4.4-16 Tetragonal, point group $\bar{4}2m$ (D_{2d}) materials, cont.

	General	Static dielectric tensor	Dissipation factor	Elastic stiffness tensor	Piezoelectric tensor	Elastooptic tensor
Material	ϱ (g/cm^3) Mohs hardness $\kappa \perp c$ $\kappa \parallel c$ (W/m K)	ε^S_{11} ε^S_{33} ε^T_{11} ε^T_{33}	$\tan\delta_1$ $\tan\delta_3$ (f (kHz))	c_{11} c_{33} c_{44} c_{66} c_{12} c_{13} (10^9 Pa)	d_{14} d_{36} (10^{-12} C/N)	p_{11} p_{12} p_{13} p_{31} p_{33} p_{44} p_{66}
Rubidium dihydrogen phosphate, RbH_2PO_4 (RbDP, RDP)	2.805 – – –	42.0 26.5 41.4 27 [4.440, 441]	0.034 0.051 (38.6 GHz) [4.440, 441]	63(7) 50.0(55) 10.6(7) 3.56(7) −6.0(6) 10.6(72) [4.7]	4.0(3) 37(1) [4.427, 428]	0.247 0.265 0.248 0.229 0.248 −0.032 −0.032 [4.409, 410]
Silver gallium selenide, $AgGaSe_2$	5.71 3–3.5 1.1 (⊥ c) 1.0 (∥ c) [4.200]	– – 10.5 12.0 [4.442]			9.0 3.7 [4.442]	
Silver gallium sulfide (silver thiogallate), $AgGaS_2$	4.58 3–3.5 1.5 (⊥ c) 1.4 (∥ c) [4.200]	10 14 [4.443][a]		87.9 75.8 24.1 30.8 58.4 59.2 [4.444, 445]		
Urea, $(NH_2)_2CO$	1.318 <2.5 – –	5–8[a] [4.26]		23.5 51.0 6.2 0.50 −0.50 7.5 [4.446]		
Zinc germanium diphosphide, $ZnGeP_2$	4.12 5.5 35 (⊥ c) 36 (∥ c) [4.200]	15 12 [4.112][a]				See [4.447, 448]

[a] These values for ε_{ij} are neither at constant stress nor at constant strain.
[b] The elastic stiffness constants were measured at constant electric field (c^E).
[c] May be only partially deuterated.
[d] *Adhav* and *Vlassopoulous* [4.369] measured arsenates with low resistivity (≈ 10^8 Ω cm), and as a result their $\tan\delta$ values are high and ε may also be high. The data for RDA in [4.369] are incorrectly labeled ADA (remarks of the compiler).

Table 4.4-16 Tetragonal, point group $\overline{4}2m$ (D_{2d}) materials, cont.

Electrooptic tensor	General optical properties	Refractive index		Nonlinear dielectric susceptibility
r^T_{41} r^T_{63} (10^{-12} m/V)	Transparency range (μm) dn_o/dT (10^{-5} K^{-1}) dn_e/dT (10^{-5} K^{-1})	n_o A B C D E F	n_e A B C D E F	d_{36} (10^{-12} m/V)
10.3 14.3 [4.449, 450]	0.22–1.5 [4.386, 436] −3.74 [4.392] −2.73 [4.392]	**1.4926** 2.249885 0.01056 0.007780475 3.688005 127.1998253 [4.386][e] –	**1.4700** 2.159913 0.009515 0.00847799 0.988431 127.692938 [4.386][e] –	0.36 [4.438]
4.5 (1.15 μm) 3.9 (1.15 μm) [4.453]	0.71–18 [4.451] 4.5 (3.39 μm) [4.452] 7.6 (3.39 μm) [4.452]	**2.7008** 4.6453 2.2057 0.1879 1.8377 1600 [4.227][f] –	**2.6800** 5.2912 1.3970 0.2845 1.9282 1600 [4.227][f] –	33 (10.6 μm) [4.12, 14]
4.0(2) 3.0(1) [4.455]	0.47–13 [4.454] For dn/dT, see [4.456, 457]	**2.4540** 3.40684 2.40065 0.09311 2.06248 950 [4.457][f] –	**2.4012** 3.60728 1.94792 0.11066 2.24544 1030.7 [4.457][f] –	17.5 (1.06 μm) [4.12, 14] 11.2 (10.6 μm) [4.12, 14]
+1.03 (633 nm) −0.75 (633 nm) [4.460]	0.2–1.8 [4.458] – –	**1.4811** 2.1823 0.0125 0.03 0 [4.461][h] – –	**1.5825** 2.51527 0.0240 0.03 0.0202 1.52 0.08771 [4.461][h]	1.18 (0.6 μm) [4.458, 459]
1.8 (5 μm) – [4.463]	0.74–12 [4.462] 21.18 [4.415] 23.01 [4.415]	**3.2324** 4.4733 5.26576 0.13381 1.49085 662.55 [4.464, 465][f] –	**3.2786** 4.63318 5.34215 0.14255 1.45785 662.55 [4.464, 465][f] –	69 (10.6 μm) [4.12, 462]

[efgh] Dispersion relations (λ (μm), $T = 20\,°\mathrm{C}$):

[e] $n^2 = A + \frac{B\lambda^2}{\lambda^2 - C} + \frac{D}{\lambda^2 - E}$.

[f] $n^2 = A + \frac{B\lambda^2}{\lambda^2 - C} + \frac{D\lambda^2}{\lambda^2 - E}$.

[g] $n^2 = A + \frac{B}{\lambda^2 - C} - D\lambda^2$.

[h] $n^2 = A + \frac{B}{\lambda^2 - C} + \frac{D(\lambda^2 - E)}{(\lambda^2 - E) + F}$.

Table 4.4-17 Tetragonal, point group $4mm$ (C_{4v}) materials

	General	Static dielectric tensor	Dissipation factor	Elastic stiffness tensor	Piezoelectric tensor	Elastooptic tensor
Material	ϱ (g/cm^3) Mohs hardness $\kappa \perp c$ $\kappa \parallel c$ (W/m K)	ε_{11}^{S} ε_{33}^{S} ε_{11}^{T} ε_{33}^{T}	$\tan\delta_1$ $\tan\delta_3$ (f (kHz))	c_{11} c_{33} c_{44} c_{66} c_{12} c_{13} (10^9 Pa)	d_{31} d_{33} d_{15} d_h (10^{-12} C/N)	p_{11} p_{12} p_{13} p_{31} p_{33} p_{44} p_{66}
Barium titanate, $BaTiO_3$[a]	6.02 – 1.34 [4.45] –	1970 109 2920 168 [4.468]	0.13 – (24 GHz) [4.469]	243 147.9 54.9 120 128 123 [4.467][b]	−33.4 +68.5 647.0 [4.467] –	See [4.443, 466]
Lead titanate, $PbTiO_3$	7.9 – – –	102(3) 33.5(10) 130–140 105–110 [4.472]	0.02–0.05 0.02–0.05 (0.1 GHz) [4.472]	$s_{11}^{E} = 7.2$ $s_{33}^{E} = 32.5$ $s_{44}^{E} = 12.2$ $s_{66}^{E} = 7.9$ $s_{12}^{E} = -2.1$ [4.470, 471][c]	−25 +117 61 [4.470, 471] –	
Lithium tetraborate, $Li_2B_4O_7$	2.44 5 – –	78.80 71.45 82.61 87.92 [4.473]		130.9(42) 55.4(11) 57.3(14) 46.4(7) 1.5(12) 30(4) [4.7]		
Potassium lithium niobate, $K_3Li_2Nb_5O_{15}$ (KLINBO)	4.3 – – –	271 83 306 115 (405 °C) [4.474][d]		220 109 68 70 74 59 [4.474][bd]	−14 57 68 [4.474][d] –	

[a] Top-seeded solution grown (TSSG) $BaTiO_3$.
[b] The elastic stiffness constants were measured at constant electric field (c^E).
[c] Units not specified in original, but probably $10^{-12}\ Pa^{-1}$.
[d] The stoichiometry was $K_{2.89}Li_{1.55}Nb_{5.11}O_{15}$ instead of $K_3Li_2Nb_5O_{15}$.

Table 4.4-17 Tetragonal, point group $4mm$ (C_{4v}) materials, cont.

Electrooptic tensor	General optical properties	Refractive index		Nonlinear dielectric susceptibility
r_{13}^{T} r_{33}^{T} r_{51}^{T} (10^{-12} m/V) at $\lambda = 633$ nm	Transparency range (μm) dn_o/dT (10^{-5} K^{-1}) dn_e/dT (10^{-5} K^{-1})	n_o A B C D E F	n_e A B C D E F	d_{31} d_{33} (10^{-12} m/V)
8 105 1300 [4.477]	0.14–10 – –	**2.3218** 1 4.195 0.04964 0 [4.478][e] – –	**2.2894** 1 4.073 0.04456 0 [4.478][e] – –	−14.4 −5.4 [4.475, 476]
13.8 (r_{13}^{S}) 5.9 (r_{33}^{S}) [4.480]	0.6–6 [4.479] – –	**2.5715** 1 5.359 0.0502 0 [4.479][e] – –	**2.5690** 1 5.365 0.0471 0 [4.479][e] – –	+35.3 −7.0 [4.479]
+3.74 +3.67 −0.11 [4.484]	0.17–3.5 [4.481, 482] ≈ 0.1 [4.482] ≈ 0.3 [4.482]	**1.5968** (1.1 μm) 2.564310 0.012337 0.013103 0.019075 [4.482][f] – –	**1.5422** (1.1 μm) 2.386510 0.010664 0.012878 0.012813 [4.482][f] – –	0.12 0.93 [4.483]
8.9 [4.485, 486] 78 [4.485, 486] 80 [4.488]	0.4–5 [4.485–487] – –	**2.208** 1 3.708 0.04601 0 [4.321][e] – –	**2.112** 1 3.349 0.03564 0 [4.321][e] – –	11.8 (0.8 μm) [4.487] 10.5 [4.485, 486]

[ef] Dispersion relations (λ (μm), $T = 20\,°\mathrm{C}$):

[e] $n^2 = A + \frac{B}{\lambda^2 - C} - D\lambda^2$.

[f] $n^2 = A + \frac{B}{\lambda^2 - C} + \frac{D(\lambda^2 - E)}{(\lambda^2 - E) + F}$.

Table 4.4-18 Orthorhombic, point group mmm (D_{2h}) materials

	General	Static dielectric tensor	Dissipation factor	Elastic stiffness tensor	Elastic compliance tensor	Piezooptic tensor
Material	ϱ (g/cm^3) Mohs hardness κ (W/m K)	ε_{11} ε_{22} ε_{33}	$\tan\delta_1$	c_{11} c_{22} c_{33} c_{44} c_{55} c_{66} c_{12} c_{13} c_{23} (10^9 Pa)	s_{11} s_{22} s_{33} s_{44} s_{55} s_{66} s_{12} s_{13} s_{23} (10^{-12} Pa^{-1})	
Magnesium silicate, Mg_2SiO_4 (forsterite)	3.217 7 5.12 [4.332]	6.2[a] [4.26]		328.5(13) 200.0(1) 235.3(7) 66.9(6) 81.3(1) 80.9(2) 68.0(16) 68.7(7) 72.8(6) [4.7]	3.37 5.84 4.93 15.0 12.3 12.4 −0.85 −0.71 −1.56 [4.7]	$q_{11}+q_{12}+q_{13}=0.16$ $q_{21}+q_{22}+q_{23}=0.20$ $q_{31}+q_{32}+q_{33}=0.27$ (589 nm) [4.489]
Yttrium aluminate, $YAlO_3$ (YAP, YALO)	5.35 8.5 11 [4.259]	16–20[a] [4.48]	0.001 [4.48]			

[a] It is not clear which relative dielectric constant ε_{ij} is specified.

Table 4.4-19 Orthorhombic, point group 222 (D_2) materials

	General	Static dielectric tensor	Elastic stiffness tensor	Piezoelectric tensor	Elastooptic tensor
Material	ϱ (g/cm^3)	ε^S_{11} ε^S_{22} ε^S_{33} ε^T_{11} ε^T_{22} ε^T_{33}	c_{11} c_{22} c_{33} c_{44} c_{55} c_{66} c_{12} c_{13} c_{23} (10^9 Pa)	d_{14} d_{25} d_{36} (10^{-12} C/N)	p_{11} p_{12} p_{13} p_{21} p_{22} p_{23} p_{31} p_{32} p_{33} p_{44} p_{55} p_{66}
Barium formate, $Ba(COOH)_2$	3.261	6.9[a] [4.490,491] 6.7[a] [4.490,491] 5.2[a] [4.490,491] 7.9 [4.331] 5.9 [4.331] 7.5 [4.331]	$s^E_{44}=78.5$ $s^E_{55}=60$ $s^E_{66}=82.5$ [4.331]	−5.6(5) −10.2(10) +1.9(1) [4.8,492]	

Part 4 | 4.4

Table 4.4-18 Orthorhombic, point group mmm (D_{2h}) materials, cont.

General optical properties	Refractive index		
Transparency range (μm) dn_X/dT (10^{-5} K^{-1}) dn_Y/dT (10^{-5} K^{-1}) dn_Z/dT (10^{-5} K^{-1})	n_X A B C	n_Y A B C	n_Z A B C
0.27–5.9 [4.87] 1.45 [4.493] – 0.98 [4.493]	**1.9111** (1 μm) 1 2.61960 0.012338 [4.494][b]	**1.9251** (1 μm) 1 2.67171 0.012605 [4.494][b]	**1.9337** (1 μm) 1 2.70381 0.012903 [4.494][b]

[b] Dispersion relation (λ (μm), $T = 20\,°\mathrm{C}$):

$$n^2 = A + \frac{B\lambda^2}{\lambda^2 - C} \,.$$

Table 4.4-19 Orthorhombic, point group 222 (D_2) materials, cont.

Electrooptic tensor	General optical properties	Refractive index			Nonlinear dielectric susceptibility
r^T_{41} r^T_{52} r^T_{63} (10^{-12} m/V)	Transparency range (μm) dn_X/dT (10^{-5} K^{-1}) dn_Y/dT (10^{-5} K^{-1}) dn_Z/dT (10^{-5} K^{-1})	n_X A B C D	n_Y A B C D	n_Z A B C D	d_{14} d_{25} d_{36} (10^{-12} m/V)
+1.81(18) −2.03(20) +0.48(10) [4.492]	0.245–2.2, 4.8–5.1 [4.495] – – –	**1.6214** 2.619 0.0177 0.039 0 [4.495][b]	**1.5819** 2.491 0.0184 0.035 0 [4.495][b]	**1.5585** 2.421 0.016 0.042 0 [4.495][b]	0.10 0.11 0.11 [4.495]

Table 4.4-19 Orthorhombic, point group 222 (D_2) materials, cont.

	General	Static dielectric tensor	Elastic stiffness tensor	Piezoelectric tensor	Elastooptic tensor
Material	ϱ (g/cm^3)	ε_{11}^S ε_{22}^S ε_{33}^S ε_{11}^T ε_{22}^T ε_{33}^T	c_{11} c_{22} c_{33} c_{44} c_{55} c_{66} c_{12} c_{13} c_{23} (10^9 Pa)	d_{14} d_{25} d_{36} (10^{-12} C/N)	p_{11} p_{12} p_{13} p_{21} p_{22} p_{23} p_{31} p_{32} p_{33} p_{44} p_{55} p_{66}
Cesium triborate, CsB_3O_5 (CBO)	3.357				
α-Iodic acid, HIO_3	4.64	7.5 12.4 8.1 [4.331][a]	57(2) 43(1) 30.0(3) 21(1) 16(1) 17.8(13) 6(3) 15(2) 11.5(5) [4.7]	−26(1) −18(1) +28(1) [4.8, 496]	0.418(16) 0.308(330) 0.313(56) 0.304(36) 0.322(31) 0.307(57) 0.527(48) 0.340(31) 0.377(103) −0.077(66) 0.107(12) 0.092(8) [4.11]
3-Methyl 4-nitropyridine 1-oxide, $C_6N_2O_3H_6$ (POM)	1.55	3.77 5.41 3.77 [4.497][a]	13.29 18.14 12.20 7.8 5.2 5.4 4.9 −2.6 10.6 [4.498]		0.45 0.53 0.49 0.53 0.57 0.42 0.36 0.70 0.64 −0.078 −0.074 −0.046 (514.5 nm) [4.498]

Table 4.4-19 Orthorhombic, point group 222 (D_2) materials, cont.

Electrooptic tensor	General optical properties	Refractive index			Nonlinear dielectric susceptibility
r^T_{41} r^T_{52} r^T_{63} (10^{-12} m/V)	Transparency range (μm) dn_X/dT (10^{-5} K^{-1}) dn_Y/dT (10^{-5} K^{-1}) dn_Z/dT (10^{-5} K^{-1})	n_X A B C D	n_Y A B C D	n_Z A B C D	d_{14} d_{25} d_{36} (10^{-12} m/V)
	0.170–3.0 [4.499] – – –	**1.5194** 2.3035 0.01378 0.01498 0.00612 [4.500][b]	**1.5505** 2.3704 0.01528 0.01581 0.00939 [4.500][b]	**1.5781** 2.4753 0.01806 0.01752 0.01654 [4.500][b]	1.5 [4.499] – –
6.6(3) 7.0(5) 6.0(3) [4.502, 503]	0.35–1.6 $E \parallel a$ 0.35–2.2 $E \parallel c$ [4.504, 505] – –	**1.8129** 2.5761 0.6973 0.05550736 0.0201 [4.505][c]	**1.9273** 2.4701 1.2054 0.05044516 0.0152 [4.505][c]	**1.9500** 2.6615 1.1316 0.05202961 0.0398 [4.505][c]	8.3 [4.501] – –
|3.6(6)| |5.1(4)| |2.6(3)| [4.507]	0.4–3.3 [4.506] – – –	**1.625** 2.4529 0.1641 0.128 0 [4.506][c]	**1.668** 2.4315 0.3556 0.1276 0.0579 [4.506][c]	**1.829** 2.5521 0.7962 0.1289 0.0941 [4.506][c]	5.3–6.0 [4.506] – –

Table 4.4-19 Orthorhombic, point group 222 (D_2) materials, cont.

	General	Static dielectric tensor	Elastic stiffness tensor	Piezoelectric tensor	Elastooptic tensor
Material	ϱ (g/cm^3)	ε^S_{11} ε^S_{22} ε^S_{33} ε^T_{11} ε^T_{22} ε^T_{33}	c_{11} c_{22} c_{33} c_{44} c_{55} c_{66} c_{12} c_{13} c_{23} (10^9 Pa)	d_{14} d_{25} d_{36} (10^{-12} C/N)	p_{11} p_{12} p_{13} p_{21} p_{22} p_{23} p_{31} p_{32} p_{33} p_{44} p_{55} p_{66}
Potassium sodium tartrate tetrahydrate, $KNa(C_4H_4O_6)\cdot 4H_2O$ (Rochelle salt)	1.767	245 – – 1100 11.1 9.2 [4.331]	40(16) 55(25) 63(23) 11.9(38) 3.1(2) 10.0(16) 24(29) 32(20) 23.8(438) [4.7]	2300 −56 11.8 [4.331]	0.35 0.41 0.42 0.37 0.28 0.34 0.36 0.35 0.36 −0.030 0.0046 −0.025 (589 nm) [4.508]
Sodium ammonium tartrate tetrahydrate, $Na(NH_4)C_4H_4O_6\cdot 4H_2O$ (Ammonium Rochelle salt)		10(1)[a] 10(1)[a] 10(1)[a] [4.509] 9.0 8.9 10.0 [4.331]	36.8 50.9 55.4 10.6 3.03 8.70 27.2 30.8 34.7 [4.331]	≈ 13 38 ≈ 7 [4.510, 511]	−0.48 −0.61 −0.68 −0.55 −0.76 −0.83 −0.44 −0.57 −0.71 0.0077 0.013 −0.0026 (589 nm) [4.512]

Table 4.4-19 Orthorhombic, point group 222 (D_2) materials, cont.

Electrooptic tensor	General optical properties	Refractive index			Nonlinear dielectric susceptibility
r^T_{41} r^T_{52} r^T_{63} (10^{-12} m/V)	Transparency range (μm) dn_X/dT (10^{-5} K^{-1}) dn_Y/dT (10^{-5} K^{-1}) dn_Z/dT (10^{-5} K^{-1})	n_X A B C D	n_Y A B C D	n_Z A B C D	d_{14} d_{25} d_{36} (10^{-12} m/V)
8.3 [4.513, 514] – –	<250 to >850 nm −6.54 [4.515, 516] −5.57 [4.515, 516] −6.00 [4.515, 516]	**1.49540** (585 nm) [4.517] **1.5622** (260 nm) [4.515, 516]	**1.49183** (585 nm) [4.517] **1.5576** (260 nm) [4.515, 516] $A = 1$ $B = 1.1851057$ $C = 0.0113636$ $D = 0$ [4.515, 516][c]	**1.49001** (585 nm) [4.517] **1.5566** (260 nm) [4.515, 516]	$\chi^{(3)}_{zzzz}$ $= 0.8 \times 10^{-22}$ V^{-2} m^2 $\chi^{(3)}_{xxxx}$ $= 0.4 \times 10^{-22}$ V^{-2} m^2 (at 576.8 and 627 nm) [4.518]
$r_{52} = 2.1$ [4.519, 520]		**1.4984** (589 nm) [4.521, 522]	**1.4996** (589 nm) [4.521, 522]	**1.4953** (589 nm) [4.521, 522]	

[a] The relative dielectric constant ε_{ij} was measured neither at constant stress nor at constant strain.

[bc] Dispersion relations (λ (μm), $T = 20\,°$C):

[b] $n^2 = A + \frac{B}{\lambda^2 - C} - D\lambda^2$.

[c] $n^2 = A + \frac{B\lambda^2}{\lambda^2 - C} - D\lambda^2$.

Table 4.4-20 Orthorhombic, point group $mm2$ (C_{2v}) materials

	General	Static dielectric tensor	Dissipation factor	Elastic stiffness tensor	Piezoelectric tensor	Elastooptic tensor
Material	ϱ (g/cm^3) Mohs hardness κ (W/m K)	ε_{11}^{S} ε_{22}^{S} ε_{33}^{S} ε_{11}^{T} ε_{22}^{T} ε_{33}^{T}	$\tan\delta_1$ $\tan\delta_2$ $\tan\delta_3$ (T (°C), f (kHz))	c_{11} c_{22} c_{33} c_{44} c_{55} c_{66} c_{12} c_{13} c_{23} (10^9 Pa)	d_{31} d_{32} d_{33} d_{15} d_{24} d_h (10^{-12} C/N)	p_{11} p_{12} p_{13} p_{21} p_{22} p_{23} p_{31} p_{32} p_{33} p_{44} p_{55} p_{66}
Ammonium sulfate, $(NH_4)_2SO_4$		– – 8[a] (−60 °C) [4.526, 527]		35.2 29.7 36.0 9.5 7.0 10.3 14.1 15.7 17.3 [4.525]		0.26 [4.523, 524] |0.27| [4.525] |0.26| [4.525] |0.23| [4.525] ≈ |0.27| [4.525] |0.25| [4.525] |0.23| [4.525] ≈ |0.26| [4.525] ≈ 0.26 [4.525] 0.02 [4.525] ≤ |0.02| [4.525] ≈ 0.01 [4.11, 525]
Barium magnesium fluoride, $BaMgF_4$ (BMF)		14 [4.528] 8 [4.528] 8.5 [4.528] 14.75(74)[a] 8.24(41)[a] 8.40(42)[a] [4.529]		104 81 130 32.1 55.1 24.7 28.7 63.7 35.8 [4.529]	+2.5(3) −4.1(4) +8.0(2) −5.3(5) −1.2(1) [4.529] –	
Barium sodium niobate, $Ba_2NaNb_5O_{15}$ ("banana")	5.41 – 3.5 [4.532]	215 205 20 238(5) 228(5) 430 [4.534][b]	0.0068(17) – 0.0024(6) (25 °C, 1 GHz) [4.296]	239 247 135 65 66 76 104 50 52 [4.535][c]	−6.8 −6.9 34 32 45 [4.533] –	$p_{66} = 0.0021$ [4.530, 531]

Table 4.4-20 Orthorhombic, point group $mm2$ (C_{2v}) materials, cont.

	Electrooptic tensor	**General optical properties**	**Refractive index**			**Nonlinear dielectric susceptibility**
	r^{T}_{13} r^{T}_{23} r^{T}_{33} r^{T}_{42} r^{T}_{51} (10^{-12} m/V)	Transparency range (μm) dn_X/dT (10^{-5} K^{-1}) dn_Y/dT (10^{-5} K^{-1}) dn_Z/dT (10^{-5} K^{-1})	n_X A B C D E F	n_Y A B C D E F	n_Z A B C D E F	d_{31} d_{32} d_{33} d_{15} d_{24} (10^{-12} m/V)
Ammonium sulfate, $(NH_4)_2SO_4$	$r_c = 0.4$ [4.536]					0.27 0.29 0.50 [4.537] – –
Barium magnesium fluoride, $BaMgF_4$		0.185–10 [4.495]	**1.4436** 2.077 0.0076 0.0079 0 [4.495][h] – –	**1.4604** 2.1238 0.0086 0 0 [4.495][h] – –	**1.4674** 2.1462 0.00736 0.009 0 [4.495][h] – –	0.022 0.033 0.009 – 0.024 [4.495]
Barium sodium niobate, $Ba_2NaNb_5O_{15}$ ("banana")	15(1) 13(1) 48(2) 92(4) 90(4) [4.534]	0.37–5 [4.534, 538, 539] −2.5 [4.534] – 8 [4.534]	**2.2573** 1 3.9495 0.04038894 0 [4.534][i] – –	**2.2571** 1 3.9495 0.04014012 0 [4.534][i] – –	**2.1694** 1 3.6008 0.03219871 0 [4.534][i] – –	−12 −12 −16.5 12 11.4 [4.534]

Table 4.4-20 Orthorhombic, point group $mm2$ (C_{2v}) materials, cont.

	General	Static dielectric tensor	Dissipation factor	Elastic stiffness tensor	Piezoelectric tensor	Elastooptic tensor
Material	ϱ (g/cm^3) Mohs hardness κ (W/m K)	ε_{11}^{S} ε_{22}^{S} ε_{33}^{S} ε_{11}^{T} ε_{22}^{T} ε_{33}^{T}	$\tan\delta_1$ $\tan\delta_2$ $\tan\delta_3$ (T (°C), f (kHz))	c_{11} c_{22} c_{33} c_{44} c_{55} c_{66} c_{12} c_{13} c_{23} (10^9 Pa)	d_{31} d_{32} d_{33} d_{15} d_{24} d_h (10^{-12} C/N)	p_{11} p_{12} p_{13} p_{21} p_{22} p_{23} p_{31} p_{32} p_{33} p_{44} p_{55} p_{66}
Calcium tartrate tetrahydrate, $Ca(C_4H_4O_6)\cdot 4H_2O$ (L-CTT)		14.3 10 20 [4.540][a]	0.9 – – (30 °C, 100 Hz) [4.540, 541]			
m-Chloronitrobenzene, $ClC_6H_4NO_2$ (CNB)						
2-Cyclooctylamino-5-nitropyridine, $C_{13}H_{19}N_3O_2$ (COANP)	1.24 – –					
N, 2-Dimethyl-4-nitrobenzenamine, $C_8H_{10}N_2O_2$ (MNMA)						

Table 4.4-20 Orthorhombic, point group $mm2$ (C_{2v}) materials, cont.

Electrooptic tensor	General optical properties	Refractive index			Nonlinear dielectric susceptibility
r^T_{13}	Transparency range (μm)	n_X	n_Y	n_Z	d_{31}
r^T_{23}	dn_X/dT (10^{-5} K^{-1})	A	A	A	d_{32}
r^T_{33}	dn_Y/dT (10^{-5} K^{-1})	B	B	B	d_{33}
r^T_{42}	dn_Z/dT (10^{-5} K^{-1})	C	C	C	d_{15}
r^T_{51}		D	D	D	d_{24}
(10^{-12} m/V)		E	E	E	(10^{-12} m/V)
		F	F	F	
	0.28–1.4 [4.542]	**1.5125**	**1.5220**	**1.5477**	<0.015
	–	1	1	1	0.20
	–	1.26	1.30	1.38	0.14
	–	0.0127273	0.0121495	0.0094521	1.73
		0 [4.542][i]	0 [4.542][i]	0 [4.542][i]	0.90 [4.542]
		–	–	–	
		–	–	–	
		1.6557	**1.6626**	**1.624**	4.6
		2.4882	2.5411	2.2469	4
		0.2384	0.2148	0.3722	7.8 [4.543]
		0.1070	0.1122	0.0810	–
		0.0091 [4.543][i]	0.0135 [4.543][i]	0.0092 [4.543][i]	–
		–	–	–	
		–	–	–	
0.63 [4.544]	0.41–... [4.545]	**1.6100**	**1.6383**	**1.7170**	11.3
–	–	2.3320	2.3994	2.5104	24
–	–	0.2215	0.2469	0.3689	10.8 [4.546, 547]
–	–	0.1686	0.1500	0.1780	–
–		0 [4.545][i]	0 [4.545][i]	0 [4.545][i]	–
		–	–	–	
		–	–	–	
8±2	0.5–2 [4.548]	**1.936**		**1.506**	13
–	–	1.6797		2.1798	–
7.5±2 [4.548]	–	1.7842		0.0736	2.6
–	–	0.1571		0.1757	12 [4.548]
–		0 [4.548][i]		0 [4.548][i]	–
		–		–	
		–		–	

Table 4.4-20 Orthorhombic, point group $mm2$ (C_{2v}) materials, cont.

	General	Static dielectric tensor	Dissipation factor	Elastic stiffness tensor	Piezoelectric tensor	Elastooptic tensor
Material	ϱ (g/cm^3) Mohs hardness κ (W/m K)	ε_{11}^{S} ε_{22}^{S} ε_{33}^{S} ε_{11}^{T} ε_{22}^{T} ε_{33}^{T}	$\tan\delta_1$ $\tan\delta_2$ $\tan\delta_3$ (T (°C), f (kHz))	c_{11} c_{22} c_{33} c_{44} c_{55} c_{66} c_{12} c_{13} c_{23} (10^9 Pa)	d_{31} d_{32} d_{33} d_{15} d_{24} d_h (10^{-12} C/N)	p_{11} p_{12} p_{13} p_{21} p_{22} p_{23} p_{31} p_{32} p_{33} p_{44} p_{55} p_{66}
Gadolinium molybdate, $Gd_2(MoO_4)_3$ (GMO)		– – 9.6 – – 10.2 [4.550]		55(4) 71(3) 101(3) 25.2(8) 25.8(2) 33.1(3) 12.8(46) 23.5(45) 27.2(76) [4.7][c]	1.5 [4.549] 0.2 [4.8] 0.5 [4.549] – – –	0.19 0.31 0.175 0.215 0.235 0.175 0.185 0.23 0.115 −0.033 −0.028 0.035 [4.551] (515–633 nm)
Lithium formate monohydrate, $LiCOOH \cdot H_2O$ (LFM)	1.46 – –	3.24 [4.552, 553] – – 4.5(5)[a] [4.554, 555] 5.0(5)[a] [4.554, 555] 6.0(3)[a] [4.554, 555]		19.15 33.55 41.98 15.02 5.31 4.89 8.62 8.40 22.77 [4.552, 553][c]	−2.1(2) −6.7(5) +9.2(7) −15.0(10) +1.2(2) [4.554–557]	
Lithium gallium oxide, $LiGaO_2$ (lithium metagallate)	4.187 7.5 –	7.0 [4.8] 6.0 [4.8] 8.3 [4.558] 7.18 [4.8] 6.18 [4.8] 8.78 [4.8]		140 120 140 57.1 47.4 69.0 14 28 31 [4.558][c]	−2.5 [4.558] −4.7 [4.558] +8.6 [4.558] −6.9 [4.558] −6.0 [4.558] +0.9 [4.8, 274]	

Table 4.4-20 Orthorhombic, point group $mm2$ (C_{2v}) materials, cont.

Electrooptic tensor	**General optical properties**	**Refractive index**			**Nonlinear dielectric susceptibility**
r^T_{13} r^T_{23} r^T_{33} r^T_{42} r^T_{51} (10^{-12} m/V)	Transparency range (μm) dn_X/dT (10^{-5} K^{-1}) dn_Y/dT (10^{-5} K^{-1}) dn_Z/dT (10^{-5} K^{-1})	n_X A B C D E F	n_Y A B C D E F	n_Z A B C D E F	d_{31} d_{32} d_{33} d_{15} d_{24} (10^{-12} m/V)
+2.15(16) −2.31(16) +0.123(15) – – [4.560, 561]		**1.8141** 1 2.2450 0.022693 0 [4.321][i] – –	**1.8145** 1 2.24654 0.0226803 0 [4.321][i] – –	**1.8637** 1 2.41957 0.0245458 0 [4.321][i] – –	−2.3 +2.3 −0.04 −4.1 +4 [4.156, 559]
−1.0(1) +3.2(2) −2.6(2) +1.0(2) +2.4(2) [4.554–557]	0.23–1.2 [4.562, 563] – – –	**1.3595** (1 μm) 1.4376 0.4045 0.01692601 0.0005 [4.564][i] – –	**1.4694** (1 μm) 1.6586 0.5006 0.023409 0.0127 [4.564][i] – –	**1.5055** (1 μm) 1.6714 0.5928 0.02534464 0.0153 [4.564][i] – –	0.13 −0.60 +0.94 [4.12] – –
	0.3–5 [4.251, 565] – – –	**1.7477** (0.5 μm) [4.566]	**1.7768** (0.5 μm) [4.566]	**1.7791** (0.5 μm) [4.566]	+0.066 −0.14 +0.57 [4.156, 565] – –

Table 4.4-20 Orthorhombic, point group $mm2$ (C_{2v}) materials, cont.

	General	Static dielectric tensor	Dissipation factor	Elastic stiffness tensor	Piezoelectric tensor	Elastooptic tensor
Material	ϱ (g/cm^3) Mohs hardness κ (W/m K)	ε_{11}^S ε_{22}^S ε_{33}^S ε_{11}^T ε_{22}^T ε_{33}^T	$\tan\delta_1$ $\tan\delta_2$ $\tan\delta_3$ (T (°C), f (kHz))	c_{11} c_{22} c_{33} c_{44} c_{55} c_{66} c_{12} c_{13} c_{23} (10^9 Pa)	d_{31} d_{32} d_{33} d_{15} d_{24} d_h (10^{-12} C/N)	p_{11} p_{12} p_{13} p_{21} p_{22} p_{23} p_{31} p_{32} p_{33} p_{44} p_{55} p_{66}
Lithium triborate, LiB_3O_5 (LBO)	2.47 6 3.5 [4.200]					
m-Nitroaniline; (3-nitrobenzenamine, meta-nitroaniline), $C_6H_4(NO_2)NH_2$ (mNA)		3.9 4.2 4.6 [4.567][a]				
Potassium acid phthalate, $KH(C_8H_4O_4)$		6.00(2) 3.87(2) 4.34(2) [4.568, 569] – – 4.34[a] [4.570]		17.6 13.3 17.0 4.86 7.59 6.23 7.4 10.4 5.0 [4.568, 569]	−15.3(0) +8.8(0) +5.5(0) −7.1(0) +4.3(0) [4.568, 569] –	−0.61 −0.25 −0.52 −0.58 −0.36 −0.60 −0.84 −0.36 −0.90 −0.26 −0.63 −0.13 (589 nm) [4.571]

Table 4.4-20 Orthorhombic, point group $mm2$ (C_{2v}) materials, cont.

Electrooptic tensor	**General optical properties**	**Refractive index**			**Nonlinear dielectric susceptibility**
r^T_{13} r^T_{23} r^T_{33} r^T_{42} r^T_{51} (10^{-12} m/V)	Transparency range (μm) dn_X/dT (10^{-5} K^{-1}) dn_Y/dT (10^{-5} K^{-1}) dn_Z/dT (10^{-5} K^{-1})	n_X A B C D E F	n_Y A B C D E F	n_Z A B C D E F	d_{31} d_{32} d_{33} d_{15} d_{24} (10^{-12} m/V)
	0.16–2.6 [4.572] −18 [4.573] −136 [4.573] −(63 + 21λ) [4.573]	**1.5656** 2.4542 0.01125 0.01135 0.01388 0 0 [4.575][j]	**1.5905** 2.5390 0.01277 0.01189 0.01849 4.3025×10^{-5} 2.9131×10^{-5} [4.575][j]	**1.6055** 2.5865 0.01310 0.01223 0.01862 4.5778×10^{-5} 3.2526×10^{-5} [4.575][j]	−0.67 0.85 0.04 [4.12, 14, 572–574] – –
7.4(7) 0.1(6) 16.7(2) [4.578] – –	0.5–2 [4.576, 577] – – –	**1.631** 2.469 0.1864 0.16 0.0199 [4.579][i] – –	**1.678** 2.6658 0.1626 0.1719 0.0212 [4.579][i] – –	**1.719** 2.8102 0.1524 0.175 0.0294 [4.579][i] – –	20 1.6 21 [4.543] – –
$r_b \approx 1.8$ [4.570]	0.3–1.7 with a narrow absorption band at 1.14 μm [4.580] – – –	**1.63** (1.1 μm) [4.580]	**1.64** (1.1 μm) [4.580]	**1.48** (1.1 μm) [4.580]	0.21 (d_{31}, 1.15 μm) 0.06 (d_{32}, 1.15 μm) 0.65 (d_{31}, 0.63 μm) 0.15 (d_{32}, 0.63 μm) – [4.580]

Table 4.4-20 Orthorhombic, point group $mm2$ (C_{2v}) materials, cont.

	General	Static dielectric tensor	Dissipation factor	Elastic stiffness tensor	Piezoelectric tensor	Elastooptic tensor
Material	ϱ (g/cm^3) Mohs hardness κ (W/m K)	ε^{S}_{11} ε^{S}_{22} ε^{S}_{33} ε^{T}_{11} ε^{T}_{22} ε^{T}_{33}	$\tan\delta_1$ $\tan\delta_2$ $\tan\delta_3$ (T (°C), f (kHz))	c_{11} c_{22} c_{33} c_{44} c_{55} c_{66} c_{12} c_{13} c_{23} (10^9 Pa)	d_{31} d_{32} d_{33} d_{15} d_{24} d_h (10^{-12} C/N)	p_{11} p_{12} p_{13} p_{21} p_{22} p_{23} p_{31} p_{32} p_{33} p_{44} p_{55} p_{66}
Potassium niobate, $KNbO_3$	4.617 – $>$3.5 [4.581]	37(2) 780(50) 24(2) 160(10) 1000(80) 55(5) (25 °C) [4.583]	0.003 0.002 0.01 (25 °C) [4.583]	226 270 280 74.3 25.0 95.5 96 [4.583] – –	+9.8(7) −19.5(20) +24.5(15) [4.582] 215(5) 159(5) [4.583] –	|0.197|[d] |0.115|[d] |0.109|[e] |0.130|[d] |0.234|[d] |0.005|[e] |0.64|[d] |0.153|[d] |0.075|[e] |0.57|[d] |0.45|[d] [4.584] –
Potassium pentaborate tetrahydrate, $KB_5O_8 \cdot 4H_2O$ (KB5)	1.74 2.5 –	5.5 4.6 4.5 [4.585] – – –		58.2 35.9 25.5 16.4 4.63 5.7 22.9 17.4 23.1 [4.585]	−0.35 −2.3 +5.5 +4.7 +20.3 [4.586] –	
Potassium titanyl arsenate, $KTiOAsO_4$ (KTA)	3.45 3 –	– – – 12(1) 12(1) 18(1) [4.587,588]				
Potassium titanate (titanyl) phosphate, $KTiOPO_4$ (KTP)	3.02 5 2 (∥ X) 3 (∥ Y) 3.3 (∥ Z) [4.590]	11.6(2) 11.0(2) 15.4(3) 11.9(2) 11.3(2) $\geq$17.5(4) [4.591]	$\lesssim$0.004 $\leq$0.004 $<$0.005 (10–10^3 MHz) $\approx$0.017 0.017 $\approx$0.35 (100 kHz) [4.591]	159 154 175 [4.589] – – – – – –		

Table 4.4-20 Orthorhombic, point group $mm2$ (C_{2v}) materials, cont.

Electrooptic tensor	General optical properties	Refractive index			Nonlinear dielectric susceptibility
r^T_{13} r^T_{23} r^T_{33} r^T_{42} r^T_{51} (10^{-12} m/V)	Transparency range (μm) dn_X/dT (10^{-5} K^{-1}) dn_Y/dT (10^{-5} K^{-1}) dn_Z/dT (10^{-5} K^{-1})	n_X A B C D E F	n_Y A B C D E F	n_Z A B C D E F	d_{31} d_{32} d_{33} d_{15} d_{24} (10^{-12} m/V)
34(2) 6(1) 63.4(10) 450(30) 120(10) [4.584]	0.4–4.5 [4.386] For temperature-dependent dispersion relations, see [4.592]	**2.2576** 1 1.44121874 0.07439136 2.54336918 0.01877036 0.02845018 [4.594][k]	**2.2195** 1 1.33660410 0.06664629 2.49710396 0.01666505 0.02517432 [4.594][k]	**2.1194** 1 1.04824955 0.06514225 2.37108379 0.01433172 0.01943289 [4.594][k]	+11 [4.12–14] −13 [4.12–14] −19.5 [4.12–14] 16 [4.593] 17 [4.593]
	0.165–1.4 [4.595] – – –	**1.4917** (0.5 μm) 1 1.1790826 0.0087815 0 [4.596, 597][i] – –	**1.4380** (0.5 μm) 1 1.0280852 0.0090222 0 [4.596, 597][i] – –	**1.4251** (0.5 μm) 1 0.9919090 0.0093289 0 [4.596, 597][i] – –	0.04 0.003 0.05 (all at 0.5 μm) – – [4.12, 598, 599]
15(1) 21(1) 40(1) – – [4.587, 588]	0.35–5.3 [4.587, 588, 600] – – –	**1.782** 1.90713 1.23552 0.0387775 0.01025 [4.601][i] – –	**1.790** 2.15912 1.00099 0.0477160 0.01096 [4.601][i] – –	**1.868** 2.14786 1.29559 0.0516153 0.01436 [4.601][i] – –	2.9 5.2 12.0 – – [4.12, 14, 602–604][l]
+9.5(5) +15.7(8) +36.3(18) 9.3(9) 7.3(7) [4.591]	0.35–4.5 [4.605, 606] 0.61 [4.607][m] 0.83 [4.607][m] 1.45 [4.607][m]	**1.7381** 3.0065 0.03901 0.04251 0.01327 [4.608][hm] – –	**1.7458** 3.0333 0.04154 0.04547 0.01408 [4.608][hm] – –	**1.8302** 3.3134 0.05694 0.05658 0.01682 [4.608][hm] – –	2.2 3.7 14.6 1.9 3.7 [4.13][l]

Table 4.4-20 Orthorhombic, point group $mm2$ (C_{2v}) materials, cont.

	General	Static dielectric tensor	Dissipation factor	Elastic stiffness tensor	Piezoelectric tensor	Elastooptic tensor
Material	ϱ (g/cm^3) Mohs hardness κ (W/m K)	ε^S_{11} ε^S_{22} ε^S_{33} ε^T_{11} ε^T_{22} ε^T_{33}	$\tan\delta_1$ $\tan\delta_2$ $\tan\delta_3$ (T (°C), f (kHz))	c_{11} c_{22} c_{33} c_{44} c_{55} c_{66} c_{12} c_{13} c_{23} (10^9 Pa)	d_{31} d_{32} d_{33} d_{15} d_{24} d_h (10^{-12} C/N)	p_{11} p_{12} p_{13} p_{21} p_{22} p_{23} p_{31} p_{32} p_{33} p_{44} p_{55} p_{66}
Rubidium titanate (titanyl) phosphate, $RbTiOPO_4$ (RTP)		23[f] (25 °C) [4.609–611]		143 142 175 33 40 57 [4.589] – – –		
Sodium nitrite, $NaNO_2$	2.168 – –	8 [4.612] 5.2 [4.612] 4.18 [4.612] 7.4 [4.614] 5.5 [4.614] 5.0 [4.614]	0.004 0.006 0.015 (3.3 GHz) [4.612]	30.6(3) 56(2) 64(2) 12(1) 9.9(1) 5.0(3) 12.5(1) 15.6(53) 14.6(48) [4.7]	−1.1 −2.8 +1.6 +9.3 −20.2 – [4.8, 614]	|0.44| [4.613] |0.37| [4.613] |0.36| [4.613] |0.39| [4.613] |0.33| [4.613] |0.27| [4.613] |0.18| [4.613] |0.19| [4.613] |0.15| [4.613] −0.050[g] [4.615] −0.30 [4.615] −0.10[g] [4.615]

[a] These values for ε_{ij} are neither at constant stress nor at constant strain.
[b] Stoichiometric crystal.
[c] c^E at constant electric field.
[d] p^E at constant electric field.
[e] $p^*_{i3} = p^E_{i3} - r^S_{i3}e_{333}/\varepsilon_0\varepsilon^S_{33}$, where r^S_{i3} is the electrooptic-tensor element $i3$ at zero strain, e_{333} is the piezoelectric-tensor element 333, and ε^S_{33} is the relative-dielectric-constant element 33 at constant strain.
[f] Element of ε unspecified, but probably ε_{11} or ε_{33}.

Table 4.4-20 Orthorhombic, point group $mm2$ (C_{2v}) materials, cont.

Electrooptic tensor	General optical properties	Refractive index			Nonlinear dielectric susceptibility
r^T_{13} r^T_{23} r^T_{33} r^T_{42} r^T_{51} (10^{-12} m/V)	Transparency range (μm) dn_X/dT (10^{-5} K^{-1}) dn_Y/dT (10^{-5} K^{-1}) dn_Z/dT (10^{-5} K^{-1})	n_X A B C D E F	n_Y A B C D E F	n_Z A B C D E F	d_{31} d_{32} d_{33} d_{15} d_{24} (10^{-12} m/V)
+9.7(9) +10.8(9) +22.5(9) +14.9(9) −7.6(9) [4.618–620]	0.35–4.5 [4.616, 617] – – –	**1.7569** 2.56666 0.53842 0.06374 0.01666 [4.617][i] – –	**1.7730** 2.34868 0.77949 0.05449 0.0211 [4.617][i] – –	**1.8540** 2.77339 0.63961 0.08151 0.02237 [4.617][i] – –	3.3 4.1 17.3 [4.603] – –
$r_{42} = -3.0(2)$ $r_{51} = -1.9(2)$ (at 546 nm) [4.623]	0.35–3.4 and 5–8 [4.621, 622] – – –	**1.3395** 1 0.727454 0.0118285 0 [4.621][i] – –	**1.4036** 1 0.978108 0.0112296 0 [4.621][i] – –	**1.6365** 1 1.616683 0.0222073 0 [4.621][i] – –	0.11 2.9 0.14 0.11 2.8 [4.624]

[g] $p^{\text{eff}}_{ijkl} = p^E_{ijkl} - r^S_{ijm} a_m a_n e_{nkl} / \boldsymbol{a} \cdot \boldsymbol{\varepsilon} \cdot \boldsymbol{a}$, where r^S is the electrooptic tensor, $\boldsymbol{a}$ is a unit acoustic-wave propagation vector, and $\boldsymbol{\varepsilon}$ is the dielectric-constant tensor at the frequency of the acoustic wave.

[hijk] Dispersion relations (λ (μm), $T = 20\,°\text{C}$):

[h] $n^2 = A + \dfrac{B}{\lambda^2 - C} - D\lambda^2$.

[i] $n^2 = A + \dfrac{B\lambda^2}{\lambda^2 - C} - D\lambda^2$.

[j] $n^2 = A + \dfrac{B}{\lambda^2 - C} - D\lambda^2 + E\lambda^4 - F\lambda^6$.

[k] $n^2 = A + \dfrac{B\lambda^2}{\lambda^2 - C} + \dfrac{D\lambda^2}{\lambda^2 - E} - F\lambda^2$.

[l] Note reversals between d_{31} and d_{32} (also between d_{15} and d_{24}) for KTA and KTP given in [4.12] and [4.13].

[m] For flux-grown crystals.

Table 4.4–21 Monoclinic, point group 2 (C_2) materials

	General	**Static dielectric tensor**	**Dissipation factor**	**Elastic stiffness tensor**	**Piezoelectric tensor**	**Elastooptic tensor**
Material	ϱ (g/cm^3) Mohs hardness κ (W/m K)	ε_{11}^{S} ε_{22}^{S} ε_{33}^{S} ε_{13}^{S} ε_{11}^{T} ε_{22}^{T} ε_{33}^{T} ε_{13}^{T}	$\tan\delta_1$ $\tan\delta_2$ $\tan\delta_3$ (f (kHz))	c_{11} c_{22} c_{33} c_{44} c_{55} c_{66} c_{12} c_{13} c_{23} c_{15} c_{25} c_{35} c_{46} (10^9 Pa)	d_{21} d_{22} d_{23} d_{14} d_{16} d_{25} d_{34} d_{36} d_h (10^{-12} C/N)	p_{11} p_{12} p_{13} p_{15} p_{21} p_{22} p_{23} p_{25} p_{31} p_{32} p_{33} p_{35} p_{44} p_{46} p_{51} p_{52} p_{53} p_{55} p_{64} p_{66}
Bismuth triborate, BiB_3O_6 (BIBO)	4.9 [4.625] – –					
Deuterated L-arginine phosphate, $(ND_xH_{2-x})_2^+(CND)(CH_2)_3$ $CH(ND_yH_{3-y})^+COO^-\cdot$ $D_2PO_4^-\cdot D_2O$ (DLAP)	≈ 1.5 3 –			17.477(111) 31.996(89) 31.575(72) – – – – – – – – – – [4.15]		
4-($N1N$-Dimethylamino)-3-acetamidonitrobenzene (N-[2-(dimethylamino)-5-nitrophenyl]-acetamide, DAN)						

Table 4.4-21 Monoclinic, point group 2 (C_2) materials, cont.

Electrooptic tensor	General optical properties	Refractive index			Nonlinear dielectric susceptibility
r^T_{12} r^T_{22} r^T_{32} r^T_{41} r^T_{43} r^T_{52} r^T_{61} r^T_{63} (10^{-12} m/V) (at $\lambda = 633$ nm)	Transparency range (μm) dn_X/dT (10^{-5} K^{-1}) dn_Y/dT (10^{-5} K^{-1}) dn_Z/dT (10^{-5} K^{-1})	n_X A B C D	n_Y A B C D	n_Z A B C D	d_{21} d_{22} d_{23} d_{25} (10^{-12} m/V)
	0.27–6.25 [4.625]	**1.9190**	**1.7585**	**1.7854**	2.3(2)
	–	3.6545	3.0740	3.1685	2.53(8)
	–	0.0511	0.0323	0.0373	1.3(1)
	–	0.0371	0.0316	0.0346	2.3(2) [4.626]
		0.0226 [4.626][a]	0.01337 [4.626][a]	0.01750 [4.626][a]	(see also [4.627])
	0.25–1.3 [4.628]	**1.4960**	**1.5584**	**1.5655**	0.48
	−3.64	2.2352	2.4313	2.4484	0.685
	−5.34	0.0118	0.0151	0.0172	−0.80
	−6.69	0.0146	0.0214	0.0229	−0.22
	(all at 532 nm [4.629])	0.00683 [4.628][a]	0.0143 [4.628][a]	0.0115 [4.628][a]	[4.12, 14, 628]
	0.485–2.27 [4.630]	**1.517**	**1.636**	**1.843**	1.1
	–	2.1390	2.3290	2.5379	3.9
	–	0.147408	0.307173	0.719557	37.5
	–	0.3681 [4.630][b]	0.3933 [4.630][b]	0.4194 [4.630][b]	1.1 [4.630, 631]
		–	–	–	

Table 4.4-21 Monoclinic, point group 2 (C_2) materials, cont.

	General	Static dielectric tensor	Dissipation factor	Elastic stiffness tensor	Piezoelectric tensor	Elastooptic tensor
Material	ϱ (g/cm^3) Mohs hardness κ (W/m K)	ε^S_{11} ε^S_{22} ε^S_{33} ε^S_{13} ε^T_{11} ε^T_{22} ε^T_{33} ε^T_{13}	tan δ_1 tan δ_2 tan δ_3 (f (kHz))	c_{11} c_{22} c_{33} c_{44} c_{55} c_{66} c_{12} c_{13} c_{23} c_{15} c_{25} c_{35} c_{46} (10^9 Pa)	d_{21} d_{22} d_{23} d_{14} d_{16} d_{25} d_{34} d_{36} d_h (10^{-12} C/N)	p_{11} p_{12} p_{13} p_{15} p_{21} p_{22} p_{23} p_{25} p_{31} p_{32} p_{33} p_{35} p_{44} p_{46} p_{51} p_{52} p_{53} p_{55} p_{64} p_{66}
Dipotassium tartrate hemihydrate, $K_2C_4H_4O_6 \cdot 0.5H_2O$ (DKT)	1.987 – –	6.44 5.80 6.49 0.005 [4.331] [d] – – – –		35.7(86) 39(17) 62(10) 9.0(5) 11.7(19) 8.3(1) 17.8(9) 22.5(99) 13.5(37) 1.8 1.2(1) 5.9(27) 0.54(18) [4.7]	−0.8 4.5 −5.3 7.9 3.4 −6.4 −12.2 −23.2 1.6 [4.632]	
Lithium sulfate monohydrate, $Li_2SO_4 \cdot H_2O$	– – ≈ 4 [4.635]	5.16 [d] [4.633, 634] 10.3 [d] [4.633, 634] 4.95 [d] [4.633, 634] – 5.6 [4.331] 10.3 [4.331] 6.5 [4.331] 0.07 [4.331]		54.9 70.5 61.8 14.0 24.2 27.0 26.3 11.4 17.1 6.5 15.7 −5.2 −26.5 [4.331] [e]	−3.6 +16.3 +1.7 +0.7 −2.0 −5.0 −2.13 −4.2 14.4 [4.632]	

Table 4.4-21 Monoclinic, point group 2 (C_2) materials, cont.

Electrooptic tensor	General optical properties	Refractive index			Nonlinear dielectric susceptibility
r^T_{12} r^T_{22} r^T_{32} r^T_{41} r^T_{43} r^T_{52} r^T_{61} r^T_{63} (10^{-12} m/V) (at $\lambda = 633$ nm)	Transparency range (μm) dn_X/dT (10^{-5} K^{-1}) dn_Y/dT (10^{-5} K^{-1}) dn_Z/dT (10^{-5} K^{-1})	n_X A B C D	n_Y A B C D	n_Z A B C D	d_{21} d_{22} d_{23} d_{25} (10^{-12} m/V)
		1.4832 (1.014 μm) [4.636]	**1.5142** (1.014 μm) [4.636]	**1.5238** (1.014 μm) [4.636]	$d_{21} = 0.11$ $d_{22} = 3.9$ $d_{14} = 0.17$ (all at 0.6943 μm) [4.637]
+8.5(4) +6.5(4) +4.5(5) 0 −1.2(7) −0.7(2) +0.41(2) +0.8(6) [4.638]		**1.4521** [4.321]	**1.4657** [4.321]	**1.4752** [4.321]	$d_{22} = 0.38 \pm 0.06$ $d_{23} = 0.27 \pm 0.04$ $d_{34} = 0.23 \pm 0.04$ [4.321]

Table 4.4-21 Monoclinic, point group 2 (C_2) materials, cont.

	General	Static dielectric tensor	Dissipation factor	Elastic stiffness tensor	Piezoelectric tensor	Elastooptic tensor
Material	ϱ (g/cm^3) Mohs hardness κ (W/m K)	ε^S_{11} ε^S_{22} ε^S_{33} ε^S_{13} ε^T_{11} ε^T_{22} ε^T_{33} ε^T_{13}	$\tan\delta_1$ $\tan\delta_2$ $\tan\delta_3$ (f (kHz))	c_{11} c_{22} c_{33} c_{44} c_{55} c_{66} c_{12} c_{13} c_{23} c_{15} c_{25} c_{35} c_{46} (10^9 Pa)	d_{21} d_{22} d_{23} d_{14} d_{16} d_{25} d_{34} d_{36} d_h (10^{-12} C/N)	p_{11} p_{12} p_{13} p_{15} p_{21} p_{22} p_{23} p_{25} p_{31} p_{32} p_{33} p_{35} p_{44} p_{46} p_{51} p_{52} p_{53} p_{55} p_{64} p_{66}
D(+)-Saccharose, $C_{12}H_{22}O_{11}$ (sucrose)	– > 2.5 –				1.47 −3.41 0.73 1.25 −2.41 −0.87 −4.21 0.423 [4.639] –	
Triglycine sulfate, $(CH_2NH_2COOH)_3 \cdot H_2SO_4$ (TGS)		9.38[d] [4.640] 20[d] [4.643] 6.00[d] [4.640] 1.17(2)[d] [4.640] 9 [4.644] 40 [4.644] 6.6 [4.644]	0.043 – – (9.6 GHz) [4.643]	41.7 32.1 33.6 9.40 9.39 6.31 17.3 18.1 20.7 4.1 −0.6 7.7 −0.6 [4.646–648]	23.5(0) 7.9(1) 25.3(0) 2.7(0) −4.5(0) 24.3(1) −3.20(1) 2.8(0) – [4.646, 647]	|0.204| [4.641, 642] |0.162| [4.641, 642] |0.175| [4.641, 642] – |0.172| [4.641, 642] |0.208| [4.641, 642] |0.150| [4.641, 642] |0.083| [4.645] |0.204| [4.641, 642] |0.169| [4.641, 642] |0.151| [4.641, 642] – |0.273| [4.645] |0.276| [4.645] – |0.075| [4.645] – – |0.075| [4.645] –

Table 4.4-21 Monoclinic, point group 2 (C_2) materials, cont.

Electrooptic tensor	General optical properties	Refractive index			Nonlinear dielectric susceptibility
r^T_{12} r^T_{22} r^T_{32} r^T_{41} r^T_{43} r^T_{52} r^T_{61} r^T_{63} (10^{-12} m/V) (at $\lambda = 633$ nm)	Transparency range (μm) dn_X/dT (10^{-5} K^{-1}) dn_Y/dT (10^{-5} K^{-1}) dn_Z/dT (10^{-5} K^{-1})	n_X A B C D	n_Y A B C D	n_Z A B C D	d_{21} d_{22} d_{23} d_{25} (10^{-12} m/V)
	0.192–1.35 [4.649] – – –	**1.5278** 1.8719 0.466 0.0214 0.0113 [4.649][c]	**1.5552** 1.9703 0.4502 0.0238 0.0101 [4.649][c]	**1.5592** 2.0526 0.3909 0.252 0.0187 [4.649][c]	See [4.649, 650]
70 [4.651, 652] – 54 [4.651, 652] – – 1 [4.654, 655] – –					$d_{23} = 0.3$ at 0.6943 μm [4.653]

[abc] Dispersion relations (λ (μm), $T = 20\,°$C):

[a] $n^2 = A + \frac{B}{\lambda^2 - C} - D\lambda^2$.

[b] $n^2 = A + \frac{B\lambda^2}{\lambda^2 - C}$.

[c] $n^2 = A + \frac{B\lambda^2}{\lambda^2 - C} - D\lambda^2$.

[d] These values for ε_{ij} are neither at constant stress nor at constant strain.

[e] The elastic stiffness constant were measured at constant electric field (c^E).

References

4.1 D. N. Nikogosyan, G. G. Gurzadyan: Kvantovaya Elektron. **13**, 2519–2520 (1986)
4.2 D. N. Nikogosyan, G. G. Gurzadyan: Sov. J. Quantum Electron. [English Transl.] **16**, 1663–1664 (1986)
4.3 V. G. Dmitriev, G. G. Gurzadyan, D. N. Nikogosyan: *Handbook of Nonlinear Optical Crystals*, 2nd edn. (Springer, Berlin, Heidelberg 1997)
4.4 V. G. Dmitriev, G. G. Gurzadyan, D. N. Nikogosyan: *Handbook of Nonlinear Optical Crystals*, 3rd edn. (Springer, Berlin, Heidelberg 1999)
4.5 D. N. Nikogosyan, G. G. Gurzadyan: Kvantovaya Elektron. **14**, 1529–1541 (1987)
4.6 D. N. Nikogosyan, G. G. Gurzadyan: Sov. J. Quantum Electron. [English Transl.] **17**, 970–977 (1987)
4.7 A. G. Every, A. K. McCurdy: *Second and Higher Order Elastic Constants*, Landolt–Börnstein, New Series III/29, ed. by D. F. Nelson, O. Madelung (Springer, Berlin, Heidelberg 1992)
4.8 W. R. Cook Jr.: *Piezoelectric, Electrostrictive, Dielectric Constants, Electromechanical Coupling Factors*, Landolt–Börnstein, New Series III/29, ed. by D. F. Nelson, O. Madelung (Springer, Berlin, Heidelberg 1993)
4.9 W. R. Cook Jr.: *Electrooptic Coefficients*, Landolt–Börnstein, New Series III/30, ed. by D. F. Nelson (Springer, Berlin, Heidelberg 1996)
4.10 D. F. Nelson: *Piezooptic, Electrooptic Constants of Crystals*, Landolt–Börnstein, New Series III/30, ed. by D. F. Nelson (Springer, Berlin, Heidelberg 1996)
4.11 K. Vedam: *Piezooptic, Elastooptic Coefficients*, Landolt–Börnstein, New Series III/30, ed. by D. F. Nelson (Springer, Berlin, Heidelberg 1996)
4.12 D. A. Roberts: IEEE J. Quantum Electron. **28**, 2057 (1992)
4.13 I. Shoji, T. Kondo, A. Kitamoto, M. Shirane, R. Ito: J. Opt. Soc. Am. B **14**, 2268 (1997)
4.14 F. Charra, G. G. Gurzadyan: *Nonlinear Dielectric Susceptibilities*, Landolt–Börnstein, New Series III/30, ed. by D. F. Nelson (Springer, Berlin, Heidelberg 2000)
4.15 G. W. Faris, L. E. Jusinski, A. P. Hickman: J. Opt. Soc. Am. B **3**, 587–599 (1993)
4.16 Opto-Technological Laboratory: Data sheet – Catalog of optical materials (Opto-Technological Laboratory, St. Petersburg 2003)
4.17 Layertec: Data sheet – Material data (Layertec GmbH, Mellingen 2003)
4.18 Röhm Plexiglas: Data sheet – Technical data (Röhm Plexiglas, Darmstadt 2003)
4.19 D. G. Cahill, R. O. Pohl: Phys. Rev. B **35**, 4067–4073 (1987)
4.20 Schott Lithotec: Data sheet – Products and applications (Schott Lithotec AG, Jena 2003)
4.21 Crystran: Data sheet – Materials (Crystran Ltd., Poole 2002)
4.22 D. E. Gray: *American Institute of Physics Handbook* (McGraw-Hill, New York 1972)
4.23 O. V. Shakin, M. F. Bryzhina, V. V. Lemanov: Fiz. Tverd. Tela **13**, 3714–3716 (1971)
4.24 O. V. Shakin, M. F. Bryzhina, V. V. Lemanov: Sov. Phys. Solid State [English Transl.] **13**, 3141–3142 (1972)
4.25 E. D. D. Schmidt, K. Vedam: J. Phys. Chem. Solids **27**, 1563–1566 (1966)
4.26 ASI Instruments: Data sheet – Dielectric constant reference guide (ASI Instruments, Inc., Houston 2003)
4.27 R. Waxler: IEEE J. Quantum Electron. **7**, 166–167 (1971)
4.28 Melles Griot: Data sheet – Optics: Windows and optical flts (Melles Griot, Rochester 2003)
4.29 W. L. Smith: Optical materials, Part I. In: *CRC Handbook of Laser Science and Technology*, Vol. 3, ed. by M. J. Weber (CRC Press, Boca Raton 1986) pp. 229–258
4.30 D. Milam, M. J. Weber: J. Appl. Phys. **47**, 2497–2501 (1976)
4.31 W. T. White III, M. A. Henesian, M. J. Weber: J. Opt. Soc. Am. B **2**, 1402–1408 (1985)
4.32 I. D. Nikolov, C. D. Ivanov: Appl. Opt. **39**, 2067–2070 (2000)
4.33 Yu. A. Repeyev, E. V. Khoroshilova, D. N. Nikogosyan: J. Photochem. Photobiol. B **12**, 259–274 (1992)
4.34 R. Adair, L. L. Chase, S. A. Payne: Phys. Rev. B **39**, 3337–3349 (1989)
4.35 G. G. Gurzadyan, R. K. Ispiryan: Int. J. Nonlin. Opt. Phys. **1**, 533–540 (1992)
4.36 R. DeSalvo, A. A. Said, D. J. Hagan, E. W. Van Stryland, M. Sheik-Bahac: IEEE J. Quantum Electron. **32**, 1324–1333 (1996)
4.37 T. Tomie, I. Okuda, M. Yano: Appl. Phys. Lett. **55**, 325–327 (1989)
4.38 P. Liu, W. L. Smith, H. Lotem, J. H. Bechtel, N. Bloembergen, R. S. Adhav: Phys. Rev. B **17**, 4620–4632 (1978)
4.39 I. H. Malitson: J. Opt. Soc. Am. **55**, 1205–1209 (1965)
4.40 A. Smakula: *Harshaw Optical Crystals* (Harshaw Chemical Co., Cleveland 1967)
4.41 A. Feldman, D. Horowitz, R. M. Waxier, M. J. Dodge: Natl. Bur. Stand. (USA) Techn. Note **993** (1979) erratum (see also [4.656])
4.42 M. D. Levenson, N. Bloembergen: Phys. Rev. B **10**, 4447–4463 (1974)
4.43 H. H. Li: J. Phys. Chem. Ref. Data **9**, 161–289 (1980)
4.44 I. H. Malitson: Appl. Opt. **2**, 1103–1107 (1963)
4.45 S. S. Ballard, J. S. Browder: Thermal properties. In: *Optical Materials, CRC Handbook of Laser Science and Technology*, Vol. 4, Subvol. 2, ed. by M. J. Weber (CRC, Boca Raton 1987) pp. 49–54
4.46 A. D. Papadopoulos, E. Anastassakis: Phys. Rev. B **43**, 9916–9923 (1991)
4.47 H. Braul, C. A. Plint: Solid State Commun. **38**, 227–230 (1981)

4.48 MarkeTech: Data sheet – Non-metallic crystals (MarkeTech International Inc., Port Townsend 2002)
4.49 K.V. Krishna Rao, V.G. Krishna Murty: Acta Crystallogr. **17**, 788–789 (1964)
4.50 K. Vedam, E.D.D. Schmidt, W.C. Schneider: *Optical Properties of Highly Transparent Solids*, ed. by S.S. Mitra, B. Bendow (Plenum, New York 1975) pp. 169–175
4.51 J.P. Szczesniak, D. Cuddeback, J.C. Corelli: J. Appl. Phys. **47**, 5356–5359 (1976)
4.52 V.M. Maevskii, A.B. Roitsin: Fiz. Tverd. Tela **31**, 294–296 (1989)
4.53 V.M. Maevskii, A.B. Roitsin: Sov. Phys. Solid State [English Transl.] **31**, 1448–1449 (1989)
4.54 D.K. Biegelsen: Phys. Rev. Lett. **32**, 1196–1199 (1974)
4.55 D.K. Biegelsen: Erratum, Phys. Rev. Lett. **33**, 51 (1974)
4.56 A.A. Berezhnoi, V.M. Fedulov, K.P. Skornyakova: Fiz. Tverd. Tela **17**, 2785–2787 (1975)
4.57 A.A. Berezhnoi, V.M. Fedulov, K.P. Skornyakova: Sov. Phys. Solid State [English Transl.] **17**, 1855–1856 (1975)
4.58 D.F. Edwards, E. Ochoa: J. Opt. Soc. Am. **71**, 607 (1981)
4.59 G.N. Ramachandran: Proc. Indian Acad. Sci. A **25**, 266–279 (1947)
4.60 J. Fontanella, R.L. Johnston, J.H. Colwell, C. Andeen: Appl. Opt. **16**, 2949–2951 (1977)
4.61 M.D. Levenson, C. Flytzanis, N. Bloembergen: Phys. Rev. B **6**, 3962–3965 (1972)
4.62 D.E. McCarthy: Appl. Opt. **2**, 591–603 (1963)
4.63 N.P. Barnes, M.S. Piltch: J. Opt. Soc. Am. **69**, 178–180 (1979)
4.64 J.J. Wynne: Phys. Rev. **178**, 1295–1303 (1969)
4.65 H.H. Li: J. Phys. Chem. Ref. Data **5**, 329–528 (1976)
4.66 P.D. Maker, R.W. Terhune: Phys. Rev. **137**, A801–A818 (1965)
4.67 A.J. Moses: *Optical Materials Properties*, Handbook of Electronic Materials, Vol. 1 (IFI/Plenum, New York 1971)
4.68 T. Radhakrishnan: Proc. Indian Acad. Sci. A **33**, 22–34 (1951)
4.69 R.F. Stephens, I.H. Malitson: Natl. Bur. Stand. J. Res. **49**, 249 (1952)
4.70 C.C. Wang, E.L. Baardsen: Phys. Rev. **185**, 1079–1082 (1969)
4.71 W.K. Burns, N. Bloembergen: Phys. Rev. B **4**, 3437–3450 (1971)
4.72 N. Bloembergen, W.K. Burns, M. Matsukoda: Opt. Commun. **1**, 195 (1969)
4.73 B. Tatian: Appl. Opt. **23**, 4477–4485 (1984)
4.74 D.F. Edwards, E. Ochoa: Appl. Opt. **19**, 4130 (1980)
4.75 J.J. Wynne, G.D. Boyd: Appl. Phys. Lett. **12**, 191–192 (1968)
4.76 W. Kucharczyk: Physica B **172**, 473–490 (1991)
4.77 A.V. Pakhnev, M.P. Shaskolškaya, S.S. Gorbach: Izv. Vyssh. Uchebn. Zaved., Fiz. **12**, 28–34 (1975)
4.78 A.V. Pakhnev, M.P. Shaskolškaya, S.S. Gorbach: Sov. Phys. J. [English Transl.] **18**, 1662 (1975)
4.79 Princeton Scientific: Data sheet – Crystalline materials (Princeton Scientific Corp., Princeton 2003)
4.80 J. Reintjes, M.B. Schulz: J. Appl. Phys. **39**, 5254–5258 (1968)
4.81 S. Musikant: *Optical Materials: An Introduction to Selection and Application* (Marcel Dekker, New York 1985)
4.82 R.W. Dixon: J. Appl. Phys. **38**, 5149–5153 (1967)
4.83 P. Billard: Acta Electron. **6**, 75–169 (1962)
4.84 M.D. Beals, L. Merker: Mater. Des. Eng. **51**, 12–13 (1960)
4.85 D.E. McCarthy: Appl. Opt. **7**, 1997 (1968)
4.86 M.O. Manasreh, D.O. Pederson: Phys. Rev. B **30**, 3482–3485 (1984)
4.87 L.G. DeShazer, S.C. Rand, B.A. Wechsler: Laser crystals. In: *Optical Materials, CRC Handbook of Laser Science and Technology*, Vol. 5, Subvol. 3, ed. by M.J. Weber (CRC, Boca Raton 1987) pp. 281–338
4.88 K.L. Ovanesyan, A.G. Petrosyan, G.O. Shirinyan, A.A. Avetisyan: Izv. Akad. Nauk SSSR, Ser. Neorg. Materiali [in Russian] **17**, 459–462 (1981)
4.89 D. Berlincourt, H. Jaffe, L.R. Shiozawa: Phys. Rev. **129**, 1009–1017 (1963)
4.90 H. Sasaki, K. Tsoubouki, N. Chubachi, N. Mikoshiba: J. Appl. Phys. **47**, 2046–2049 (1976)
4.91 R. Weil, M.J. Sun: Proc. Int. Symp. Cadmium Telluride Materials Gamma-Ray Detectors, Tech. Dig., Strasbourg 1971, pages XIX-1 to XIX-6
4.92 E.H. Turner, I.P. Kaminow, C. Schwab: Phys. Rev. B **9**, 2524–2529 (1974)
4.93 R.C. Hanson, J.R. Hallberg, C. Schwab: Appl. Phys. Lett. **21**, 490–492 (1972)
4.94 A. Boese, E. Mohler, R. Pitka: J. Mater. Sci. **9**, 1754–1758 (1974)
4.95 B. Prevot, C. Carabatos, C. Schwab, B. Hennion, F. Moussa: Solid State Commun. **13**, 1725–1727 (1973)
4.96 D.K. Biegelsen, J.C. Zesch, C. Schwab: Phys. Rev. B **14**, 3578–3582 (1976)
4.97 L.M. Belyaev, G.S. Belikova, G.F. Dobrzhanskii, G.B. Netesov, Yu.V. Shaldin: Fiz. Tverd. Tela **6**, 2526–2528 (1964)
4.98 L.M. Belyaev, G.S. Belikova, G.F. Dobrzhanskii, G.B. Netesov, Yu.V. Shaldin: Sov. Phys. Solid State [English Transl.] **6**, 2007–2008 (1965)
4.99 R.C. Hanson, K. Helliwell, C. Schwab: Phys. Rev. B **9**, 2649–2654 (1974)
4.100 G. Arlt, P. Quadflieg: Phys. Status Solidi **25**, 323–330 (1968)
4.101 Ioffe Institute: Database (Ioffe Institute, St. Petersburg 2003)
4.102 K.S. Champlin, R.J. Erlandson, G.H. Glover, P.S. Hauge, T. Lu: Appl. Phys. Lett. **11**, 348–349 (1967)
4.103 T.E. Walsh: RCA Rev. **27**, 323–335 (1966)
4.104 A. Feldman, R.M. Waxler: J. Appl. Phys. **53**, 1477–1483 (1982)
4.105 D.R. Nelson, E.H. Turner: J. Appl. Phys. **39**, 3337–3343 (1968)

4.106 J. E. Kiefer, A. Yariv: Appl. Phys. Lett. **15**, 26–27 (1969)
4.107 G. H. Sherman, P. D. Coleman: J. Appl. Phys. **44**, 238 (1973)
4.108 R. C. Miller, W. A. Nordland: Phys. Rev. B **5**, 4931–4934 (1972)
4.109 J. Pastrnak, L. Roskovcova: Phys. Status Solidi **14**, K5–K8 (1966)
4.110 N. P. Barnes, M. S. Piltch: J. Opt. Soc. Am. **67**, 628 (1977)
4.111 A. G. DeBell, E. L. Dereniak, J. Harvey, J. Nissley, J. Palmer, A. Selvarajan, W. L. Wolfe: Appl. Opt. **18**, 3114–3115 (1979)
4.112 E. H. Turner, E. Buehler, H. Kasper: Phys. Rev. B **9**, 558–561 (1974)
4.113 D. S. Chemla, P. Kupecek, C. Schwartz, C. Schwab, A. Goltzene: IEEE J. Quantum Electron. **7**, 126 (1971)
4.114 R. C. Miller, W. A. Nordland, S. C. Abrahams, J. L. Bernstein, C. Schwab: J. Appl. Phys. **44**, 3700 (1973)
4.115 E. Mohler, B. Thomas: Phys. Status Solidi (b) **79**, 509–517 (1977)
4.116 A. Feldman, D. Horowitz: J. Opt. Soc. Am. **59**, 1406–1408 (1969)
4.117 T. G. Okroashvili: Opt. Spektrosk. **47**, 798–800 (1979)
4.118 T. G. Okroashvili: Opt. Spectrosc. [English Transl.] **47**, 442–443 (1979)
4.119 J. J. Wynne, N. Bloembergen: Phys. Rev. **188**, 1211 (1969)
4.120 J. J. Wynne, N. Bloembergen: Erratum, Phys. Rev. B **2**, 4306 (1970)
4.121 B. O. Seraphin, H. E. Bennett: *Semiconductors and Semimetals 3*, ed. by R. K. Willard, R. C. Beer (Academic Press, New York 1967)
4.122 R. A. Soref, H. W. Moos: J. Appl. Phys. **35**, 2152 (1964)
4.123 R. K. Chang, J. Ducuing, N. Bloembergen: Phys. Rev. Lett. **15**, 415–418 (1965)
4.124 W. D. Johnston Jr., I. P. Kaminow: Phys. Rev. **188**, 1209–1211 (1969)
4.125 E. Yablonovitch, C. Flytzanis, N. Bloembergen: Phys. Rev. Lett. **29**, 865–868 (1972)
4.126 M. Sugie, K. Tada: Jpn. J. Appl. Phys. **15**, 421–430 (1976)
4.127 M. Bertolotti, V. Bogdanov, A. Ferrari, A. Jascow, N. Nazorova, A. Pikhtin, L. Schirone: J. Opt. Soc. Am. B **7**, 918–922 (1990)
4.128 B. F. Levine, C. G. Bethea: Appl. Phys. Lett. **20**, 272–274 (1972)
4.129 A. H. Kachare, W. G. Spitzer, J. E. Fredrickson: J. Appl. Phys. **47**, 4209 (1976)
4.130 Yu. Berezashvili, S. Machavariani, A. Natsvlishvili, A. Chirakadze: J. Phys. D **22**, 682–686 (1989)
4.131 M. M. Choy, R. L. Byer: Phys. Rev. B **14**, 1693–1705 (1976)
4.132 D. F. Parsons, P. D. Coleman: Appl. Opt. **10**, 1683–1685 (1971)
4.133 Yu. X. Ilisavskii, L. A. Kutakova: Fiz. Tverd. Tela **23**, 3299–3307 (1981)
4.134 Yu. X. Ilisavskii, L. A. Kutakova: Sov. Phys. Solid State [English Transl.] **23**, 1916–1920 (1981)
4.135 L. G. Meiners: J. Appl. Phys. **59**, 1611–1613 (1986)
4.136 D. N. Nichols, D. S. Rimai, R. J. Sladek: Solid State Commun. **36**, 667–669 (1980)
4.137 A. A. Blistanov, V. S. Bondarenko, N. V. Perelomova, F. N. Strizhevskaya, V. V. Tchakalova, M. P. Shaskolskaya: *Acoustic Crystals* [in Russian] (Nauka, Moscow 1982)
4.138 E. Käräjämäki, R. Laiho, T. Levola: Physica **429**, 3–18 (1982)
4.139 G. R. Mariner, K. Vedam: Appl. Opt. **20**, 2878–2879 (1981)
4.140 A. V. Novoselova (Ed.): *Physical-Chemical Properties of Semiconductors - Handbook* [in Russian] (Nauka, Moscow 1979)
4.141 M. Yamada, K. Yamamoto, K. Abe: J. Phys. D **10**, 1309–1313 (1977)
4.142 S. Adachi, C. Hamaguchi: J. Phys. Soc. Jpn. **43**, 1637–1645 (1977)
4.143 C. K. N. Patel, R. E. Slusher, P. A. Fleury: Phys. Rev. Lett. **17**, 1011–1014 (1966)
4.144 H. McKenzie, D. J. Hagan, H. A. Al-Attar: IEEE J. Quantum Electron. **22**, 1328 (1986)
4.145 C. C. Lee, H. Y. Fan: Phys. Rev. B **9**, 3502 (1974)
4.146 O. G. Lorimor, W. G. Spitzer: J. Appl. Phys. **36**, 1841–1844 (1965)
4.147 T. S. Moss: *Optical Properties of Semiconductors* (Academic Press, New York 1959) p. 224
4.148 R. Braunstein, N. Ockman: *Interactions of Coherent Optical Radiation with Solids*, Final Report, Office of Naval Research (Department of Navy, Washington, DC 1964) ARPA Order No. 306-362
4.149 K. Tada, N. Suzuki: Jpn. J. Appl. Phys. **19**, 2295–2296 (1980)
4.150 Y. Tsay, B. Bendow, S. S. Mitra: Phys. Rev. B **5**, 2688–2696 (1972)
4.151 J. Stone, M. S. Whalen: Appl. Phys. Lett. **41**, 1140–1142 (1982)
4.152 A. N. Pikhtin, A. D. Yaskov: Sov. Phys. Semicond. **12**, 622–626 (1978)
4.153 I. I. Adrianova, A. A. Berezhnoi, K. K. Dubenskii, V. A. Sokolov: Fiz. Tverd. Tela **12**, 2462–2464 (1970)
4.154 I. I. Adrianova, A. A. Berezhnoi, K. K. Dubenskii, V. A. Sokolov: Sov. Phys. Solid State [English Transl.] **12**, 1972–1973 (1970)
4.155 C. Kojima, T. Shikama, S. Kuninobu, A. Kawabata, T. Tanaka: Jpn. J. Appl. Phys. **8**, 1361–1362 (1969)
4.156 R. C. Miller, W. A. Nordland: Phys. Rev. B **2**, 4896–4902 (1970)
4.157 A. Chergui, J. L. Deiss, J. B. Grun, J. L. Loison, M. Robino, R. Besermann: Appl. Surf. Sci. **96**, 874 (1996)
4.158 D. T. F. Marple: J. Appl. Phys. **35**, 539 (1964)
4.159 Alkor Technologies: Data sheet – Optical crystals (Alkor Technologies, St. Petersburg 2002)
4.160 E. A. Kraut, B. R. Tittman, L. J. Graham, T. C. Lim: Appl. Phys. Lett. **17**, 271–272 (1970)

4.161 R. E. Aldrich, S. L. Hou, M. L. Harvill: J. Appl. Phys. **42**, 493–494 (1971)
4.162 J. Zelenka: Czech. J. Phys. B **28**, 165–169 (1978)
4.163 M. Krzesinska: Ultrasonics **24**, 88–92 (1986)
4.164 V. V. Kucha, V. I. Mirgorodskii, S. V. Peshin, A. T. Sobolev: Pis̕ma Zh. Tekh. Fiz. **10**, 124–126 (1984)
4.165 V. V. Kucha, V. I. Mirgorodskii, S. V. Peshin, A. T. Sobolev: Sov. Tech. Phys. Lett. [English Transl.] **10**, 51–52 (1984)
4.166 A. Reza, G. Babonas, D. Senuliene: Liet. Fiz. Rinkinys. **26**, 41–47 (1986)
4.167 A. Reza, G. Babonas, D. Senuliene: Sov. Phys. Collection [English Transl.] **26**, 33–38 (1986)
4.168 P. I. Ropot: Opt. Spektrosk. **70**, 371–375 (1991)
4.169 P. I. Ropot: Opt. Spectrosc. [English Transl.] **70**, 217–220 (1991)
4.170 H. Schweppe, P. Quadfleg: IEEE Trans. Sonics Ultrason. **21**, 56–57 (1974)
4.171 Y. R. Reddy, L. Sirdeshmukh: Phys. Status Solidi (a) **103**, K157–K160 (1987)
4.172 G. A. Babonas, A. A. Reza, E. I. Leonov, V. I. Shandaris: Zh. Tekh. Fiz. **55**, 1203–1205 (1985)
4.173 G. A. Babonas, A. A. Reza, E. I. Leonov, V. I. Shandaris: Sov. Phys. Tech. Phys. [English Transl.] **30**, 689–690 (1985)
4.174 R. Bechmann: Proc. R. Phys. Soc. (London) B **64**, 323–337 (1951)
4.175 A. D. Prasad Rao, P. da R. Andrade, S. P. S. Porto: Phys. Rev. B **9**, 1077–1084 (1974)
4.176 T. S. Narasimhamurty: Proc. Indian Acad. Sci. A **40**, 167–175 (1954)
4.177 M. Tanaka, M. Yamada, C. Hamaguchi: J. Phys. Soc. Jpn. **38**, 1708–1714 (1975)
4.178 Eksma: Data sheet – Optics and optomechanics (Eksma Co., Vilnius 2003)
4.179 P. C. Leung, G. Andermann, W. G. Spitzer, C. A. Mead: J. Phys. Chem. Solids **27**, 849–855 (1966)
4.180 V. M. Skorikov, I. S. Zakharov, V. V. Volkov, E. A. Spirin: Inorgan. Mater. **38**, 172–178 (2002)
4.181 F. Vachss, L. Hesselink: Opt. Commun. **62**, 159–165 (1987)
4.182 P. Bayvel, M. McCall, R. V. Wright: Opt. Lett. **13**, 27–29 (1988)
4.183 P. Bayvel: Sensors Actuators **16**, 247–254 (1989)
4.184 H. J. Simon, N. Bloembergen: Phys. Rev. **171**, 1104 (1968)
4.185 H.-J. Weber: Acta Crystallogr. A **35**, 225–232 (1979)
4.186 S. Chandrasekhar: Proc. R. Soc. A **259**, 531 (1961)
4.187 K. L. Vodopyanov, L. A. Kulevskii, V. G. Voevodin, A. I. Gribenyukov, K. R. Allakhverdiev, T. A. Kerimov: Opt. Commun. **83**, 322–326 (1991)
4.188 V. I. Sokolov, V. K. Subashiev: Fiz. Tverd. Tela **14**, 222–228 (1972)
4.189 V. I. Sokolov, V. K. Subashiev: Sov. Phys. Solid State [English Transl.] **14**, 178–183 (1972)
4.190 G. B. Abdullaev, L. A. Kulevsky, A. M. Prokhorov, A. D. Saleev, E. Yu. Salaev, V. V. Smirnov: Pisma Zh. Eksp. Teor. Fiz. **16**, 130 (1972)
4.191 G. B. Abdullaev, L. A. Kulevsky, A. M. Prokhorov, A. D. Saleev, E. Yu. Salaev, V. V. Smirnov: JETP Lett. [English Transl.] **16**, 90–92 (1972)
4.192 P. J. Kupecek, E. Batifol, A. Kuhn: Opt. Commun. **11**, 291–295 (1974)
4.193 K. L. Vodopyanov, I. A. Kulevskii: Opt. Commun. **118**, 375 (1995)
4.194 G. A. Akundov, A. A. Agaeve, V. M. Salmanov, Yu. P. Sharorov, I. D. Yaroshetskii: Sov. Phys. Semicond. **7**, 826–827 (1973)
4.195 C. A. Arguello, D. L. Rousseau, S. P. S. Porto: Phys. Rev. **181**, 1351–1363 (1969)
4.196 C. F. Cline, H. L. Dunegan, G. W. Henderson: J. Appl. Phys. **38**, 1944 (1967)
4.197 S. B. Austerman, D. A. Berlincourt, H. H. A. Krueger: J. Appl. Phys. **34**, 339–341 (1963)
4.198 A. A. Reza, G. A. Babonas: Fiz. Tverd. Tela **16**, 1414–1418 (1974)
4.199 A. A. Reza, G. A. Babonas: Sov. Phys. Solid State [English Transl.] **16**, 909–911 (1974)
4.200 J. D. Beasley: Appl. Opt. **33**, 1000–1003 (1994)
4.201 E. F. Tokarev, G. S. Pado, L. A. Chernozatonskii, V. V. Drachev: Fiz. Tverd. Tela **15**, 1593–1595 (1973)
4.202 E. F. Tokarev, G. S. Pado, L. A. Chernozatonskii, V. V. Drachev: Sov. Phys. Solid State [English Transl.] **15**, 1064–1065 (1973)
4.203 I. A. Dankov, G. S. Pado, I. B. Kobyakov, V. V. Berdnik: Fiz. Tverd. Tela **21**, 2570–2575 (1979)
4.204 I. A. Dankov, G. S. Pado, I. B. Kobyakov, V. V. Berdnik: Sov. Phys. Solid State [English Transl.] **21**, 1481–1483 (1979)
4.205 I. B. Kobyakov, V. M. Arutyunova: Zh. Tekh. Fiz. [in Russian] **58**, 983 (1988)
4.206 K. Vedam, T. A. Davis: Phys. Rev. **181**, 1196–1201 (1969)
4.207 A. S. Barker, M. Ilegems: Phys. Rev. B **7**, 743–750 (1973)
4.208 E. K. Sichel, J. I. Pankove: J. Phys. Chem. Solids **38**, 330 (1977)
4.209 V. A. Savastenko, A. U. Sheleg: Phys. Status Solidi (a) **48**, K135 (1978)
4.210 M. Gospodinov, P. Sveshtarov, N. Petkov, T. Milenov, V. Tasev, A. Nikolov: Bulg. J. Phys. **16**, 520–522 (1989)
4.211 S. A. Geidur: Opt. Spektrosk. **49**, 193–195 (1980)
4.212 S. A. Geidur: Opt. Spectrosc. [English Transl.] **49**, 105–106 (1980)
4.213 S. Singh, J. R. Potopowicz, L. G. Van Uitert, S. H. Wemple: Appl. Phys. Lett. **19**, 53–56 (1971)
4.214 G. Arlt, G. R. Schodder: J. Acoust. Soc. Am. **37**, 384 (1965)
4.215 E. E. Tokarev, I. B. Kobyakov, I. P. Kuz̕mina, A. N. Lobachev, G. S. Pado: Fiz. Tverd. Tela **17**, 980–986 (1975)
4.216 E. E. Tokarev, I. B. Kobyakov, I. P. Kuz̕mina, A. N. Lobachev, G. S. Pado: Sov. Phys. Solid State [English Transl.] **17**, 629–632 (1975)
4.217 R. J. Morrow, H. W. Newkirk: Rev. Int. Hautes Temp. Refract. **6**(2), 99–104 (1969)

4.218 D. A. Belogurov, T. G. Okroashvili, Yu. V. Shaldin, V. A. Maslov: Opt. Spektrosk. **54**, 298–301 (1983)
4.219 D. A. Belogurov, T. G. Okroashvili, Yu. V. Shaldin, V. A. Maslov: Opt. Spectrosc. [English Transl.] **54**, 177–179 (1983)
4.220 H. W. Newkirk, D. K. Smith, J. S. Kahn: Am. Mineral. **51**, 141–151 (1966)
4.221 J. Jerphagnon, H. W. Newkirk: Appl. Phys. Lett. **18**, 245–247 (1971)
4.222 M. Bass (Ed.): *Handbook of Optics*, Vol. 2 (McGraw-Hill, New York 1995)
4.223 A. A. Davydov, L. A. Kulevsky, A. M. Prokhorov, A. D. Savelev, V. V. Smirnov: Pisma Zh. Eksp. Teor. Fiz. **15**, 725 (1972)
4.224 A. A. Davydov, L. A. Kulevsky, A. M. Prokhorov, A. D. Savelev, V. V. Smirnov: JETP Lett. [English Transl.] **15**, 513 (1972)
4.225 S. H. Wemple, M. DiDomenico Jr.: Appl. Solid State Sci. **3**, 263–283 (1972)
4.226 A. Penzkofer, M. Schaffner, X. Bao: Opt. Quantum Electron. **22**, 351 (1990)
4.227 G. C. Bhar: Appl. Opt. **15**, 305 (1976)
4.228 R. C. Miller: Appl. Phys. Lett. **5**, 17–19 (1964)
4.229 R. Weil, D. Neshmit: J. Opt. Soc. Am. **67**, 190–195 (1977)
4.230 S. J. Czyzak, W. M. Baker, R. C. Crane, J. B. Howe: J. Opt. Soc. Am. **47**, 240 (1957)
4.231 X. C. Long, R. A. Myers, S. R. J. Brueck, R. Ramer, K. Zheng, S. D. Hersee: Appl. Phys. Lett. **67**, 1349–1351 (1995)
4.232 I. Ishidate, K. Inone, M. Aoki: Jpn. J. Appl. Phys. **19**, 1641–1645 (1980)
4.233 P. M. Lundquist, W. P. Lin, G. K. Wong, M. Razeghi, J. B. Ketterson: Appl. Phys. Lett. **66**, 1883–1885 (1995)
4.234 I. P. Kaminow, E. H. Turner: Appl. Opt. **5**, 1612–1628 (1966)
4.235 I. P. Kaminow, E. H. Turner: Proc. IEEE **54**, 1374–1390 (1966)
4.236 O. W. Madelung, E. Mollwo: Z. Phys. **249**, 12–30 (1971)
4.237 L. B. Kobyakov: Kristallografiya **11**, 419–421 (1966)
4.238 L. B. Kobyakov: Sov. Phys. Crystallogr. [English Transl.] **11**, 369–371 (1966)
4.239 I. A. Dankov, I. B. Kobyakov, S. Yu. Davydov: Fiz. Tverd. Tela **24**, 3613–3620 (1982)
4.240 I. A. Dankov, I. B. Kobyakov, S. Yu. Davydov: Sov. Phys. Solid State [English Transl.] **24**, 2058–2063 (1982)
4.241 N. Uchida, S. Saito: J. Appl. Phys. **43**, 971–976 (1972)
4.242 S. Haussühl: Acta Crystallogr. A **34**, 547–550 (1978)
4.243 N. A. Zakharov, A. V. Egorov, N. S. Kozlova, O. G. Portnov: Fiz. Tverd. Tela **30**, 3166–3168 (1988)
4.244 N. A. Zakharov, A. V. Egorov, N. S. Kozlova, O. G. Portnov: Sov. Phys. Solid State [English Transl.] **30**, 1823–1824 (1988)
4.245 S. Haussühl: Acustica **23**, 165–169 (1970)
4.246 S. Haussühl: Phys. Status Solidi **29**, K159–K162 (1968)
4.247 Ya. V. Burak, K. Ya. Borman, I. S. Girnyk: Fiz. Tverd. Tela **26**, 3692–3694 (1984)
4.248 Ya. V. Burak, K. Ya. Borman, I. S. Girnyk: Sov. Phys. Solid State [English Transl.] **26**, 2223–2224 (1984)
4.249 J. Wang, H. Li, L. Zhang, R. Wang: Acta Acust. (China) **11**, 338 (1986)
4.250 A. W. Warner, D. A. Pinnow, J. G. Bergman Jr, G. R. Crane: J. Acoust. Soc. Am. **47**, 791–794 (1970)
4.251 R. C. Miller, S. C. Abrahams, R. L. Barns, J. L. Bernstein, W. A. Nordland, E. H. Turner: Solid State Commun. **9**, 1463–1465 (1971)
4.252 A. A. Abdullaev, A. V. Vasil'eva, G. F. Dobrzhanskii, Yu. N. Polivanov: Sov. J. Quantum Electron. **7**, 56–59 (1977)
4.253 G. Nath, H. Nehmanesch, M. Gsänger: Appl. Phys. Lett. **17**, 286–288 (1970)
4.254 S. V. Bogdanov (Ed.): *Lithium Iodate – Growth, Properties and Applications* [in Russian] (Nauka, Novosibirsk 1980)
4.255 D. J. Gettemy, W. C. Harker, G. Lindholm, N. P. Barnes: IEEE J. Quantum Electron. **24**, 2231 (1988)
4.256 O. G. Vlokh, I. A. Velichko, L. A. Lazko: Kristallografiya **20**, 430–432 (1975)
4.257 O. G. Vlokh, I. A. Velichko, L. A. Lazko: Sov. Phys. Crystallogr. [English Transl.] **20**, 263–264 (1975)
4.258 K. Kato: IEEE J. Quantum Electron. **21**, 119 (1985)
4.259 M. J. Weber: Insulating crystal lasers. In: *CRC Handbook of Lasers with Selected Data on Optical Technology*, ed. by R. J. Pressley (Chemical Rubber Co., Cleveland 1971) pp. 371–417
4.260 T. Goto, O. L. Anderson, I. Ohno, S. Yamamoto: J. Geophys. Res. **94**, 7588 (1989)
4.261 R. M. Waxler, E. N. Farabaugh: J. Res. Natl. Bur. Stand. A **74**, 215–220 (1970)
4.262 E. M. Voronkova, B. N. Grechushnikov, G. I. Distler, I. P. Petrov: *Optical Materials for the Infrared Technique* [in Russian] (Nauka, Moscow 1965)
4.263 N. N. Khromova: Izv. Akad. Nauk SSSR, Ser. Neorg. Materiali [in Russian] **23**, 1500 (1987)
4.264 D. F. Nelson, P. D. Lazay, M. Lax: Phys. Rev. B **6**, 3109–3120 (1972)
4.265 Z. P. Chang, G. R. Barsch: IEEE Trans. Sonics Ultrason. **23**, 127–135 (1976)
4.266 I. M. Sil'vestrova, Yu. V. Pisarevskii, O. V. Zvereva, A. A. Shternberg: Kristallografiya **32**, 792–794 (1987)
4.267 I. M. Sil'vestrova, Yu. V. Pisarevskii, O. V. Zvereva, A. A. Shternberg: Sov. Phys. Crystallogr. [English Transl.] **32**, 467–468 (1987)
4.268 I. M. Sil'vestrova, O. A. Aleshko-Ozhevskii, Yu. V. Pisarevskii, A. A. Shternberg, G. S. Mironova: Fiz. Tverd. Tela **29**, 3454–3456 (1987)
4.269 I. M. Sil'vestrova, O. A. Aleshko-Ozhevskii, Yu. V. Pisarevskii, A. A. Shternberg, G. S. Mironova: Sov. Phys. Solid State [English Transl.] **29**, 1979–1980 (1987)
4.270 H. A. A. Sidek, G. A. Saunders, H. Wang, B. Xu, J. Han: Phys. Rev. B **36**, 7612 (1987)
4.271 A. C. DeFranzo, B. G. Pazol: Appl. Opt. **32**, 2224–2234 (1993)
4.272 E. Carvallo: Compte Rendus **126**, 950 (1898)
4.273 W. L. Bond: J. Appl. Phys. **36**, 1674–1677 (1965)

4.274 H. Jaffe, D.A. Berlincourt: Proc. IEEE **53**, 1372–1386 (1965)
4.275 J. Sapriel, R. Lancon: Proc. IEEE **61**, 678–679 (1973)
4.276 J. Sapriel: Appl. Phys. Lett. **19**, 533–535 (1971)
4.277 I.M. Silvestrova, V.A. Kuznetsov, N.A. Moiseeva, E.P. Efremova, Yu.V. Pisarevskii: Fiz. Tverd. Tela [in Russian] **28**, 180 (1986)
4.278 W.G. Cady: *Piezoelectricity* (McGraw-Hill, New York 1946)
4.279 P.H. Carr: J. Acoust. Soc. Am. **41**, 75–83 (1967)
4.280 R. Bechmann: Phys. Rev. **110**, 1060–1061 (1958)
4.281 T.S. Narasimhamurty: J. Opt. Soc. Am. **59**, 682–686 (1969)
4.282 H. Wagner: Z. Phys. **193**, 218–234 (1966)
4.283 P. Grosse: *Die Festkörpereigenschaften von Tellur*, Springer Tracts in Modern Physics, No. 48 (Springer, Berlin, Heidelberg 1969) p. 65
4.284 G. Arlt, P. Quadfleg: Phys. Status Solidi **32**, 687–689 (1969)
4.285 I.S. Grigoriev, E.Z. Melikhov (Eds.): *Physical Quantities Handbook* [in Russian] (Emergoatomizdat, Moscow 1991)
4.286 D. Souilhac, D. Billeret, A. Gundjan: Appl. Opt. **28**, 3993–3996 (1989)
4.287 W.L. Bond, G.D. Boyd, H.L. Carter Jr.: J. Appl. Phys. **38**, 4090–4091 (1967)
4.288 G.D. Boyd, T.J. Bridges, E.G. Burkhardt: IEEE J. Quantum Electron. **4**, 515 (1968)
4.289 E.H. Turner: IEEE J. Quantum Electron. **3**, 695 (1967)
4.290 C. Chen, Z. Xu, D. Deng, J. Zhang, G.K.L. Wong, B. Wu, N. Ye, D. Tang: Appl. Phys. Lett. **68**, 2930–2932 (1996)
4.291 J. Jerphagnon, S.K. Kurtz: Phys. Rev. B **1**, 1739 (1970)
4.292 F. Hache, A. Zeboulon, G. Gallot, G.M. Gale: Opt. Lett. **20**, 1556–1558 (1995)
4.293 A.J. Rogers: Proc. R. Soc. London A **353**, 177–192 (1977)
4.294 J.H. McFee, G.D. Boyd, P.H. Schmidt: Appl. Phys. Lett. **17**, 57–59 (1970)
4.295 D. Eimerl, L. Davis, S. Velsko, E.K. Graham, A. Zalkin: J. Appl. Phys. **62**, 1968–1983 (1987)
4.296 J.R. Teague, R.R. Rice, R. Gerson: J. Appl. Phys. **46**, 2864–2866 (1975)
4.297 T. Yamada, N. Niizeki, H. Toyoda: Jpn. J. Appl. Phys. **6**, 151–155 (1967)
4.298 R.T. Smith, F.S. Welsh: J. Appl. Phys. **42**, 2219–2230 (1971)
4.299 R.A. Graham: Ferroelectrics **10**, 65–69 (1976)
4.300 T. Yamada, H. Iwasaki, N. Niizeki: Jpn. J. Appl. Phys. **8**, 1127–1132 (1969)
4.301 I. Tomeno, N. Hirano: J. Phys. Soc. Jpn. **50**, 1809 (1981)
4.302 I. Tomeno, N. Hirano: J. Phys. Soc. Jpn. **51**, 339 (1982)
4.303 L.P. Avakyants, D.F. Kiselev, N.N. Shchitov: Fiz. Tverd. Tela **18**, 2129–2130 (1976)
4.304 L.P. Avakyants, D.F. Kiselev, N.N. Shchitov: Sov. Phys. Solid State [English Transl.] **18**, 1242–1243 (1976)
4.305 C. Chen, B. Wu, A. Jiang, G. You: Sci. Sin. B **28**, 235 (1985)
4.306 L.J. Bromley, A. Guy, D.C. Hanna: Opt. Commun. **67**, 316–320 (1988)
4.307 G.G. Gurzadyan, A.S. Oganesyan, A.V. Petrosyan, R.O. Sharkhatunyan: Zh. Tekh. Fiz. **61**, 152–154 (1991)
4.308 G.G. Gurzadyan, A.S. Oganesyan, A.V. Petrosyan, R.O. Sharkhatunyan: Sov. Phys. Tech. Phys. [English Transl.] **36**, 341–342 (1991)
4.309 R.C. Eckardt, H. Masuda, Y.X. Fan, R.L. Byer: IEEE J. Quantum Electron. **26**, 922–933 (1990)
4.310 L. Bohaty, J. Liebertz: Z. Kristallogr. **192**, 91–95 (1990)
4.311 D. Zhang, Y. Kong, J.-Y. Zhang: Opt. Commun. **184**, 485–491 (2000)
4.312 M.V. Hobden, J. Warner: Phys. Lett. **22**, 243 (1966)
4.313 D.H. Jundt, M.M. Fejer, R.L. Byer: IEEE J. Quantum Electron. **26**, 135–138 (1990)
4.314 J.E. Midwinter: J. Appl. Phys. **39**, 3033–3038 (1968)
4.315 K.F. Hulme, P.H. Davies, V.M. Cound: J. Phys. C **2**, 855–857 (1969)
4.316 J.D. Zook, D. Chen, G.N. Otto: Appl. Phys. Lett. **11**, 159–161 (1967)
4.317 A.L. Aleksandrovskii, G.I. Ershova, G.Kh. Kitaeva, S.P. Kulik, I.I. Naumova, V.V. Tarasenko: Kvantovaya Elektron. **18**, 254–256 (1991)
4.318 A.L. Aleksandrovskii, G.I. Ershova, G.Kh. Kitaeva, S.P. Kulik, I.I. Naumova, V.V. Tarasenko: Sov. J. Quantum Electron. [English Transl.] **21**, 225–227 (1991)
4.319 Y. Chang, J. Wen, H. Wang, B. Li: Chin. Phys. Lett. **9**, 427–430 (1992)
4.320 R.J. Holmes, Y.S. Kim, C.D. Brandle, D.M. Smyth: Ferroelectrics **51**, 41–45 (1983)
4.321 S. Singh: Nonlinear optical materials. In: *CRC Handbook of Lasers with Selected Data on Optical Technology*, ed. by R.J. Pressley (Chemical Rubber Co., Cleveland 1971) pp. 489–525
4.322 R.C. Miller, A. Savage: Appl. Phys. Lett. **9**, 169–171 (1966)
4.323 E.H. Turner: Meeting Opt. Soc. Am., Tech. Dig. Paper A13, San Francisco 1966
4.324 C. OHara, N.M. Shorrocks, R.W. Whatmore, O. Jones: J. Phys. D **15**, 1289–1299 (1982)
4.325 Ya.M. Olikh: Fiz. Tverd. Tela **25**, 2222–2225 (1983)
4.326 Ya.M. Olikh: Sov. Phys. Solid State [English Transl.] **25**, 1282–1283 (1983)
4.327 Ya.M. Olikh: Phys. Status Solidi (a) **80**, K81 (1983)
4.328 I.I. Zubrinov, V.I. Semenov, D.V. Sheloput: Fiz. Tverd. Tela **15**, 2871–2873 (1973)
4.329 I.I. Zubrinov, V.I. Semenov, D.V. Sheloput: Sov. Phys. Solid State [English Transl.] **15**, 1921–1922 (1974)
4.330 V.V. Lomanov: *Acoustooptical Radioelectronic System Devices* [in Russian] (Nauka, Leningrad 1988) pp. 48–61
4.331 W.P. Mason: *Piezoelectric Crystals and their Applications to Ultrasonics* (Van Nostrand, New York 1950)
4.332 Y.S. Touloukian, R.W. Powell, C.Y. Ho, P.G. Klemens: *Thermal Conductivity, Non-Metallic Solids*, Thermal Properties of Matter, Vol. 2 (IFI/Plenum, New York 1970)

4.333 I. S. Zheludev, M. M. Tagieva: Kristallografiya **7**, 589–592 (1962)
4.334 I. S. Zheludev, M. M. Tagieva: Sov. Phys. Crystallogr. [English Transl.] **7**, 473–475 (1963)
4.335 J. D. Feichtner, R. Johannes, G. W. Roland: Appl. Opt. **9**, 1716–1717 (1970)
4.336 W. B. Gandrud, G. D. Boyd, J. H. McFee, F. H. Wehmeier: Appl. Phys. Lett. **16**, 59–61 (1970)
4.337 I. V. Kityk: Fiz. Tverd. Tela **33**, 1826–1833 (1991)
4.338 I. V. Kityk: Sov. Phys. Solid State [English Transl.] **33**, 1026–1030 (1991)
4.339 K. F. Hulme, O. Jones, P. H. Davies, M. V. Hobden: Appl. Phys. Lett. **10**, 133–135 (1967)
4.340 J. Warner: Brit. J. Appl. Phys. II **1**, 949–950 (1968)
4.341 R. F. Lucy: Appl. Opt. **11**, 1329 (1972)
4.342 D. S. Chemla, P. J. Kupecek, C. A. Schwartz: Opt. Commun. **7**, 225–228 (1973)
4.343 J. D. Feichtner, G. W. Roland: Appl. Opt. **11**, 993–998 (1972)
4.344 M. D. Ewbank, P. R. Newman, N. L. Mota, S. M. Lee, W. L. Wolfe, A. G. DeBell, W. A. Harrison: J. Appl. Phys. **51**, 3848–3852 (1980)
4.345 R. C. Y. Auyeung, D. M. Zielke, B. J. Feldman: Appl. Phys. B **48**, 293 (1989)
4.346 D. R. Suhre: Appl. Phys. B **52**, 367–370 (1991)
4.347 I. P. Kaminow, E. H. Turner: *CRC Handbook of Lasers with Selected Data on Optical Technology*, ed. by R. J. Pressley (Chemical Rubber Co., Cleveland 1971) pp. 447–459
4.348 F. Pockels: Abhandl. Ges. Wiss. Göttingen **39**, 1–204 (1894)
4.349 M. J. Weber (Ed.): *Handbook of Laser Science and Technology*, Vol. 3 (CRC, Boca Raton 1986)
4.350 S. Chang, H. R. Carleton: *Basic Optical Properties of Materials*, National Bureau of Standards Special Publication 574, ed. by A. Feldman (National Bureau of Standards, Boulder 1980) pp. 213–216
4.351 M. H. Grimsditch, A. K. Ramdas: Phys. Rev. B **22**, 4094–4096 (1980)
4.352 H. J. Eichler, H. Fery, J. Knof, J. Eichler: Z. Phys. B **28**, 297–306 (1977)
4.353 J. R. DeVore: J. Opt. Soc. Am. **41**, 416–419 (1951)
4.354 Bluelightning Optics: Data sheet – Product (Bluelightning Optics Co., Ltd., Beijing 2002)
4.355 G. C. Coquin, D. A. Pinnow, A. W. Warner: J. Appl. Phys. **42**, 2162–2168 (1971)
4.356 K. Wu, W. Hua: Acta. Acust. (China) **12**, 64 (1987)
4.357 I. L. Chistyi, L. Csillag, V. F. Kitaeva, N. Kroo, N. N. Sobolev: Phys. Status Solidi (a) **47**, 609–615 (1978)
4.358 B. W. Woods, S. A. Payne, J. E. Marion, R. S. Hughes, L. E. Davis: J. Opt. Soc. Am. B **8**, 970–977 (1991)
4.359 P. Blanchfield, G. A. Saunders: J. Phys. C **12**, 4673 (1979)
4.360 Y. Ohmachi, N. Uchida: J. Appl. Phys. **41**, 2307–2311 (1970)
4.361 G. Arlt, H. Schweppe: Solid State Commun. **6**, 783–784 (1968)
4.362 A. W. Warner, D. L. White, W. A. Bonner: J. Appl. Phys. **11**, 4489–4495 (1972)
4.363 Y. Ohmachi, N. Uchida: Rev. Electr. Commun. Lab. (Tokyo) **20**, 529–541 (1972)
4.364 Almaz Optics Inc.: Data sheet – Optical components (Almaz Optics Inc., Marlton 2002)
4.365 M. D. Ewbank, F. R. Newman: J. Appl. Phys. **53**, 1150–1153 (1982)
4.366 G. H. Sherman, P. D. Coleman: IEEE J. Quantum Electron. **9**, 403–409 (1973)
4.367 Y. Fujii, S. Yoshida, S. Misawa, S. Mackawa, T. Sakudo: Appl. Phys. Lett. **31**, 815–816 (1977)
4.368 N. Uchida: Phys. Rev. **4**, 3736 (1971)
4.369 R. S. Adhav, A. D. Vlassopoulos: IEEE J. Quantum Electron. **10**, 688 (1974)
4.370 E. N. Volkova, A. N. Izrailenko: Kristallografiya **28**, 1217–1219 (1983)
4.371 E. N. Volkova, A. N. Izrailenko: Sov. Phys. Crystallogr. [English Transl.] **28**, 716–717 (1983)
4.372 W. P. Mason, B. T. Matthias: Phys. Rev. **88**, 477–479 (1952)
4.373 A. S. Vasilevskaya, A. S. Sonin: Kristallografiya **14**, 713–716 (1969)
4.374 A. S. Vasilevskaya, A. S. Sonin: Sov. Phys. Crystallogr. [English Transl.] **14**, 611–613 (1970)
4.375 Clevite Corp.: *Reference Data on Linear Electro-Optic Effects* (Clevite Corp., Cleveland 1967)
4.376 H. Jaffe: Meeting Opt. Soc. Am., Tech. Dig., Cleveland 1950
4.377 G. K. Afanaseva: Kristallografiya **13**, 1024 (1968)
4.378 G. K. Afanaseva: Sov. Phys. Crystallogr. [English Transl.] **13**, 1024 (1968)
4.379 R. S. Adhav: J. Acoust. Soc. Am. **43**, 835 (1968)
4.380 F. Hoff, B. Stadnik: Electron. Lett. **2**, 293 (1966)
4.381 T. S. Narasimhamurty, K. Veerabhadra Rao, H. E. Pettersen: J. Mater. Sci. **8**, 577–580 (1973)
4.382 T. Hailing, G. A. Saunders, W. A. Lambson: Phys. Rev. B **26**, 5786 (1982)
4.383 R. S. Adhav: 76th Annual Meeting Acoust. Soc. Am., Tech. Dig. Paper No. 79, Cleveland 1968
4.384 B. A. Strukov, T. P. Spiridonov, N. N. Tkachev, K. A. Minaeva, M. Yu. Kozhevnikov: Izv. Akad. Nauk SSSR Ser. Fiz. **53**, 1320–1326 (1989)
4.385 B. A. Strukov, T. P. Spiridonov, N. N. Tkachev, K. A. Minaeva, M. Yu. Kozhevnikov: Bull. Acad. Sci. USSR, Phys. Ser. [English Transl.] **53**, 86–92 (1989)
4.386 D. Eimerl: Ferroelectrics **72**, 95–139 (1987)
4.387 A. S. Sonin, A. S. Vasilevskaya: *Electrooptic Crystals* [in Russian] (Atomizdat, Moscow 1971)
4.388 A. S. Vasilevskaya, E. N. Volkova, V. A. Koptsik, L. N. Rashkovich, T. A. Regulskaya, I. S. Rez, A. S. Sonin, V. S. Suvorov: Kristallografiya **12**, 518 (1967)
4.389 A. S. Vasilevskaya, E. N. Volkova, V. A. Koptsik, L. N. Rashkovich, T. A. Regulskaya, I. S. Rez, A. S. Sonin, V. S. Suvorov: Sov. Phys. Crystallogr. [English Transl.] **12**, 446 (1967)

4.390 V. S. Suvorov, A. S. Sonin, I. S. Rez: Zh. Eksp. Teor. Fiz. **53**, 491 (1967)
4.391 V. S. Suvorov, A. S. Sonin, I. S. Rez: Sov. Phys. JETP [English Transl.] **26**, 23 (1967)
4.392 N. P. Barnes, D. J. Gettemy, R. S. Adhav: J. Opt. Soc. Am. **72**, 895 (1982)
4.393 F. M. Johnson, J. A. Duardo: Laser Focus **11**, 31 (1967)
4.394 F. Zernike: J. Opt. Soc. Am. **55**, 91 (1965)
4.395 R. L. Byer, H. Kildal, R. S. Feigelson: Appl. Phys. Lett. **19**, 237–240 (1971)
4.396 G. D. Boyd, E. Buehler, F. G. Storz, J. H. Wernick: IEEE J. Quantum Electron. **8**, 419–426 (1972)
4.397 G. C. Ghosh, G. C. Bhar: IEEE J. Quantum Electron. **18**, 143 (1982)
4.398 K. Kato: IEEE J. Quantum Electron. **10**, 616 (1974)
4.399 R. S. Adhav: J. Opt. Soc. Am. **59**, 414–418 (1969)
4.400 R. S. Adhav: J. Appl. Phys. **46**, 2808 (1975)
4.401 M. P. Zaitseva, Yu. I. Kokorin, A. M. Sysoev, I. S. Rez: Kristallografiya **27**, 146–151 (1982)
4.402 M. P. Zaitseva, Yu. I. Kokorin, A. M. Sysoev, I. S. Rez: Sov. Phys. Crystallogr. [English Transl.] **27**, 86–89 (1982)
4.403 A. T. Anistratov, V. G. Martynov, L. A. Shabanova, I. S. Rez: Fiz. Tverd. Tela **21**, 1354 (1979)
4.404 A. T. Anistratov, V. G. Martynov, L. A. Shabanova, I. S. Rez: Sov. Phys. Solid State [English Transl.] **21** (1979)
4.405 C. W. Fairall, W. Reese: Phys. Rev. B **6**, 193–199 (1972)
4.406 R. S. Adhav: J. Appl. Phys. **39**, 4091 (1968)
4.407 T. R. Sliker, S. R. Burlage: J. Appl. Phys. **34**, 1837–1840 (1963)
4.408 L. A. Shuvalov, A. V. Mnatsakanyan: Kristallografiya **11**, 222 (1966)
4.409 L. P. Avakyants, D. F. Kiselev, N. V. Perelomova, V. I. Sugrei: Fiz. Tverd. Tela **25**, 580–582 (1983)
4.410 L. P. Avakyants, D. F. Kiselev, N. V. Perelomova, V. I. Sugrei: Sov. Phys. Solid State [English Transl.] **25**, 329–330 (1983)
4.411 Y. Mori, I. Kuroda, S. Nakajima, T. Sasaki, S. Nakai: Appl. Phys. Lett. **67**, 1818–1820 (1995)
4.412 N. Umemura, K. Kato: Appl. Opt. **36**, 6794 (1997)
4.413 G. D. Boyd, H. M. Kasper, J. H. McFee, F. G. Storz: IEEE J. Quantum Electron. **8**, 900–908 (1972)
4.414 G. D. Boyd, H. Kasper, J. M. McFee: IEEE J. Quantum Electron. **7**, 563–573 (1971)
4.415 G. C. Bhar, G. Ghosh: J. Opt. Soc. Am. **69**, 730–733 (1979)
4.416 R. S. Adhav: J. Appl. Phys. **39**, 4095–4098 (1968)
4.417 R. Danelyus, A. Piskarskas, V. Sirutkaitis, A. Stabinis, Ya. Yasevichyute: *Parametric Generators of Light and Picosecond Spectroscopy* [in Russian] (Mokslas, Vilnius 1983)
4.418 J. H. Ott, T. R. Sliker: J. Opt. Soc. Am. **54**, 1442–1444 (1964)
4.419 M. I. Zugrav, A. Dumitrica, A. Dumitras, N. Comaniciu: Cryst. Res. Technol. **17**, 475–480 (1982)
4.420 I. P. Kaminow: Phys. Rev. A **138**, 1539–1543 (1965)
4.421 D. A. Berlincourt, D. R. Curran, H. Jaffe: *Physical Acoustics*, Vol. 1A, ed. by W. P. Mason (Academic Press, New York 1964) pp. 1169–1270
4.422 S. Haussühl: Z. Kristallogr. **120**, 401 (1964)
4.423 I. P. Kaminow, G. O. Harding: Phys. Rev. **129**, 1562–1566 (1963)
4.424 R. S. Adhav: J. Phys. D **2**, 171–175 (1969)
4.425 E. N. Volkova, B. M. Berezhnoi, A. N. Izrailenko, A. V. Mishchenko, L. N. Rashkovich: Izv. Akad. Nauk SSSR, Ser. Fiz. **35**, 1858–1861 (1971)
4.426 E. N. Volkova, B. M. Berezhnoi, A. N. Izrailenko, A. V. Mishchenko, L. N. Rashkovich: Bull. Acad. Sci. USSR, Phys. Ser. [English Transl.] **35**, 1690–1693 (1971)
4.427 L. A. Shuvalov, I. S. Zheludev, A. V. Mnatsakanyan, Ts.-Zh. Ludupov, I. Fiala: Izv. Akad. Nauk SSSR, Ser. Fiz. **31**, 1919–1922 (1967)
4.428 L. A. Shuvalov, I. S. Zheludev, A. V. Mnatsakanyan, Ts.-Zh. Ludupov, I. Fiala: Bull. Acad. Sci. USSR, Phys. Ser. [English Transl.] **31**, 1963–1966 (1967)
4.429 R. S. Adhav: J. Phys. D **2**, 177–182 (1969)
4.430 K. S. Aleksandrov, A. T. Anistratov, A. R. Zamkov, I. S. Rez: Fiz. Tverd. Tela **19**, 1863–1866 (1977)
4.431 K. S. Aleksandrov, A. T. Anistratov, A. R. Zamkov, I. S. Rez: Sov. Phys. Solid State [English Transl.] **19**, 1090–1091 (1977)
4.432 W. L. Smith: Appl. Opt. **6**, 1798 (1977)
4.433 K. Veerabhadra Rao, T. S. Narasimhamurty: J. Phys. C **11**, 2343–2347 (1978)
4.434 F. Zernike: J. Opt. Soc. Am. **54**, 1215 (1964)
4.435 F. Zernike: Erratum, J. Opt. Soc. Am. **55**, 210 (1965)
4.436 K. Kato, A. J. Alcock, M. C. Richardson: Opt. Commun. **11**, 5 (1974)
4.437 K. Kato: Opt. Commun. **13**, 93 (1975)
4.438 J. E. Pearson, G. A. Evans, A. Yariv: Opt. Commun. **4**, 366–367 (1972)
4.439 K. Kato: IEEE J. Quantum Electron. **10**, 622 (1974)
4.440 I. S. Zheludev, Ts.-Zh. Ludupov: Kristallografiya **10**, 764–766 (1965)
4.441 I. S. Zheludev, Ts.-Zh. Ludupov: Sov. Phys. Crystallogr. [English Transl.] **10**, 645–646 (1966)
4.442 H. Horinaka, H. Nozuchi, H. Sonomura, T. Miyauchi: Jpn. J. Appl. Phys. **22**, 546 (1983)
4.443 V. M. Cound, P. H. Davies, K. F. Hulme, D. Robertson: J. Phys. C **3**, L83–L84 (1970)
4.444 M. H. Grimsditch, G. D. Holah: Phys. Rev. B **12**, 4377 (1975)
4.445 M. H. Grimsditch, G. D. Holah: J. Phys. (Paris) **36**, Suppl. C, C3–C185 (1975)
4.446 A. Yoshihara, E. R. Bernstein: J. Chem. Phys. **77**, 5319 (1982)
4.447 G. Ambrazevicius, G. Babonas: Liet. Fiz. Sbornik. **18**, 765–774 (1978)
4.448 G. Ambrazevicius, G. Babonas: Sov. Phys. Collection [English Transl.] **18**, 52–59 (1978)
4.449 Yu. K. Shaldin, D. A. Belogurov: Opt. Spektrosk. **35**, 693–701 (1973)
4.450 Yu. K. Shaldin, D. A. Belogurov: Opt. Spectrosc. [English Transl.] **35**, 403–407 (1973)

4.451 R.L. Byer, M.M. Choy, R.L. Herbst, D.S. Chemla, R.S. Feigelson: Appl. Phys. Lett. **24**, 65–68 (1974)
4.452 N.P. Barnes, D.J. Gettemy, J.R. Hietanen, R.A. Iannini: Appl. Opt. **28**, 5162–5168 (1989)
4.453 H. Horinaka, H. Sonomura, T. Migauchi: Jpn. J. Appl. Phys. **21**, 1485–1488 (1982)
4.454 D.S. Chemla, P.J. Kupecek, D.S. Robertson, R.C. Smith: Opt. Commun. **3**, 29 (1971)
4.455 M.G. Cohen, M. DiDomenico Jr., S.H. Wemple: Phys. Rev. B **1**, 4334–4337 (1970)
4.456 G.C. Bhar, D.K. Ghosh, P.S. Ghosh, D. Schmitt: Appl. Opt. **22**, 2492 (1983)
4.457 J.J. Zondy, D. Touahri: J. Opt. Soc. Am. B **14**, 1331 (1997)
4.458 J.M. Halbout, S. Blit, W. Donaldson, C.L. Tang: IEEE J. Quantum Electron. **15**, 1176–1180 (1979)
4.459 K. Betzler, H. Hesse, P. Loose: J. Mol. Struct. **47**, 393–396 (1978)
4.460 L. Bohaty: Z. Kristallogr. **163**, 307–309 (1983)
4.461 M.J. Rosker, K. Cheng, C.L. Tang: IEEE J. Quantum Electron. **21**, 1600 (1985)
4.462 G.D. Boyd, E. Buehler, F.G. Storz: Appl. Phys. Lett. **18**, 301–304 (1971)
4.463 S.R. Sashital, R.R. Stephens, J.F. Lotspeich: J. Appl. Phys. **59**, 757–760 (1986)
4.464 G.C. Bhar, L.K. Samanta, D.K. Ghosh, S. Das: Kvantovaya Elektron. **14**, 1361 (1987)
4.465 G.C. Bhar, L.K. Samanta, D.K. Ghosh, S. Das: Sov. J. Quantum Electron. [English Transl.] **17**, 860 (1987)
4.466 K. Tada, K. Kikuchi: Jpn. J. Appl. Phys. **19**, 1311–1315 (1980)
4.467 A. Schaefer, H. Schmitt, A. Dörr: Ferroelectrics **69**, 253–266 (1986)
4.468 D. Berlincourt, H. Jaffe: Phys. Rev. **111**, 143–148 (1958)
4.469 W.J. Merz: Phys. Rev. **76**, 1221–1225 (1949)
4.470 V.G. Gavrilyachenko, E.G. Fesenko: Kristallografiya **16**, 640–641 (1971)
4.471 V.G. Gavrilyachenko, E.G. Fesenko: Sov. Phys. Crystallogr. [English Transl.] **16**, 549–550 (1971)
4.472 A.V. Turik, N.B. Shevchenko, V.G. Gavrilyachenko, E.G. Fesenko: Phys. Status Solidi (b) **94**, 525–528 (1979)
4.473 Molecular Technology: Data sheet – World of crystals (Molecular Technology GmbH, Berlin 2002)
4.474 M. Adachi, A. Kawabata: Jpn. J. Appl. Phys. **17**, 1969–1973 (1978)
4.475 R.C. Miller, D.A. Kleinman, A. Savage: Phys. Rev. Lett. **11**, 146–149 (1963)
4.476 R.C. Miller, W.A. Nordland: Opt. Commun. **1**, 400–402 (1970)
4.477 M. Zgonik, P. Bernasconi, M. Duelli, R. Schlesser, P. Güter, M.H. Garrett, D. Rytz, Y. Zhu, X. Wu: Phys. Rev. B **50**, 5941–5949 (1994)
4.478 S.H. Wemple, M. DiDomenico Jr., I. Camlibel: J. Phys. Chem. Solids **29**, 1797–1803 (1968)
4.479 S. Singh, J.P. Remeika, J.R. Potopowicz: Appl. Phys. Lett. **20**, 135–137 (1972)
4.480 E.H. Turner: unpublished work (quoted in [4.225, 347])
4.481 R. Komatsu, T. Sugawara, K. Sassa, N. Sarukura, Z. Liu, S. Izumida, Y. Segawa, S. Uda, T. Fukuda, K. Yamanouchi: Appl. Phys. Lett. **70**, 3492–3494 (1997)
4.482 T. Sugawara, R. Komatsu, S. Uda: Solid State Commun. **107**, 233 (1998)
4.483 S.I. Furusawa, O. Chikagawa, S. Tange, T. Ishidate, H. Orihara, Y. Ishibashi, K. Miwa: J. Phys. Soc. Jpn. **60**, 2691–2693 (1991)
4.484 L. Bohaty, S. Haussühl, J. Liebertz: Cryst. Res. Technol. **24**, 1159–1163 (1989)
4.485 L.G. Van Uitert, S. Singh, H.J. Levinstein, J.E. Geusic, W.A. Bonner: Appl. Phys. Lett. **11**, 161–163 (1967)
4.486 L.G. Van Uitert, S. Singh, H.J. Levinstein, J.E. Geusic, W.A. Bonner: Erratum, Appl. Phys. Lett. **12**, 224 (1968)
4.487 J.J.E. Reid: Appl. Phys. Lett. **62**, 19–21 (1993)
4.488 M. Adachi, T. Shiosaki, A. Kawabata: Ferroelectrics **27**, 89–92 (1980)
4.489 J.L. Kirk, K. Vedam: J. Phys. Chem. Solids **33**, 1251–1255 (1972)
4.490 R. Kiriyama: Science (Tokyo) **17**, 239–240 (1947)
4.491 R. Kiriyama: Chem. Abstr. **45**, 2278c (1951)
4.492 M.S. Madhava, S. Haussühl: Z. Kristallogr. **141**, 25–30 (1975)
4.493 M.J. Dodge: Refractive index. In: *Optical Materials*, CRC Handbook of Laser Science and Technology, Vol. 4, Subvol. 2, ed. by M.J. Weber (CRC, Boca Raton 1987) pp. 21–47
4.494 K.W. Martin, L.G. DeShazer: Appl. Opt. **12**, 941–943 (1973)
4.495 P.S. Bechthold, S. Haussühl: Appl. Phys. **14**, 403–410 (1977)
4.496 S. Haussühl: Acta Crystallogr. A **24**, 697–698 (1968)
4.497 P. Güter: Ferroelectrics **75**, 5–23 (1987)
4.498 J. Sapriel, R. Hierle, J. Zyss, M. Bossier: Appl. Phys. Lett. **55**, 2594–2596 (1989)
4.499 Y. Wu, T. Sasaki, S. Nakai, A. Yokotani, H. Tang, C. Chen: Appl. Phys. Lett. **62**, 2614–2615 (1993)
4.500 K. Kato: IEEE J. Quantum Electron. **31**, 169–171 (1995)
4.501 G.R. Crane: J. Chem. Phys. **62**, 3571 (1975)
4.502 E.N. Volkova, V.A. Dianova, A.L. Zuev, A.N. Izrailenko, A.S. Lipatov, V.N. Parygin, L.N. Rashkovich, L.E. Chirkov: Kristallografiya **16**, 346–349 (1971)
4.503 E.N. Volkova, V.A. Dianova, A.L. Zuev, A.N. Izrailenko, A.S. Lipatov, V.N. Parygin, L.N. Rashkovich, L.E. Chirkov: Sov. Phys. Crystallogr. [English Transl.] **16**, 284–286 (1971)
4.504 S.K. Kurtz, T.T. Perry, J.G. Bergman Jr.: Appl. Phys. Lett. **12**, 186–188 (1968)
4.505 H. Naito, H. Inaba: Optoelectronics **4**, 335 (1972)
4.506 J. Zyss, D.S. Chemla, J.F. Nicoud: J. Chem. Phys. **74**, 4800–4811 (1981)
4.507 M. Sigelle, R. Hierle: J. Appl. Phys. **52**, 4199–4204 (1981)
4.508 T.S. Narasimhamurty: Phys. Rev. **186**, 945–948 (1969)

4.509 Y. Takagi, Y. Makita: J. Phys. Soc. Jpn. **13**, 272–277 (1958)
4.510 V.V. Gladkii, V.K. Magataev, V.A. Kirikov: Fiz. Tverd. Tela **19**, 1102–1106 (1977)
4.511 V.V. Gladkii, V.K. Magataev, V.A. Kirikov: Sov. Phys. Solid State [English Transl.] **19**, 641–644 (1977)
4.512 M.S. Khan, T.S. Narasimhamurty: Solid State Commun. **43**, 941–943 (1982)
4.513 A.T. Anistratov, S.X. Melhikova: Kristallografiya **17**, 149–152 (1972)
4.514 A.T. Anistratov, S.X. Melhikova: Sov. Phys. Crystallogr. [English Transl.] **17**, 119–121 (1972)
4.515 N.A. Romanyuk, A.M. Kostetskii: Fiz. Tverd. Tela **18**, 1489–1491 (1976)
4.516 N.A. Romanyuk, A.M. Kostetskii: Sov. Phys. Solid State [English Transl.] **18**, 867–869 (1976)
4.517 J. Valasek: Phys. Rev. **20**, 639–664 (1922)
4.518 G.B. Hadjichristov, P.P. Kircheva, N. Kirov: J. Mol. Struct. **382**, 33–37 (1996)
4.519 I.V. Berezhnoi, R.O. Vlokh: Fiz. Tverd. Tela **30**, 2223–2225 (1988)
4.520 I.V. Berezhnoi, R.O. Vlokh: Sov. Phys. Solid State [English Transl.] **30**, 1282–1284 (1988)
4.521 N.R. Ivanov, D. Khusravov, L.A. Shuvalov, N.M. Shchagina: Izv. Akad. Nauk SSSR, Ser. Fiz. **43**, 1691–1701 (1979)
4.522 N.R. Ivanov, D. Khusravov, L.A. Shuvalov, N.M. Shchagina: Bull. Acad. Sci. USSR, Phys. Ser. [English Transl.] **43**, 121–129 (1979)
4.523 V.G. Martynov, K.S. Aleksandrov, A.T. Anistratov: Fiz. Tverd. Tela **15**, 2922–2926 (1973)
4.524 V.G. Martynov, K.S. Aleksandrov, A.T. Anistratov: Sov. Phys. Solid State [English Transl.] **15**, 1950–1952 (1974)
4.525 Y. Luspin, G. Hauret: C. R. Hebd. Seances Acad. Sci. (Paris), Ser. B **274**, 995 (1972)
4.526 A.T. Anistratov, V.G. Martynov: Kristallografiya **15**, 308–312 (1970)
4.527 A.T. Anistratov, V.G. Martynov: Sov. Phys. Crystallogr. [English Transl.] **15**, 256–260 (1970)
4.528 M. Eibschütz, H.J. Guggenheim, S.H. Wemple, I. Camlibel, M. DiDomenico Jr.: Phys. Lett. A **29**, 409–410 (1969)
4.529 K. Recker, F. Wallrafen, S. Haussühl: J. Cryst. Growth **26**, 97–100 (1974)
4.530 R.O. Vlokh, I.P. Skab: Fiz. Tverd. Tela **34**, 3250–3255 (1992)
4.531 R.O. Vlokh, I.P. Skab: Sov. Phys. Solid State [English Transl.] **34**, 1739–1741 (1992)
4.532 J.D. Barry, C.J. Kennedy: IEEE J. Quantum Electron. **11**, 575–579 (1975)
4.533 T. Yamada, H. Iwasaki, N. Niizeki: J. Appl. Phys. **41**, 4141–4147 (1970)
4.534 S. Singh, D.A. Draegert, J.E. Geusic: Phys. Rev. B **2**, 2709–2724 (1970)
4.535 A.W. Warner, G.A. Coquin, J.L. Fink: J. Appl. Phys. **40**, 4353 (1969)
4.536 J. Fousek, C. Konak: Ferroelectrics **12**, 185–187 (1976)
4.537 K. Suzuki, K. Inoue, M. Shibuya: J. Phys. Soc. Jpn. **43**, 1457–1458 (1977)
4.538 J.E. Geusic, H.J. Levinstein, J.J. Rubin, S. Singh, L.G. Van Uitert: Appl. Phys. Lett. **11**, 269–271 (1967)
4.539 J.E. Geusic, H.J. Levinstein, J.J. Rubin, S. Singh, L.G. Van Uitert: Erratum, Appl. Phys. Lett. **12**, 224 (1968)
4.540 S.K. Gupta, H.B. Gon, K.V. Rao: Ferroelectr. Lett. **7**, 15–19 (1987)
4.541 S.K. Gupta, H.B. Gon, K.V. Rao: J. Mater. Sci. Lett. **6**, 4–6 (1987)
4.542 C. Medrano, P. Günter, H. Arend: Phys. Status Solidi **143**, 749–754 (1987)
4.543 A. Carenco, J. Jerphagnon, A. Perigaud: J. Chem. Phys. **66**, 3806–3813 (1977)
4.544 M.Y. Antipin, T.V. Timofeeva, R.D. Clark, V.N. Nesterov, F.M. Dolgushin, J. Wu, A.J. Leyderman: Mater. Chem. **11**, 351–358 (2001)
4.545 C. Bosshard, K. Sutter, P. Günter, G. Chapuis, R.J. Twieg, D. Dobrowolsky: Proc. SPIE **1017**, 207–211 (1988)
4.546 P. Günter, C. Bosshard, K. Sutter, H. Arend, G. Chapuis, R.J. Twieg, D. Dobrowolski: Appl. Phys. Lett. **50**, 486–488 (1987)
4.547 C. Bosshard, K. Sutter, P. Günter: Ferroelectrics **92**, 387–393 (1989)
4.548 K. Sutter, C. Bosshard, M. Ehrensperger, P. Günter, R.J. Twieg: IEEE J. Quantum Electron. **24**, 2362–2366 (1988)
4.549 C. Scheiding, G. Schmidt, H.D. Kürsten: Krist. Tech. **8**, 311–321 (1973)
4.550 L.E. Cross, A. Fouskova, S.E. Cummins: Phys. Rev. Lett. **21**, 812–814 (1968)
4.551 J. Sapriel, R. Vacher: J. Appl. Phys. **48**, 1191–1194 (1977)
4.552 M.P. Zaitseva, L.A. Shabanova, B.I. Kidyarov, Yu.I. Kokorin, S.I. Burkov: Kristallografiya **28**, 741–744 (1983)
4.553 M.P. Zaitseva, L.A. Shabanova, B.I. Kidyarov, Yu.I. Kokorin, S.I. Burkov: Sov. Phys. Crystallogr. [English Transl.] **28**, 439–440 (1983)
4.554 A.L. Aleksandrovskii, A.N. Izrailenko, L.N. Razhkovich: Kvantovaya Elektron. **1**, 1261–1264 (1974)
4.555 A.L. Aleksandrovskii, A.N. Izrailenko, L.N. Razhkovich: Sov. J. Quantum Electron. [English Transl.] **4**, 699–700 (1974)
4.556 A.L. Aleksandrovskii: Prib. Tekh. Eksp. **1**, 205–206 (1974)
4.557 A.L. Aleksandrovskii: Instrum. Exp. Tech., USSR [English Transl.] **17**, 234–235 (1974)
4.558 S. Nanamatsu, K. Doi, M. Takahashi: Jpn. J. Appl. Phys. **11**, 816–822 (1972)
4.559 R.C. Miller, W.A. Nordland, K. Nassau: Ferroelectrics **2**, 97–99 (1971)
4.560 Yu.V. Shaldin, D.A. Belogurov, T.M. Prokhortseva: Fiz. Tverd. Tela **15**, 1383–1387 (1973)
4.561 Yu.V. Shaldin, D.A. Belogurov, T.M. Prokhortseva: Sov. Phys. Solid State [English Transl.] **15**, 936–938 (1973)

4.562 S. Singh, W. A. Bonner, J. R. Potopowicz, L. G. Van Uitert: Appl. Phys. Lett. **17**, 292–294 (1970)
4.563 F. B. Dunning, F. K. Tittel, R. F. Stebbings: Opt. Commun. **7**, 181 (1973)
4.564 H. Naito, H. Inaba: Optoelectronics **5**, 256 (1973)
4.565 R. C. Miller, W. A. Nordland, E. D. Kolb, W. L. Bond: J. Appl. Phys. **41**, 3008–3011 (1970)
4.566 P. V. Lenzo, E. G. Spencer, J. P. Remeika: Appl. Opt. **4**, 1036–1037 (1965)
4.567 K. Araki, T. Tanaka: Jpn. J. Appl. Phys. **11**, 472–479 (1972)
4.568 L. M. Belyaev, G. S. Belikova, A. B. Gilvarg, I. M. Silvestrova: Kristallografiya **14**, 645–651 (1969)
4.569 L. M. Belyaev, G. S. Belikova, A. B. Gilvarg, I. M. Silvestrova: Sov. Phys. Crystallogr. [English Transl.] **14**, 544–549 (1970)
4.570 A. Miniewicz, A. Samoc, J. Sworakowski: J. Mol. Electron. **4**, 25–29 (1988)
4.571 M. S. Khan, T. S. Narasimhamurty: J. Mater. Sci. Lett. **1**, 268–270 (1982)
4.572 C. Chen, Y. Wu, A. Jiang, B. Wu, G. You, R. Li, S. Lin: J. Opt. Soc. Am. B **6**, 616–621 (1989)
4.573 S. P. Velsko, M. Webb, L. Davis, C. Huang: IEEE J. Quantum Electron. **27**, 2182–2192 (1991)
4.574 S. Lin, Z. Sun, B. Wu, C. Chen: J. Appl. Phys. **67**, 634 (1990)
4.575 K. Kato: IEEE J. Quantum Electron. **30**, 2950–2952 (1994)
4.576 B. L. Davydov, L. G. Koreneva, E. A. Lavrovsky: Radiotekh. Elektron. **19**, 1313 (1974)
4.577 B. L. Davydov, L. G. Koreneva, E. A. Lavrovsky: Radio Eng. Electron. Phys. [English Transl.] **19**(6), 130 (1974)
4.578 J. L. Stevenson: J. Phys. D **6**, L13–L16 (1973)
4.579 S. K. Kurtz, J. Jerphagnon, M. M. Choy: *Elastic, Piezoelectric, Pyroelectric, Piezooptic, Electrooptic Constant, and Nonlinear Dielectric Susceptibilities of Crystals*, Landolt–Börnstein, New Series III/11, ed. by K. H. Hellwege (Springer, Berlin, Heidelberg 1979) pp. 671–743
4.580 G. S. Belyaev, G. S. Belikova, A. G. Gilvarg, M. P. Golovei, I. N. Kalinkina, G. I. Kosourov: Opt. Spectr. (USSR) [English Transl.] **29**, 522 (1970)
4.581 Y. Uematsu, T. Fukuda: Jpn. J. Appl. Phys. **12**, 841–844 (1973)
4.582 P. Güter: Jpn. J. Appl. Phys. **16**, 1727–1728 (1977)
4.583 E. Wiesendanger: Ferroelectrics **6**, 263–281 (1974)
4.584 M. Zgonik, R. Schlesser, I. Biaggio, I. Voit, J. Tscherry, P. Güter: J. Appl. Phys. **74**, 1287–1297 (1993)
4.585 W. R. Cook Jr., H. Jaffe: Acta Crystallogr. **10**, 705–707 (1957)
4.586 T. Krajewski, Z. Tylczynski, T. Breczewski: Acta Phys. Polonica A **47**, 455–466 (1975)
4.587 J. D. Bierlein, H. Vanherzeele, A. A. Ballman: Appl. Phys. Lett. **54**, 783–785 (1989)
4.588 J. D. Bierlein, H. Vanherzeele, A. A. Ballman: Erratum, Appl. Phys. Lett. **61**, 3193 (1992)
4.589 V. V. Aleksandrov, T. S. Velichkina, V. I. Voronkova, L. V. Koltsova, I. A. Yakovlev, V. K. Yanovskii: Solid State Commun. **69**, 877 (1989)
4.590 J. D. Bierlein, H. Vanherzeele: J. Opt. Soc. Am. B **6**, 622–633 (1989)
4.591 J. D. Bierlein, C. B. Arweiler: Appl. Phys. Lett. **49**, 917–919 (1986)
4.592 D. H. Jundt, P. Güter, B. Zysset: Nonlin. Opt. **4**, 341–345 (1993)
4.593 Y. Uematsu: Jpn. J. Appl. Phys. **13**, 1362–1368 (1974)
4.594 B. Zysset, I. Biaggio, P. Güter: J. Opt. Soc. Am. B **9**, 380–386 (1992)
4.595 J. A. Paisner, M. L. Spaeth, D. C. Gerstenberger, I. W. Ruderman: Appl. Phys. Lett. **32**, 476–478 (1978)
4.596 W. R. Cook, L. H. Hubby Jr.: J. Opt. Soc. Am. **66**, 72 (1976)
4.597 F. B. Dunning, R. G. Stickel Jr.: Appl. Opt. **15**, 3131 (1976)
4.598 H. J. Dewey: IEEE J. Quantum Electron. **12**, 303 (1976)
4.599 Y. Wu, C. Chen: Wuli Xuebao (Acta Phys. Sin.) **35**, 1–6 (1986)
4.600 A. H. Kung: Appl. Phys. Lett. **65**, 1082–1084 (1994)
4.601 D. L. Fenimore, K. L. Schepler, U. B. Ramabadran, S. R. McPherson: J. Opt. Soc. Am. B **12**, 794–796 (1995)
4.602 L. T. Cheng, L. K. Cheng, J. D. Bierlein, F. C. Zumsteg: Appl. Phys. Lett. **63**, 2618–2620 (1993)
4.603 L. K. Cheng, L. T. Cheng, J. Galperin, P. A. M. Hotsenpiller, J. D. Bierlein: J. Cryst. Growth **137**, 107–115 (1994)
4.604 K. Kato: IEEE J. Quantum Electron. **30**, 881–883 (1994)
4.605 A. C. Aleksandrovsky, S. A. Akhmanov, V. A. Dyakov, N. I. Zheludev, V. I. Pryalkyn: Kvantovaya Elektron. **12**, 1333 (1985)
4.606 A. C. Aleksandrovsky, S. A. Akhmanov, V. A. Dyakov, N. I. Zheludev, V. I. Pryalkyn: Sov. J. Quantum Electron. [English Transl.] **15**, 885 (1985)
4.607 W. Wiechmann, S. Kubota, T. Fukui, H. Masuda: Opt. Lett. **18**, 1208–1210 (1993)
4.608 K. Kato: IEEE J. Quantum Electron. **QE-27**, 1137–1140 (1991)
4.609 V. K. Yanovskii, V. I. Voronkova, A. P. Leonov, S. Yu. Stefanovich: Fiz. Tverd. Tela **27**, 2516–2517 (1985)
4.610 V. K. Yanovskii, V. I. Voronkova, A. P. Leonov, S. Yu. Stefanovich: Sov. Phys. Solid State [English Transl.] **27**, 1508–1509 (1985)
4.611 V. K. Yanovskii, V. I. Voronkova: Phys. Status Solidi (a) **93**, 665–668 (1980)
4.612 E. Nakamura: J. Phys. Soc. Jpn. **17**, 961–966 (1962)
4.613 H. Shimizu, M. Tsukamoto, Y. Ishibashi, Y. Takagi: J. Phys. Soc. Jpn. **38**, 195–201 (1975)
4.614 K. Hamano, K. Negishi, M. Marutake, S. Nomura: Jpn. J. Appl. Phys. **2**, 83–90 (1963)
4.615 G. Hauret, J. P. Chapelle, L. Taurel: Phys. Status Solidi (a) **11**, 255–261 (1972)
4.616 F. C. Zumsteg, J. D. D. Bierlein, T. E. Gier: J. Appl. Phys. **47**, 4980 (1976)

4.617 Yu.S. Oseledchik, A.I. Pisarevsky, A.L. Prosvirin, V.N. Lopatko, L.E. Kholodenkov, E.F. Titkov, A.A. Demidovich, A.P. Shkadarevich: Proc. of Laser Optics Conf. [in Russian], (Leningrad Univ. Press, Leningrad 1990)
4.618 J.Y. Wang, Y.G. Liu, J.Q. Wei, L.P. Shi, M. Wang: Z. Kristallogr. **191**, 231–238 (1990)
4.619 J.Y. Wang, Y.G. Liu, J.Q. Wei, L.P. Shi, M. Wang: Guisuanyan Xuebao **18**, 165–170 (1990)
4.620 J.Y. Wang, Y.G. Liu, J.Q. Wei, L.P. Shi, M. Wang: Chem. Abstr. **114** (1991) No. 154040x
4.621 K. Iio: J. Phys. Soc. Jpn. **34**, 138 (1973)
4.622 D.S. Chemla, E. Batifol, R.L. Byer, R.L. Herbst: Opt. Commun. **11**, 57 (1974)
4.623 A.R. Johnston, T. Nakamura: J. Appl. Phys. **40**, 3656–3658 (1969)
4.624 K. Inoue, T. Ishidate: Ferroelectrics **7**, 105–106 (1974)
4.625 B. Teng, J. Wang, Z. Wang, H. Jiang, X. Hu, R. Song, H. Liu, Y. Liu, J. Wei, Z. Shao: J. Cryst. Growth **224**, 280–283 (2001)
4.626 H. Hellwig, J. Liebertz, L. Bohaty: J. Appl. Phys. **88**, 240 (2000)
4.627 Z. Lin, Z. Wang, C. Chen, M.-H. Lee: J. Appl. Phys. **90**, 5585 (2001)
4.628 D. Eimerl, S. Velsko, L. Davis, F. Wang, G. Loiacono, G. Kennedy: IEEE J. Quantum Electron. **25**, 179–193 (1989)
4.629 C.E. Barker, D. Eimerl, S.P. Velsko: J. Opt. Soc. Am. B **8**, 2481–2492 (1991)
4.630 P. Kerkoc, M. Zgonik, K. Sutter, C. Bosshard, P. Günter: J. Opt. Soc. Am. B **7**, 313–319 (1990)
4.631 I. Biaggio, P. Kerkoc, L.-S. Wu, P. Günter, B. Zysset: J. Opt. Soc. Am. B **9**, 507–517 (1992)
4.632 R. Bechmann: *Elastic, Piezoelectric, Piezooptic and Electrooptic Constants of Crystals*, Landolt-Börnstein, New Series III/1 (Springer, Berlin, Heidelberg 1966) pp. 40–123
4.633 C.S. Brown, R.C. Kell, R. Taylor, L.A. Thomas: Proc. Inst. Elec. Eng. (London) B **109**, 99–114 (1962)
4.634 C.S. Brown, R.C. Kell, R. Taylor, L.A. Thomas: IRE Trans. Compon. Parts **9**, 193–211 (1962)
4.635 B.M. Suleiman, A. Lundén, E. Karawacki: Solid State Ionics **136–137**, 325–330 (2000)
4.636 M.V. Hobden: J. Appl. Phys. **38**, 4365 (1967)
4.637 A. Sonin, A.A. Filimonov, S.V. Suvorov: Sov. Phys. Solid State **10**, 1481 (1968)
4.638 L. Bohaty, S. Haussühl, G. Nohl: Z. Kristallogr. **139**, 33–38 (1974)
4.639 W.F. Holman: Ann. Phys. **29**, 160–178 (1909)
4.640 G. Brosowki, G. Luther, H.E. Muser: Phys. Status Solidi (a) **14**, K15–K17 (1972)
4.641 B.A. Strukov, K.A. Minaeva: Prib. Tekh. Eksp. **29**, 157–160 (1986)
4.642 B.A. Strukov, K.A. Minaeva: Instrum. Exp. Tech. (USSR) [English Transl.] **29**, Part II, 1413–1416 (1986)
4.643 P.P. Craig: Phys. Lett. **20**, 140–142 (1966)
4.644 F. Jona, G. Shirane: *Ferroelectric Crystals* (MacMillan, New York 1962)
4.645 J. Berdowski, A. Opilski, J. Szuber: J. Tech. Phys. (Poland) **18**, 211–217 (1977)
4.646 V.P. Konstantinova, I.M. Silvestrova, K.S. Aleksandrov: Kristallografiya **4**, 69–73 (1959)
4.647 V.P. Konstantinova, I.M. Silvestrova, K.S. Aleksandrov: Sov. Phys. Crystallogr. [English Transl.] **4**, 63–67 (1960)
4.648 S. Haussühl, J. Albers: Ferroelectrics **15**, 73 (1977)
4.649 J.M. Halbout, C.L. Tang: IEEE J. Quantum Electron. **18**, 410–415 (1982)
4.650 M.J. Rosker, C.L. Tang: IEEE J. Quantum Electron. **20**, 334 (1984)
4.651 N.R. Ivanov, S.Ya. Benderskii, L.A. Shuvalov: Kristallografiya **22**, 115–125 (1977)
4.652 N.R. Ivanov, S.Ya. Benderskii, L.A. Shuvalov: Sov. Phys. Crystallogr. [English Transl.] **22**, 64–69 (1977)
4.653 A.S. Sonin, V.S. Suvorov: Sov. Phys. Solid State **9**, 1437 (1967)
4.654 N.R. Ivanov, L.A. Shuvalov: Kristallografiya **11**, 760–765 (1966)
4.655 N.R. Ivanov, L.A. Shuvalov: Sov. Phys. Crystallogr. [English Transl.] **11**, 648–651 (1967)
4.656 A. Feldman, R.M. Waxler: Phys. Rev. Lett. **45**, 126–129 (1980)

4.5. Ferroelectrics and Antiferroelectrics

Ferroelectrics

Ferroelectric crystals (especially oxides in the form of ceramics) are important basic materials for technological applications in capacitors and in piezoelectric, pyroelectric, and optical devices. In many cases their nonlinear characteristics turn out to be very useful, for example in optical second-harmonic generators and other nonlinear optical devices. In recent decades, ceramic thin-film ferroelectrics have been utilized intensively as parts of memory devices. Liquid crystal and polymer ferroelectrics are utilized in the broad field of fast displays in electronic equipment.

This chapter surveys the nature of ferroelectrics, making reference to the data presented in the Landolt–Börnstein data collection *Numerical Data and Functional Relationships in Science and Technology*, Vol. III/36, *Ferroelectrics and Related Substances* (LB III/36). The data in the figures in this chapter have been taken mainly from the Landolt–Börnstein collection. The Landolt–Börnstein volume mentioned above consists of three subvolumes: Subvolume A [5.1,2], covering oxides; Subvolume B [5.3], covering inorganic crystals other than oxides; and Subvolume C [5.4], covering organic crystals, liquid crystals, and polymers.

Matter consists of electrons and nuclei. Most of the electrons generally are tightly bound to the nuclei, but some of the electrons are only weakly bound or are freely mobile in a lattice of ions. The physical properties of matter can be considered as being split into two categories. The properties in the first category are determined directly by the electrons and by the interaction of the electrons with lattice vibrations. Examples are the metallic, magnetic, superconductive, and semiconductive properties. The properties in the second category are only indirectly related to the electrons and can be discussed as being due to interaction between atoms, ions, or molecules. In this category we have, for example, the dielectric, elastic, piezoelectric, and pyroelectric properties; we have the dispersion relations of the lattice vibrations; and we have most of the properties of liquid crystals and polymers. The important properties of ferroelectrics are linked to all the latter properties, and they exhibit diverse types of phase transitions together with anomalies in these properties. These specific modifications convey information about cooperative interactions among ions, atoms, or molecules in the condensed phase of matter.

4.5.1 Definition of Ferroelectrics and Antiferroelectrics

A ferroelectric crystal is defined as a crystal which belongs to the pyroelectric family (i. e. shows a spontaneous electric polarization) and whose direction of spontaneous polarization can be reversed by an electric field. An antiferroelectric crystal is defined as a crystal whose structure can be considered as being composed of two sublattices polarized spontaneously in antiparallel directions and in which a ferroelectric phase can

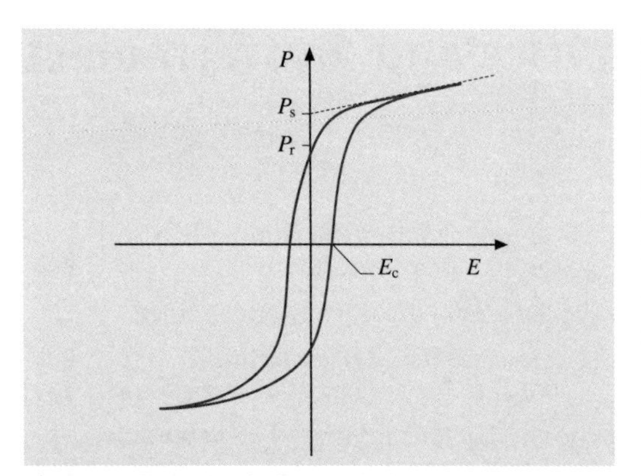

Fig. 4.5-1 Ferroelectric hysteresis loop. P_s, spontaneous polarization; P_r, remanent polarization; E_c, coercive field

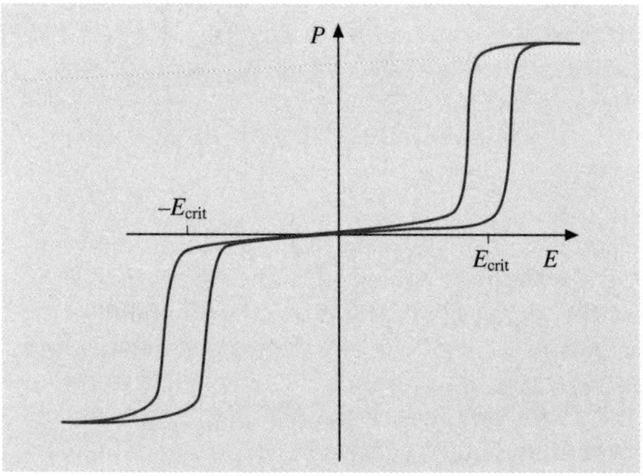

Fig. 4.5-2 Antiferroelectric hysteresis loop. E_{crit}, critical field

be induced by applying an electric field. Experimentally, the reversal of the spontaneous polarization in ferroelectrics is observed as a single hysteresis loop (Fig. 4.5-1), and the induced phase transition in antiferroelectrics as a double hysteresis loop (Fig. 4.5-2), when a low-frequency ac field of a suitable strength is applied.

The spontaneous polarization in ferroelectrics and the sublattice polarizations in antiferroelectrics are analogous to their magnetic counterparts. As described above, however, these polarizations are a necessary but not sufficient condition for ferroelectricity or antiferroelectricity. In other words, ferroelectricity and antiferroelectricity are concepts based not only upon the crystal structure, but also upon the dielectric behavior of the crystal. It is a common dielectric characteristic of ferroelectrics and antiferroelectrics that, in a certain temperature range, the dielectric polarization is observed to be a two-valued function of the electric field.

The definition of ferroelectric liquid crystals needs some comments; see remark *l* in Sect. 4.5.3.1.

4.5.2 Survey of Research on Ferroelectrics

The ferroelectric effect was discovered in 1920 by *Valasek*, who obtained hysteresis curves for Rochelle salt analogous to the *B–H* curves of ferromagnetism [5.5], and studied the electric hysteresis and piezoelectric response of the crystal in some detail [5.6]. For about 15 years thereafter, ferroelectricity was considered as a very specific property of Rochelle salt, until Busch and Scherrer discovered ferroelectricity in KH_2PO_4 and its sister crystals in 1935. During World War II, the anomalous dielectric properties of $BaTiO_3$ were discovered in ceramic specimens independently by Wainer and Solomon in the USA in 1942, by Ogawa in Japan in 1944, and by Wul and Goldman in Russia in 1946. Since then, many ferroelectrics have been discovered and research activity has rapidly increased. In recent decades, active studies have been made on ferroelectric liquid crystals and high polymers, after ferroelectricity had been considered as a characteristic property of solids for more than 50 years.

Figures 4.5-3, 4.5-4, and 4.5-5 demonstrate how ferroelectric research has developed. Figure 4.5-3 indicates the number of ferroelectrics discovered each year for oxide (Fig. 4.5-3a) and nonoxide ferroelectrics (Fig. 4.5-3b). Figure 4.5-4 gives the total number of ferroelectrics known at the end of each year. At present more than 300 ferroelectric substances are known. Figure 4.5-5 indicates the number of research papers on ferroelectrics and related substances published each year.

Advanced experimental methods (e.g. inelastic neutron scattering and hyper-Raman scattering) have been applied effectively to studies of ferroelectrics, and several new concepts (e.g. soft modes of lattice vibrations and the dipole glass) have been introduced to understand the nature of ferroelectrics. Ferroelectric crystals have been widely used in capacitors and piezoelectric devices. Steady developments in crystal growth and in the preparation of ceramics and ceramic thin

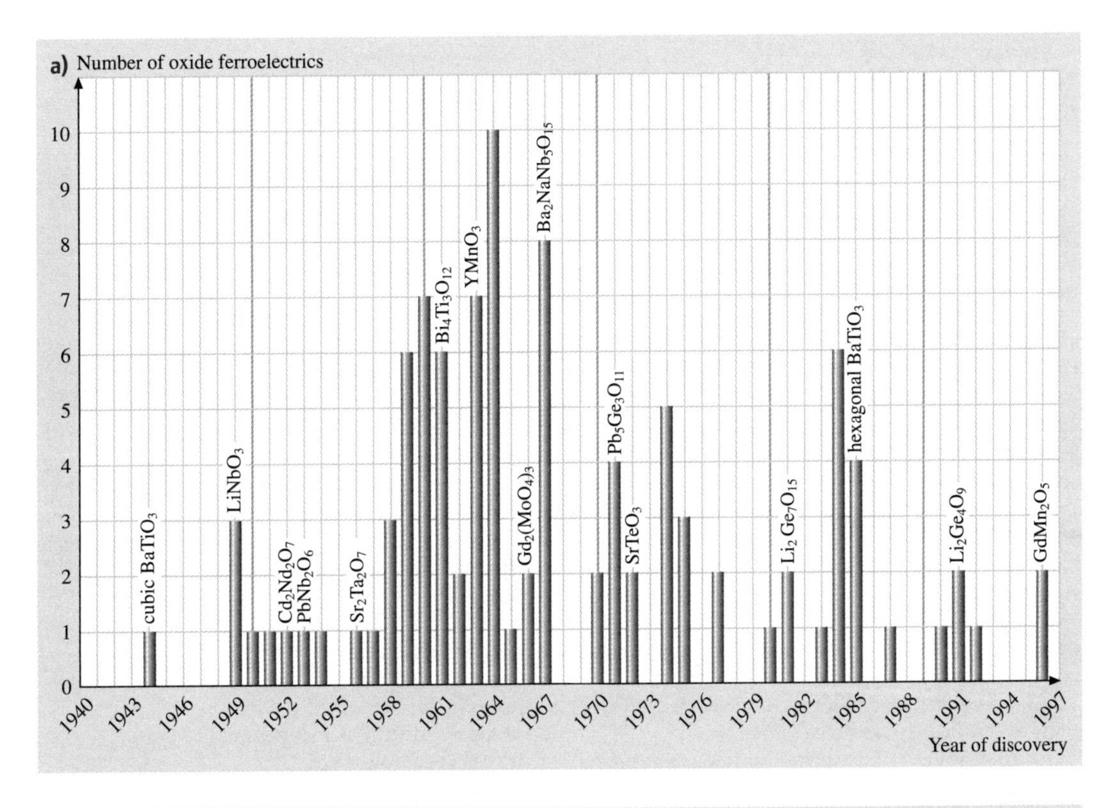

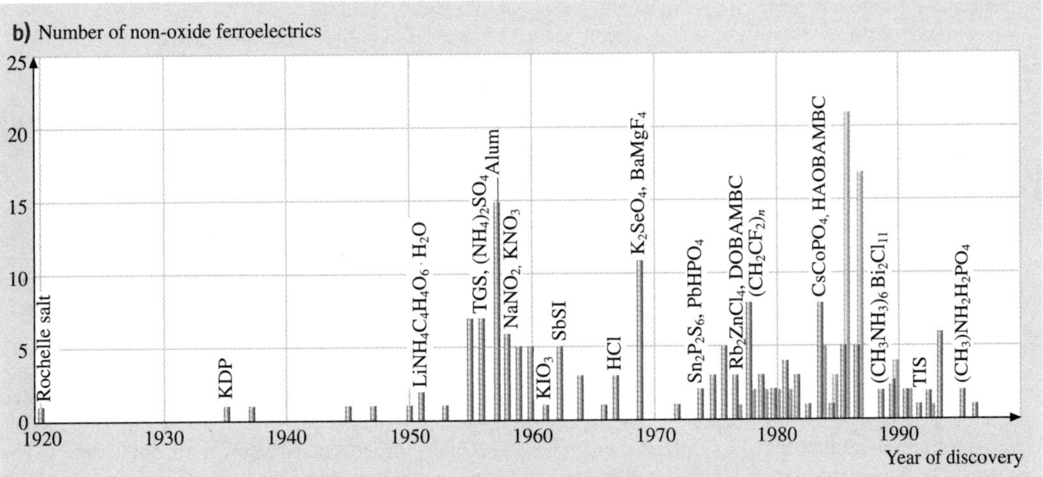

Fig. 4.5-3a,b Number of ferroelectric substances discovered in each year. Representative ferroelectrics are indicated at their year of discovery. **(a)** Oxide ferroelectrics. Only pure compounds are taken into account. **(b)** Nonoxide ferroelectrics. *Gray bars* stand for nonoxide crystals, counting each pure compound as one unit. *Brown bars* stand for liquid crystals and polymers, counting each group of homologues (cf. Chaps. 71 and 72 in [5.4]) as one unit. The figure was prepared by Prof. K. Deguchi using data from LB III/36

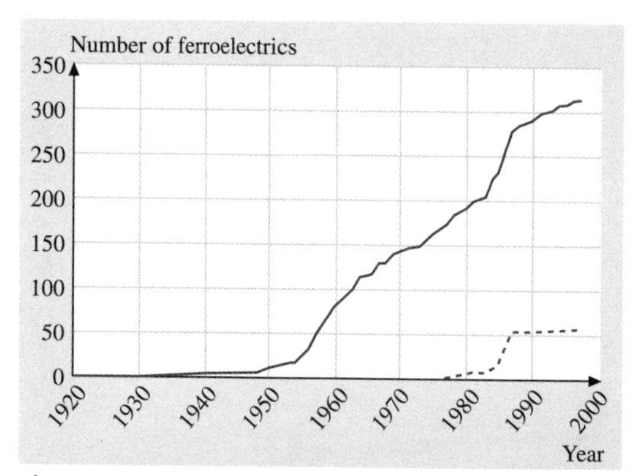

Fig. 4.5-4 Number of ferroelectric substances known at the end of each year. The *solid line* represents all ferroelectrics, including liquid crystals and polymers. For liquid crystals and polymers, each group of homologues is counted as one substance. The *dashed line* represents ferroelectric liquid crystals and polymers alone. Figure prepared by Prof. K. Deguchi

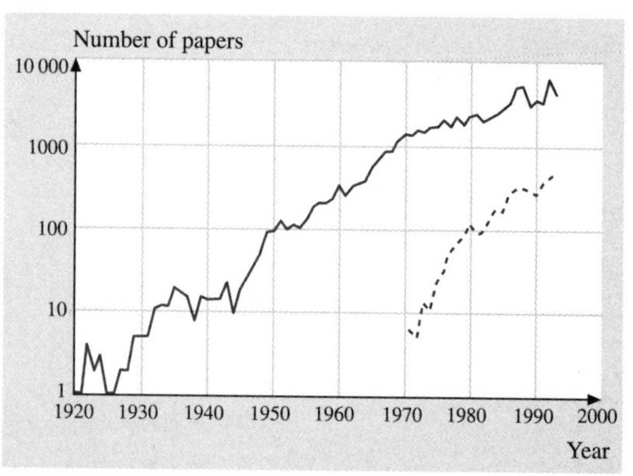

Fig. 4.5-5 Number of research papers on ferroelectrics and related substances published in each year. The *solid line* indicates the number of papers concerning all ferroelectrics (crystals + liquid crystals + polymers). The *dashed line* indicates the number of papers concerning liquid crystals and polymers alone. Prepared by Prof. K. Deguchi

films have opened the way to various other applications (e.g. second-harmonic generation and memory devices). Liquid crystals and high-polymer ferroelectrics are very useful as fast display elements.

Corresponding to this intense development, many textbooks, monographs, and review articles have been published on ferroelectric research during recent years. Some of them, arranged according to the various research fields, are listed here:

- Introduction to ferroelectrics: [5.7–10]
- Applications in general: [5.10, 11]
- Piezoelectricity: [5.12–14]
- Structural phase transitions: [5.15–17]
- Incommensurate phases: [5.18]
- Soft-mode spectroscopy: [5.19]
- Inelastic neutron scattering studies of ferroelectrics: [5.20]
- Raman and Brillouin scattering: [5.21]
- Ceramic capacitors: [5.22]
- Dipole glasses: [5.23]
- Relaxors: [5.24]
- Second-harmonic generation (SHG): [5.25–28]
- Ferroelectric ceramics: [5.29]
- Thin films: [5.30–35]
- Acoustic surface waves (ASWs): [5.36–39]
- Ferroelectric transducers and sensors: [5.40]
- Memory applications: [5.35]
- Ferroelectric liquid crystals: [5.41–47]
- Ferroelectric polymers: [5.48–51]

4.5.3 Classification of Ferroelectrics

Ferroelectricity is caused by a cooperative interaction of molecules or ions in condensed matter. The transition to ferroelectricity is characterized by a phase transition. Depending on the mechanism of how the molecules or ions interact in the material, we can classify the ferroelectric phase transitions and also the ferroelectric materials themselves into three categories: (I) order–disorder type, (II) displacive type, and (III) indirect type. In the order–disorder type (I), the spontaneous polarization is caused by orientational order of dipolar molecules, which is best visualized by the Ising model. The dielectric constants of order–disorder type ferroelectrics increase markedly in the vicinity of the Curie point. In the displacive type (II), the spontaneous polarization results from softening of the transverse optical modes of the lattice vibrations at the origin of the Brillouin zone; again, a marked increase of the dielectric constants is observed near the Curie temperature. The indirect type (III) is further classified into III_{op} and III_{ac}. In type III_{op}, the phase transition is originally caused by softening of the optical modes at the Brillouin zone boundary; a coupling of the soft modes with the electric polarization (through a complex mechanism) causes the spontaneous polarization (e.g. substance 18, GMO, in Sect. 4.5.4.1); only a very slight anomaly of the dielectric constants is observed above the Curie point (see Fig. 4.5-6). In type III_{ac}, softening of the acoustic modes (decrease of the tangent of the dispersion relation) takes place at the origin of the Brillouin zone, and a piezoelectric coupling between the soft modes and the polarization results in the spontaneous polarization (e.g. substance 40, LAT, in Sect. 4.5.4.3). No dielectric-constant anomaly appears above the Curie temperature when the crystal is clamped so that elastic deformation is prohibited (see Fig. 4.5-7). Most of the nonoxide ferroelectrics are of the order–disorder type and most of the oxide ferroelectrics are of the displacive type, while a few ferroelectrics are of the indirect type.

It might be expected that the dielectric dispersion would be dissipative and occur at relatively low frequency (e.g. in the microwave region) in the order–disorder ferroelectrics, while the dispersion would be of the resonance type and occur in the millimeter or infrared region in the displacive ferroelectrics. Actually, however, the situation is more complex owing to phonon–phonon coupling and cluster formation near the Curie temperature. Phonon–phonon coupling is inevitable in displacive ferroelectrics because their ferroelectricity is closely related to the anharmonic term of the ion potential, as first pointed out by *Slater* (see [5.7]). Accordingly, the resonance-type oscillation tends to be overdamped. High dielectric constants

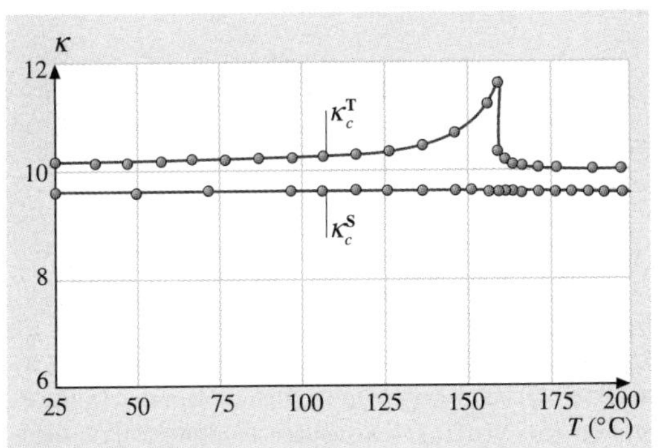

Fig. 4.5-6 $Gd_2(MoO_4)_3$. κ_c^T, and κ_c^S versus T. κ_c^T is the free dielectric constant κ_c measured at 1 kHz, and κ_c^S is the clamped dielectric constant κ_c measured at 19 MHz

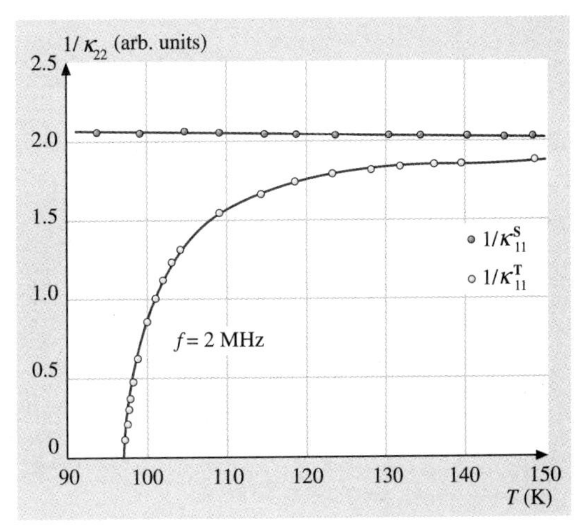

Fig. 4.5-7 $LiNH_4C_4H_4O_6 \cdot H_2O$. $1/\kappa_{22}^S$, and $1/\kappa_{22}^T$ versus T. $\kappa_{22} = \kappa_b$. $f = 2$ MHz. κ_{22}^S is κ_{22} of the clamped crystal. κ_{22}^T is κ_{22} of the free crystal

tend to induce the formation of clusters in which unit cell polarizations are aligned, similarly to the domains in the ferroelectric phase, but the boundaries of these clusters fluctuate thermally. Inelastic neutron, hyper-Raman, and hyper-Rayleigh scattering, etc. indicate the existence of such clusters in several displacive ferroelectrics. It is known that a ferromagnetic domain wall can move to follow an ac magnetic field and contribute to the magnetic susceptibility, while a ferroelectric domain wall usually cannot follow an ac electric field and in practice does not contribute to the dielectric constant. The cluster boundaries, however, fluctuate thermally and hence will be able to follow an ac electric field and contribute to the dielectric constant. This contribution will be dissipative in its character. Accordingly, it is possible that dielectric dispersion occurs as a combination of an overdamped-resonance dispersion and a dissipative dispersion in most displacive ferroelectrics.

In second-order or nearly second-order phase transitions, the dielectric dispersion is observed to show a critical slowing-down: a phenomenon in which the response of the polarization to a change of the electric field becomes slower as the temperature approaches the Curie point. Critical slowing-down has been observed in the GHz region in several order–disorder ferroelectrics (e.g. Figs. 4.5-8 and 4.5-9) and displacive ferroelectrics (e.g. Fig. 4.5-10). The dielectric constants at the Curie point in the GHz region are very small in order–disorder

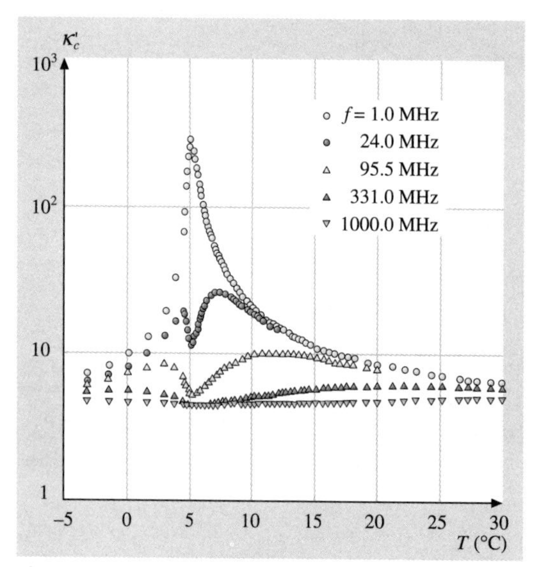

Fig. 4.5-8 $Ca_2Sr(CH_3CH_2COO)_6$. κ_c' versus T. Parameter: f

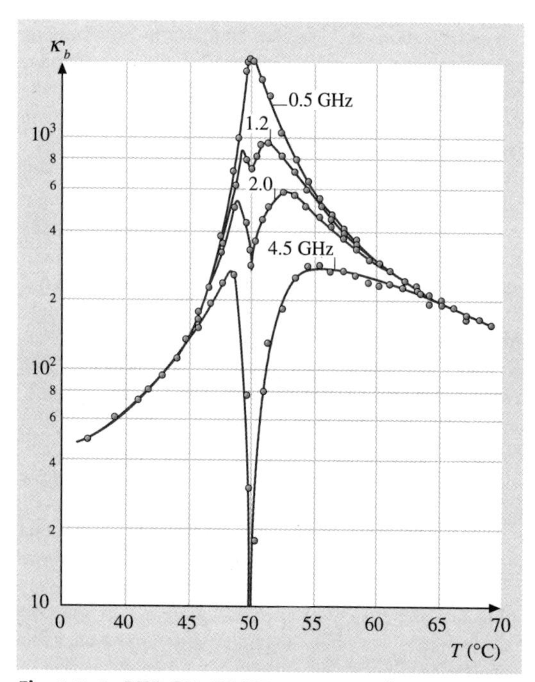

Fig. 4.5-9 $(NH_2CH_2COOH)_3 \cdot H_2SO_4$. κ_b' versus T relation, showing critical slowing-down. Parameter: f

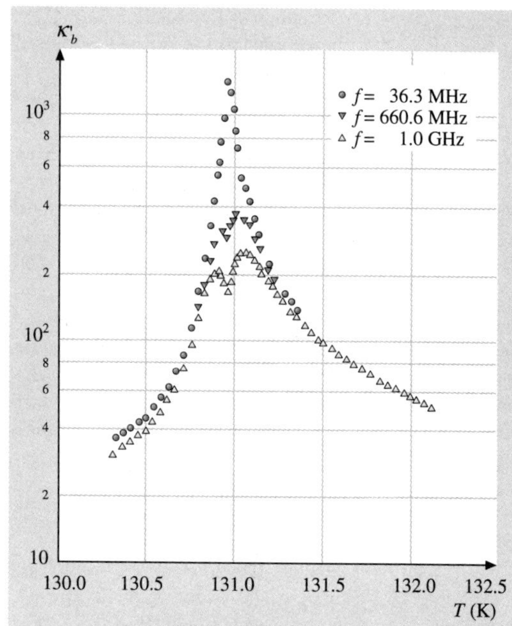

Fig. 4.5-10 $(CH_3NHCH_2COOH)_3 \cdot CaCl_2$. κ_b' versus T. Parameter: f. Critical slowing-down takes place

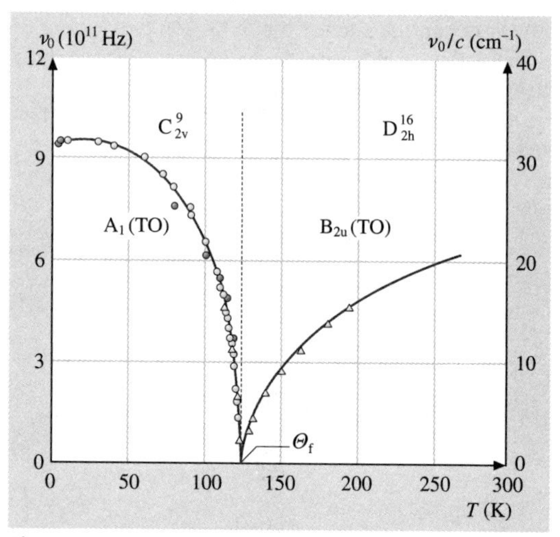

Fig. 4.5-11 $(CH_3NHCH_2COOH)_3 \cdot CaCl_2$. ν_0 versus T. ν_0 is the phonon mode frequency. *Triangles*: measured by millimeter spectroscopy. *Brown circles*: measured from far-infrared spectra. *Gray circles*: measured from electric-field-induced Raman spectra. In the paraelectric phase ($T > \Theta_f$), ν_0 decreases as the temperature decreases toward Θ_f, that is, the phonon mode softens

ferroelectrics, as seen in Figs. 4.5-8 and 4.5-9. The dielectric constant at the Curie point at 1 GHz is not so small in displacive ferroelectrics, as seen in Fig. 4.5-10, where the critical slowing-down seems to be related to cluster boundary motion, and the dielectric constant at the Curie point at 1 GHz contains a contribution from the soft phonon as observed in millimeter spectroscopy (Fig. 4.5-11).

4.5.3.1 The 72 Families of Ferroelectrics

In the Landolt–Börnstein data collection, ferroelectric and antiferroelectric substances are classified into 72 families according to their chemical composition and their crystallographic structure. Some substances which are in fact neither ferroelectric nor antiferroelectric but which are important in relation to ferroelectricity or antiferroelectricity, for instance as an end material of a solid solution, are also included in these families as related substances. This subsection surveys these 72 families of ferroelectrics presented in Landolt–Börnstein Vol. III/36 (LB III/36). Nineteen of these families concern oxides [5.1, 2], 30 of them concern inorganic crystals other than oxides [5.3], and 23 of them concern organic crystals, liquid crystals, and polymers [5.4]. Table 4.5-1 lists these families and gives some information about each family. Substances classified in LB III/36 as miscellaneous crystals (outside the families) are not included.

In the following, remarks are made on 13 of the families, labeled by the letters a – m in Table 4.5-1. The corresponding family numbers are repeated in the headings.

a. Perovskite-Type Family (Family Number 1). The name of this group is derived from the mineral perovskite ($CaTiO_3$). The perovskite-type oxides are cubic (e.g. $CaTiO_3$ above 1260 °C and $BaTiO_3$ above 123 °C) or pseudocubic with various small lattice distortions (e.g. $CaTiO_3$ below 1260 °C and $BaTiO_3$ below 123 °C). Ceramics made from solid solutions of perovskite-type oxides are the most useful ferroelectrics in high-capacitance capacitors, piezoelectric elements, and infrared sensors. Ceramic thin films are useful in memory devices.

The pure compounds are divided into simple perovskite-type oxides and complex perovskite-type oxides. Simple perovskite-type oxides have the chemical formula $A^{1+}B^{5+}O_3$ or $A^{2+}B^{4+}O_3$. Complex perovskite-type oxides have chemical formulas expressed by $(A^{1+}_{1/2}A'^{3+}_{1/2})BO_3$, $A^{2+}(B^{2+}_{1/2}B'^{6+}_{1/2})O_3$, $A^{2+}(B^{3+}_{1/2}B'^{5+}_{1/2})O_3$, $A^{2+}(B^{2+}_{1/3}B'^{5+}_{2/3})O_3$, $A^{2+}(B^{3+}_{2/3}B'^{6+}_{1/3})O_3$, $A(B, B', B'')O_3$, or $(A, A')(B, B')O_3$. Among the complex perovskite-type oxides, most of the $Pb(B, B')O_3$-type oxides show a diffuse phase transition such that the transition point is smeared out over a relatively wide temperature range and exhibits a characteristic dielectric relaxation; these materials therefore are called "relaxors".

b. $LiNbO_3$ Family (Family Number 2). This family contains $LiNbO_3$ and $LiTaO_3$. Their chemical formulas are similar to those of the simple perovskite oxides, but their structures are trigonal, unlike the perovskite oxides.

c. Stibiotantalite Family (Family Number 5). The members of this family are isomorphous with the mineral stibiotantalite $Sb(Ta, Nb)O_4$. They have the common chemical formula ABO_4, where A stands for Sc, Sb, or Bi and B for Ta, Nb, or Sb.

d. Tungsten Bronze-Type Family (Family Number 6). The tungsten bronzes are a group of compounds having the chemical formula M_xWO_3, where M stands for an alkali metal, an alkaline earth metal, Ag, Tl, etc. (e.g. Na_xWO_3, where $x = 0.1–0.95$). Most of them exhibit a bronze-like luster. The tungsten bronze-type oxides consist of crystals isomorphous with tungsten bronze, including a simple type (e.g. $Pb_{1/2}NbO_3$) and a complex type (e.g. $Ba_2NaNb_5O_{15}$). Single crystals (not ceramics) are used for technological applications.

e. Pyrochlore-Type Family (Family Number 7). The members of this family are isomorphous with the mineral pyrochlore, $CaNaNb_2O_6F$. Most of the members have the general chemical formula $A_2B_2O_7$ or $A_2B_2O_6$ (anion-deficient compounds), where A stands for Cd, Pb, Bi, etc. and B for Nb, Ta, etc.

f. $Sr_2Nb_2O_7$ Family (Family Number 8). This family contains high-temperature ferroelectrics such as $Nb_2Ti_2O_7$ and $La_2Ti_2O_7$. Their Curie points are higher than 1500 °C.

g. Layer-Structure Family (Family Number 9). The common chemical formula of these oxides is $(Bi_2O_2)(A_{n-1}B_nO_{3n-1})$, where A stands for Ca, Sr, Ba, Pb, Bi, etc., B stands for Ti, Nb, Ta, Mo, W, etc., and n varies from 1 to 9. The crystal structure is a repeated stacking of a layer of $(Bi_2O_2)^{2+}$ and a layer of $(A_{n-1}B_nO_{3n-1})^{2-}$, which can be approximately repre-

Inorganic Crystals Oxides [5.1,2]		Inorganic Crystals other than Oxides [5.3]		Organic Crystals, Liquid Crystals, and Polymers [5.4]	
Family Nr.	Name	Family Nr.	Name	Family Nr.	Name
1	Perovskite-type family (90, 40; 11) a	20	SbSI family (11, 4; 1)	50	$SC(NH_2)_2$ family (1, 1; 1)
2	$LiNbO_3$ family (2, 2; 1) b	21	TlS family (1, 1; 0)	51	CCl_3CONH_2 family (1, 1; 0)
3	$YMnO_3$ family (6, 6; 0)	22	$TlInS_2$ family (5, 2; 0)	52	$Cu(HCOO)_2 \cdot 4H_2O$ family (1, 1; 0)
4	$SrTeO_3$ family (1, 1; 1)	23	Ag_3AsS_3 family (2, 1; 0)	53	$N(CH_3)_4HgCl_3$ family (6, 5; 0)
5	Stibiotantalite family (7, 6; 0) c	24	$Sn_2P_2S_6$ family (2, 2; 0)	54	$(CH_3NH_3)_2AlCl_5 \cdot 6H_2O$ family (3, 2; 0)
6	Tungsten bronze-type family (141, 21; 2) d	25	$KNiCl_3$ family (3, 3; 0)	55	$[(CH_3)_2NH_2]_2CoCl_4$ family (3, 2; 0)
7	Pyrochlore-type family (19, 2; 0) e	26	$BaMnF_4$ family (6, 6; 1)	56	$[(CH_3)_2NH_2]_3Sb_2Cl_9$ family (5, 5; 0)
8	$Sr_2Nb_2O_7$ family (6, 5; 1) f	27	HCl family (2, 2; 1)	57	$(CH_3NH_3)_5Bi_2Cl_{11}$ family (2, 2; 0)
9	Layer-structure family (36, 16; 0) g	28	$NaNO_2$ family (2, 2; 1)	58	DSP ($Ca_2Sr(CH_3CH_2COO)_6$) family (3, 3; 1)
10	$BaAl_2O_4$-type family (1, 1; 0)	29	$CsCd(NO_2)_3$ family (2, 2; 0)	59	$(CH_2ClCOO)_2H \cdot NH_4$ family (2, 2; 0)
11	$LaBGeO_5$ family (1, 1; 0)	30	KNO_3 family (4, 3; 1)	60	TGS ($(NH_2CH_2COOH)_3 \cdot H_2SO_4$) family (3, 3; 1)
12	$LiNaGe_4O_9$ family (2, 2; 0)	31	$LiH_3(SeO_3)_2$ family (7, 5; 0)	61	$NH_2CH_2COOH \cdot AgNO_3$ family (1, 1; 0)
13	$Li_2Ge_7O_{15}$ family (1, 1; 1)	32	KIO_3 family (3, 3; 0)	62	$(NH_2CH_2COOH)_2 \cdot HNO_3$ family (1, 1; 0)
14	$Pb_5Ge_3O_{11}$ family (1,1; 0)	33	KDP (KH_2PO_4) family (12, 12; 3)	63	$(NH_2CH_2COOH)_2 \cdot MnCl_2 \cdot 2H_2O$ family (1, 1; 0)
15	$5PbO \cdot 2P_2O_5$ family (1, 1; 0) (exact chemical formula unknown)	34	$PbHPO_4$ family (2, 2; 1)	64	$(CH_3NHCH_2COOH)_3 \cdot CaCl_2$ family (2, 2; 1)
16	$Ca_3(VO_4)_2$ family (2, 2; 0)	35	$KTiOPO_4$ family (23, 15; 0)	65	$(CH_3)_3NCH_2COO \cdot H_3PO_4$ family (3, 3; 0)
17	GMO ($Gd_2(MoO_4)_3$) family (5, 5; 1)	36	$CsCoPO_4$ family (5, 5; 0)	66	$(CH_3)_3NCH_2COO \cdot CaCl_2 \cdot 2H_2O$ family (1, 1; 1)
18	Boracite-type family (28, 14; 1) h	37	$NaTh_2(PO_4)_3$ family (2, 2; 0)	67	Rochelle salt ($NaKC_4H_4O_6 \cdot 4H_2O$) family (3, 2; 1)
19	$Rb_3MoO_3F_3$ family (4.4; 0)	38	$Te(OH)_6 \cdot 2NH_4H_2PO_4 \cdot (NH_4)_2HPO_4$ family (1, 1; 0)	68	$LiNH_4C_4H_4O_6 \cdot H_2O$ family (3, 2; 1) k
		39	$(NH_4)_2SO_4$ family (22, 21; 1)	69	$C_5H_6NBF_4$ family (1, 1; 0)
		40	NH_4HSO_4 family (9, 4; 1)	70	$3C_6H_4(OH)_2 \cdot CH_3OH$ family (1, 1; 0)
		41	NH_4LiSO_4 family (9, 6; 1)	71	Liquid crystal family (97, 90; 2) l
		42	$(NH_4)_3H(SO_4)_2$ family (2, 2; 1)	72	Polymer family (5, 5; 1) m
		43	Langbeinite-type family (16, 5; 1) i		
		44	Lecontite ($NaNH_4SO_4 \cdot 2H_2O$) family (2, 2; 0)		
		45	Alum family (16, 15; 0) j		
		46	GASH ($C(NH_2)_3Al(SO_4)_2 \cdot 6H_2O$) family (9, 9; 0)		
		47	Colemanite ($Ca_2B_6O_{11} \cdot 5H_2O$) family (1, 1; 0)		
		48	$K_4Fe(CN)_6 \cdot 3H_2O$ family (4, 4; 0)		
		49	$K_3BiCl_6 \cdot 2KCl \cdot KH_3F_4$ family (1, 1; 0)		

◄ **Table 4.5-1** The 72 families of ferroelectric materials. The number assigned to each family corresponds to the number used in LB III/36. The numbers in parentheses (N_{Sub}, N_{F+A}; n) after the family name serve the purpose of conveying some information about the size and importance of the family. The numbers indicate the following: N_{Sub}, the number of pure substances (ferroelectric, antiferroelectric, and related substances) which are treated as members of this family in LB III/36; N_{F+A}, the number of ferroelectric and antiferroelectric substances which are treated as members of this family in LB III/36; n, the number of representative substances from this family whose properties are surveyed in Sect. 4.5.4. For some of these families, additional remarks are needed: for instance, because the perovskite-type oxide family has many members and consists of several subfamilies; because the liquid crystal and polymer families have very specific properties compared with crystalline ferroelectrics; and because the traditional names of some families are apt to lead to misconceptions about their members. Such families are marked by letters a–m following the parentheses, and remarks on these families are given under the corresponding letter in the text in Sect. 4.5.3.1

sented by a chain of n perovskite-type units of ABO_3 perpendicular to the layer.

h. Boracite-Type Family (Family Number 18). Boracite is a mineral, $Mg_3B_7O_{13}Cl$. The boracite-type family contains crystals isomorphous with the mineral, and has the chemical formula $M_3^{2+}B_7O_{13}X^{1-}$, where M^{2+} stands for a divalent cation of Mg, Cr, Mn, Fe, Co, Ni, Cu, Zn, or Cd, and X^{1-} stands for an anion of Cl, Br, or I.

i. Langbeinite-Type Family (Family Number 43). This family consists of crystals which are basically isomorphous with $K_2Mg_2(SO_4)_3$ (langbeinite), and have the common chemical formula $M_2^{1+}M_2^{2+}(SO_4)_3$, where M^{1+} stands for a monovalent ion of K, Rb, Cs, Tl, or NH_4, and M^{2+} stands for a divalent ion of Mg, Ca, Mn, Fe, Co, Ni, Zn, or Cd.

j. Alum Family (Family Number 45). The alums are compounds with the chemical formula $M^{1+}M^{3+}(SO_4)_2 \cdot 12\,H_2O$, where M^{1+} is a monovalent cation and M^{3+} is a trivalent cation. Ferroelectricity has been found for the monovalent cations of NH_4, CH_3NH_3, etc. and the trivalent cations of Al, V, Cr, Fe, In, and Ga. This family contains a few isomorphous selenates, $M^{1+}M^{3+}(SeO_4)_2 \cdot 12\,H_2O$.

k. $LiNH_4C_4H_4O_6 \cdot H_2O$ Family (Family Number 68). This family contains two ferroelectrics, $LiNH_4C_4H_4O_6 \cdot H_2O$ and $LiTlC_4H_4O_6 \cdot H_2O$. The polar directions of the two ferroelectrics are different from each other.

l. Liquid Crystals (Family Number 71). Ferroelectric and antiferroelectric liquid crystals are very useful as fast display elements.

A. *Ferroelectric liquid crystals (family number 71A).* Ferroelectric liquid crystals are defined as liquid crystals which exhibit a ferroelectric hysteresis loop like that shown in Fig. 4.5-1. Unlike ferroelectric crystals, however, ferroelectric liquid crystals generally have no spontaneous polarization in the bulk state. The chiral smectic phase denoted by Sm C* (e.g. of DOBAMBC) consists of many layers, each of which has a spontaneous polarization parallel to the layer plane, but the spontaneous polarization varies helically in different directions from layer to layer, so that the bulk has no spontaneous polarization as a whole. A sufficiently strong electric field causes a transition from the helical phase to a polar phase. Under an alternating electric field, the helical structure does not have a chance to build up owing to the delay in the transition. Instead, a direct transition occurs between the induced polar phases, resulting in a hysteresis loop of the type shown in Fig. 4.5-1. Accordingly, the hysteresis loop may be regarded as one in which the linear part of the antiferroelectric hysteresis shown in Fig. 4.5-2 is eliminated. It should be noted, however, that the helical structure disappears and two stable states with parallel and antiparallel polarizations appear, similar to the domain structure of a ferroelectric crystal, when a liquid crystal is put in a cell which is thinner than the helical pitch [5.52].

B. *Antiferroelectric liquid crystals (family number 71B).* The phase denoted by Sm C_A^* (e.g. of MHPOBC) exhibits a double hysteresis of the type shown in Fig. 4.5-2, and a liquid crystal showing this phase is called antiferroelectric [5.53].

m. Polymers (Family Number 72). Polyvinylidene fluoride $(CH_2CF_2)_n$ and its copolymers with trifluoroethylene $(CHFCF_2)_n$, etc. are ferroelectric. Ferroelectric polymers are usually prepared as thin films in which crystalline and amorphous regions coexist. The ferroelectric hysteresis loop originates from reversal of the spontaneous polarization in the crystalline regions. The electric-field distribution is expected to be complex in the thin film because of the interposition of the amorphous regions. A ferroelectric hysteresis loop can be observed when the ratio of the volume of the crystalline regions to the total volume is relatively large (e.g. more than 50%). The coercive field is larger (e.g. > 50 MV/m) than that of solid ferroelectrics (usually a few MV/m or smaller). The ferroelectric properties depend sensitively upon the details of sample preparation, for example the use of melt quenching or melt extrusion, the annealing temperature, or the details of the poling procedure. Polymer ferroelectrics are useful for soft transducers.

4.5.4 Physical Properties of 43 Representative Ferroelectrics

This section surveys the characteristic properties of 43 representative ferroelectrics with the aim of demonstrating the wide variety in the behavior of ferroelectrics. The 43 representative ferroelectrics are selected from 29 of the above-mentioned families. The presentation is mainly graphical. Most of the figures are reproduced from LB III/36, where the relevant references can be found. Table 4.5-2 summarizes the meaning of the symbols frequently used in the figure captions and indicates the units in which the data are given.

To facilitate to get more information on each representative substance in LB II/36, the number assigned to the substance in LB III/36 is given in parenthesis following its chemical formula, e.g., $KNbO_3$ (LB number 1A-2).

4.5.4.1 Inorganic Crystals Oxides [5.1, 2]

Perovskite-Type Family

$KNbO_3$ (LB Number 1A-2). This crystal is ferroelectric below about 418 °C. Further phase transitions take place at about 225 °C and about −10 °C, retaining ferroelectric activity. The crystal has large electromechanical coupling constants and is useful in lead-free piezoelectric elements and SAW (surface acoustic wave) filters in communications technology (Fig. 4.5-13, 4.5-14).

Table 4.5-2 Symbols and units frequently used in the figure captions

Symbol	Meaning
$\boldsymbol{a}, \boldsymbol{b}, \boldsymbol{c}$	unit cell vector, units Å
$\boldsymbol{a}^*, \boldsymbol{b}^*, \boldsymbol{c}^*$	unit cell vector in reciprocal space, units Å^{-1}
E_c	coercive field, units V/m
f	frequency, units Hz = 1/s
P_s	spontaneous polarization, units C/m^2
T	temperature, units K or °C
κ	dielectric constant (or relative permittivity) = $\varepsilon/\varepsilon_0$, where ε is the permittivity of the material and ε_0 is the permittivity of a vacuum. κ is a dimensionless number
κ', κ''	real and imaginary parts of the complex dielectric constant $\kappa^* = \kappa' + i\kappa''$. κ' and κ'' are both dimensionless numbers
κ_{ij}	component of dielectric-constant tensor; dimensionless numbers
$\kappa_a, \kappa_b, \kappa_c$	κ measured along a, b, c axes; dimensionless numbers
$\kappa_{(hkl)}$	κ measured perpendicular to the (hkl) plane; dimensionless number
$\kappa_{[uvw]}$	κ measured parallel to the [uvw] direction; dimensionless number
$\kappa^{\mathbf{T}}$	κ of free crystal, i. e. κ measured at constant stress **T**; dimensionless number
$\kappa^{\mathbf{S}}$	κ of clamped crystal, i. e. κ measured at constant strain **S**; dimensionless number
Θ_f	ferroelectric transition temperature, units K or °C

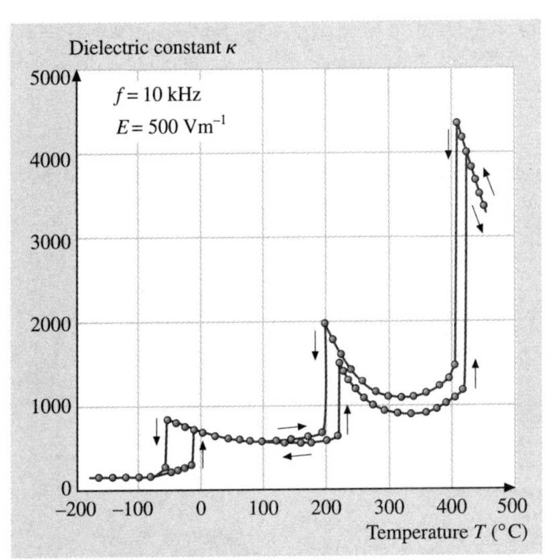

Fig. 4.5-12 $KNbO_3$. Dielectric constant κ versus temperature T

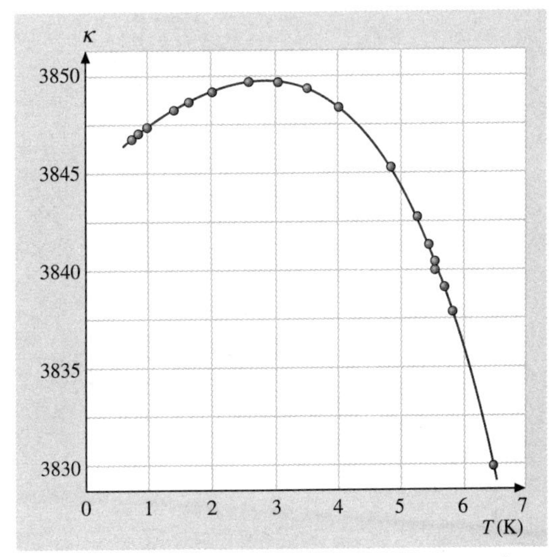

Fig. 4.5-14 $KTaO_3$. Real part of dielectric constant κ' versus T. $f = 1\,kHz$

$KTaO_3$ (LB Number 1A-5). $KTaO_3$ is cubic at all temperatures, and its dielectric constant becomes very large at low temperatures without a phase transition (Fig. 4.5-14). It is generally believed that this behavior is related to the zero-point lattice vibrations. Replacement of Nb by Ta generally lowers drastically the ferroelectric Curie temperature , as seen by comparing Fig. 4.5-14 with Fig. 4.5-12. (This effect is well demonstrated later in Figs. 4.5-39 and 4.5-40).

$SrTiO_3$ (LB Number 1A-8). This crystal is cubic at room temperature and slightly tetragonal below 105 K. The phase transition at 105 K is caused by softening of the lattice vibration mode at the (1/2, 1/2, 1/2) Brillouin zone corner (Fig. 4.5-15) without an appreciable dielectric anomaly. At very low temperatures, the dielectric constants become extraordinarily high without a phase transition (Fig. 4.5-16), owing to softening of the optical phonon at the origin of the Brillouin zone (Fig. 4.5-17). It is generally believed that the absence of a low-temperature transition is related to zero-point lat-

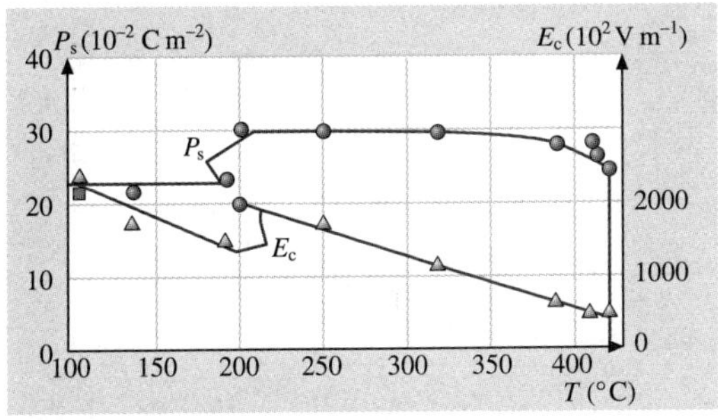

Fig. 4.5-13 $KNbO_3$. P_s and E_c versus T. Measurements were made by applying the electric field parallel to the pseudocubic [100] direction

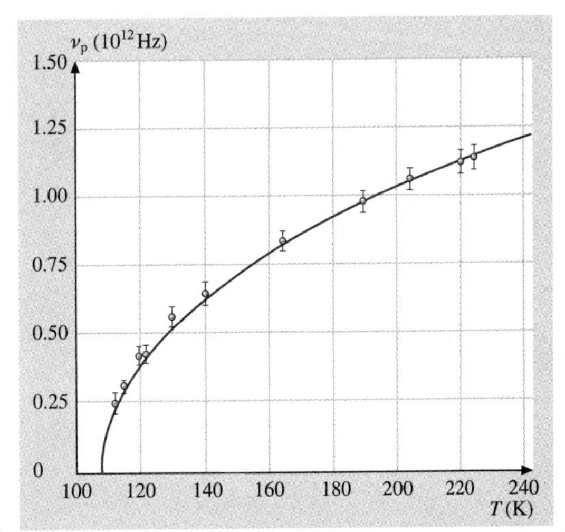

Fig. 4.5-15 $SrTiO_3$. ν_P versus T. ν_P is the frequency of the R_{25} (Γ_{25}) optical phonon

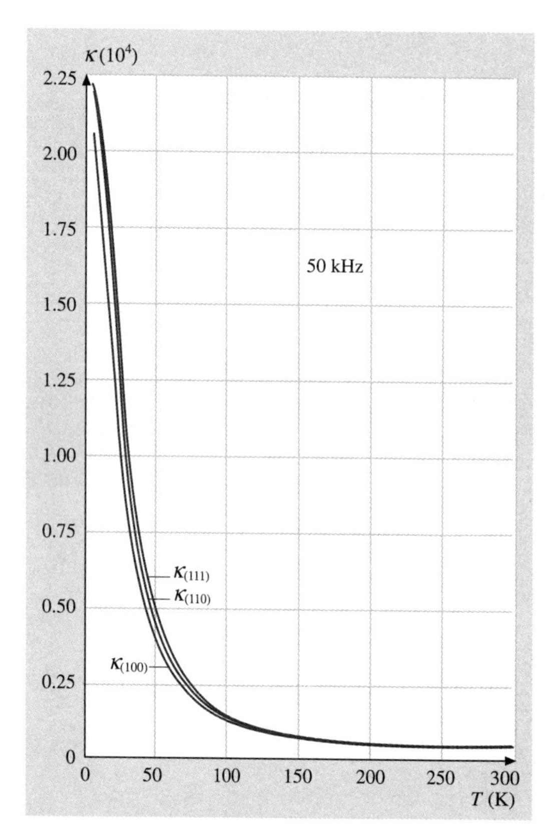

Fig. 4.5-16 $SrTiO_3$. $\kappa_{(111)}$, $\kappa_{(110)}$, and $\kappa_{(100)}$ versus T at $f = 50$ kHz. $\kappa_{(111)}$, $\kappa_{(110)}$, and $\kappa_{(100)}$ are slightly different from each other in the tetragonal phase

tice vibrations as in $KTaO_3$. Solid solutions of $SrTiO_3$ are useful as ceramic thin films in memory devices.

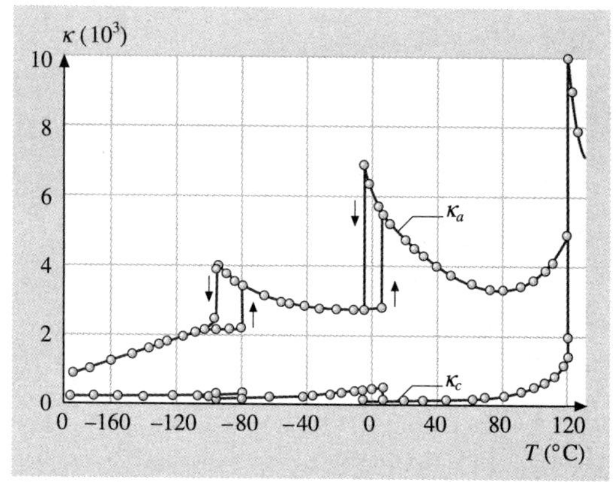

Fig. 4.5-18 $BaTiO_3$. κ_a and κ_b versus T. κ_a and κ_b are the values of κ along the a and b axes, respectively, of the tetragonal phase

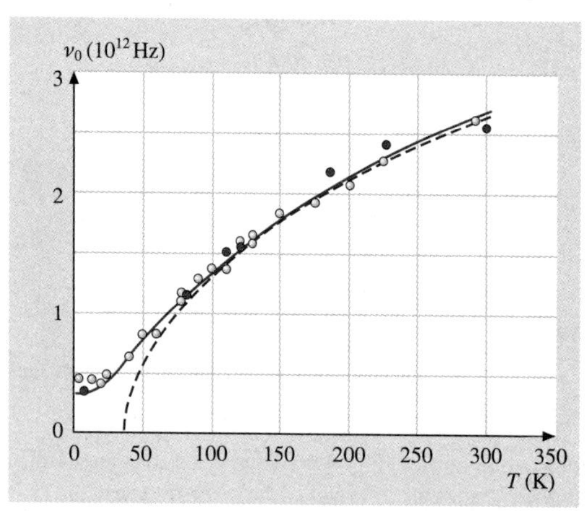

Fig. 4.5-17 $SrTiO_3$. ν_0 versus T. ν_0 is the frequency of the soft phonon at $q = 0$. *Open circles*: inelastic neutron scattering. *Filled circles*: Raman scattering. The *solid curve* is $194.4\,\kappa^{1/2}$ and the *dashed curve* is $0.677(T - T_0)^{1/2}$, with $T_0 = 38$ K

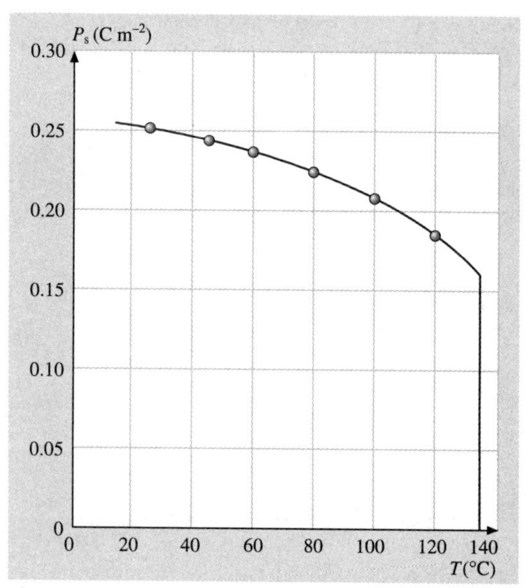

Fig. 4.5-19 $BaTiO_3$. P_s versus T in the tetragonal phase. P_s is parallel to the c axis

$BaTiO_3$ (LB Number 1A-10). $BaTiO_3$ is the most extensively studied ferroelectric crystal. It is ferroelectric below about 123 °C, where the crystal symmetry changes from cubic to tetragonal. Further phase transitions take place from tetragonal to orthorhombic at about 5 °C and to rhombohedral at about −90 °C (Figs. 4.5-18 to 4.5-20). It is believed that the ferroelectric transition

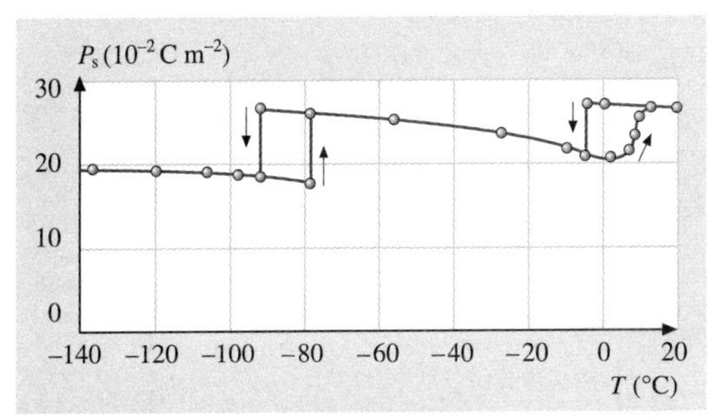

Fig. 4.5-20 $BaTiO_3$. $P_{s[001]}$ versus T below 20 °C, where $P_{s[001]}$ is the component of the spontaneous polarization parallel to the c axis in the tetragonal phase

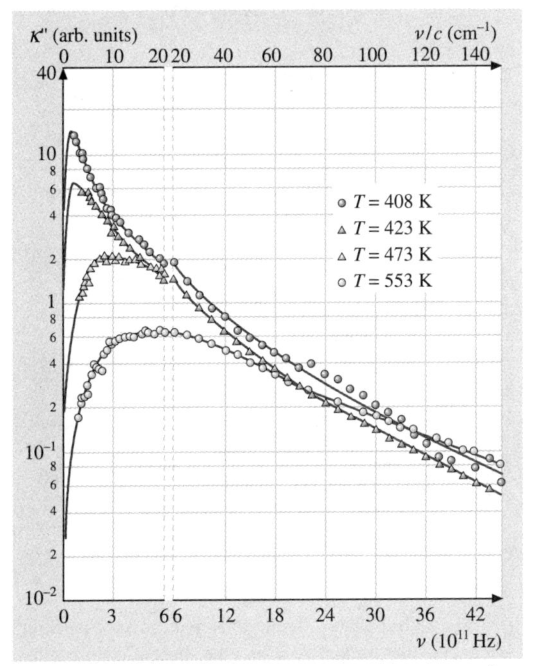

Fig. 4.5-21 $BaTiO_3$. κ'' versus ν. Parameter: T. κ'' is the imaginary part of κ obtained from the hyper-Raman spectrum. *Curves*: calculations based upon the classical dispersion oscillator model. c: light velocity

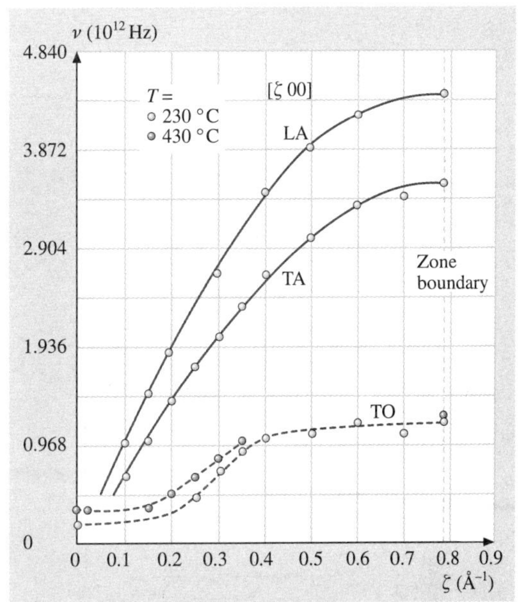

Fig. 4.5-22 $BaTiO_3$. Phonon dispersion relation determined by neutron scattering along the [100] direction in the cubic phase. ν is the phonon frequency. LA, longitudinal acoustic branch; TA, transverse acoustic branch; TO, transverse optical branch. The frequency of the TO branch is lower (softer) at 230 °C than at 430 °C, indicating mode softening

is caused by softening of an optical mode at the center of the Brillouin zone. Hyper-Raman scattering studies support this model (Figs. 4.5-21 and 4.5-22). The results of neutron scattering studies favor this model (Fig. 4.5-23), but the measurement was not easy owing to the intense elastic peaks (Fig. 4.5-24a). These elastic peaks indicate marked cluster formation in the vicinity of the Curie point. As discussed in Sect. 4.5.3, there seem to be two components contributing to the dielectric constant measured in the vicinity of the Curie point: a component due to the soft mode and another one due to the motion of the cluster boundaries. Presumably the two components suggested by infrared and hyper-Raman scattering data correspond to these two components. For theoretical studies of the phase transitions, readers should refer to [5.7]. Solid solutions of $BaTiO_3$ are the most useful ferroelectrics in ceramic condensers and as thin films in memory devices.

When the temperature is raised above 1460 °C, cubic $BaTiO_3$ performs another phase transition to a hexagonal structure. This hexagonal phase can be quenched

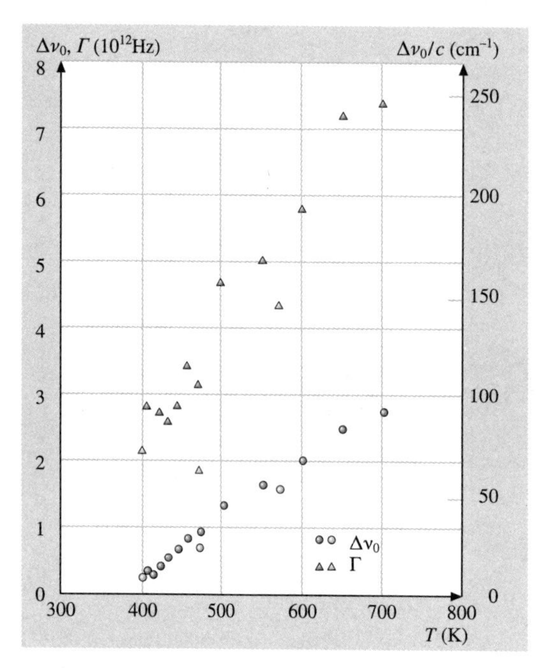

Fig. 4.5-23 $BaTiO_3$. $\Delta\nu_0$ and Γ versus T, obtained from hyper-Raman scattering in the cubic phase. $\Delta\nu_0$ and Γ are the optical mode frequency and damping constant, respectively. The different symbols (*brown* and *gray*) show results from different authors. $\Delta\nu_0$ decreases as the temperature decreases to the Curie point, showing the presence of mode softening. c: ligth velocity

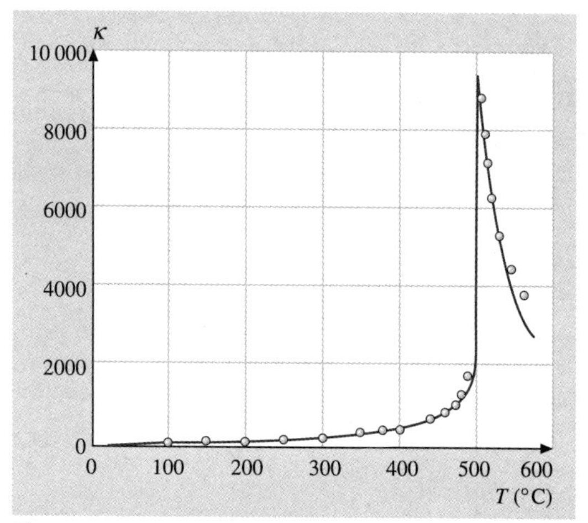

Fig. 4.5-25 $PbTiO_3$. κ versus T

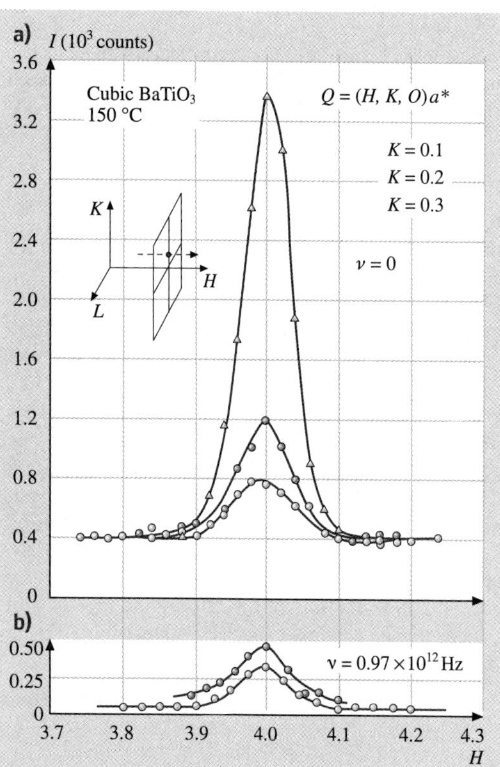

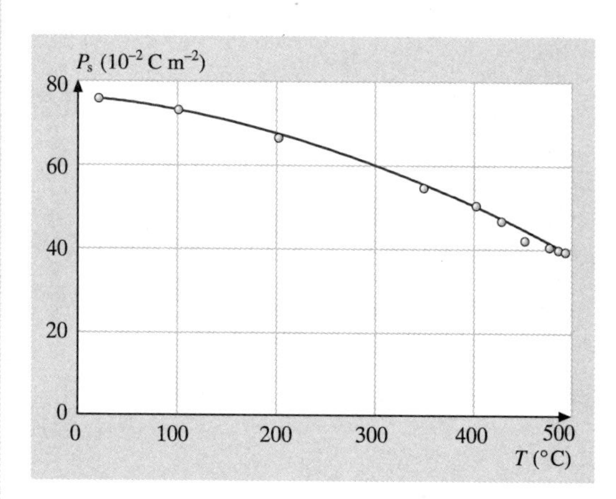

Fig. 4.5-26 $PbTiO_3$. P_s versus T

Fig. 4.5-24a,b $BaTiO_3$. Triple-axis neutron spectrometer scans at constant frequency across the sheet of diffuse scattering at 150 °C. The path of the scans is shown in the *inset*. (**a**) shows the elastic scan ($\nu = 0$), where the high background level is due to nuclear incoherent scattering. (**b**) Ineleastic scan ($\nu = 0.97 \times 10^{12}$ Hz)

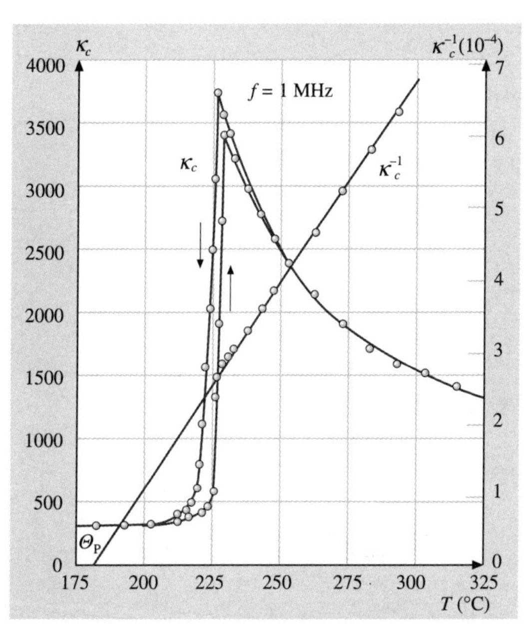

Fig. 4.5-27 $PbZrO_3$ (single crystal). κ_c and κ_c^{-1} versus T

to room temperature by relatively rapid cooling. This hexagonal $BaTiO_3$ also shows ferroelectric activity below −199 °C.

$PbTiO_3$ (LB Number 1A-11). This crystal is ferroelectric below about 500 °C (Figs. 4.5-25, 4.5-26). The spontaneous polarization is large (Fig. 4.5-26) and thus the pyroelectric coefficient is large, which makes the crystal useful in infrared sensors. Solid solutions with other perovskite-type oxides provide good dielectric and piezoelectric materials (see PZT, PLZT, and $(PbTiO_3)_x(Pb(Sc_{1/2}Nb_{1/2})O_3)_{1-x}$ below).

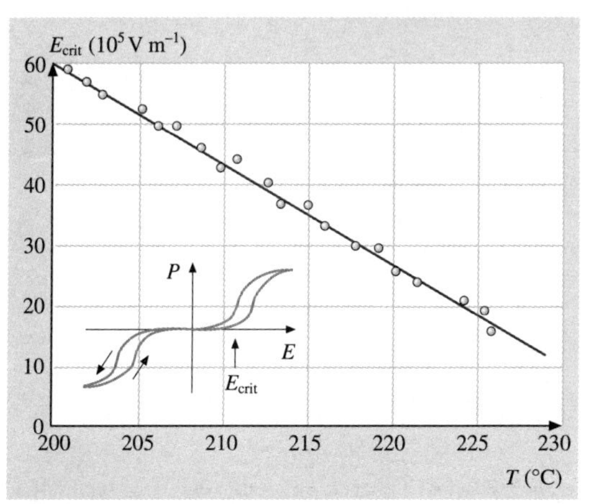

Fig. 4.5-28 $PbZrO_3$. E_{crit} versus T. E_{crit} is the critical field of the antiferroelectric hysteresis loop

$PbZrO_3$ (LB Number 1A-15). This crystal is antiferroelectric below about 230 °C (Fig. 4.5-27), exhibiting a typical antiferroelectric hysteresis loop (Fig. 4.5-28). Solid solutions of this substance are very important in technological applications (see PZT and PLZT below).

$Pb(Mg_{1/3}Nb_{2/3})O_3$ (LB Number 1B-d4). This crystal exhibits a broad ferroelectric phase transition with an average transition temperature of −8 °C determined by the maximum of the low-frequency dielectric constant. A marked frequency dispersion of the

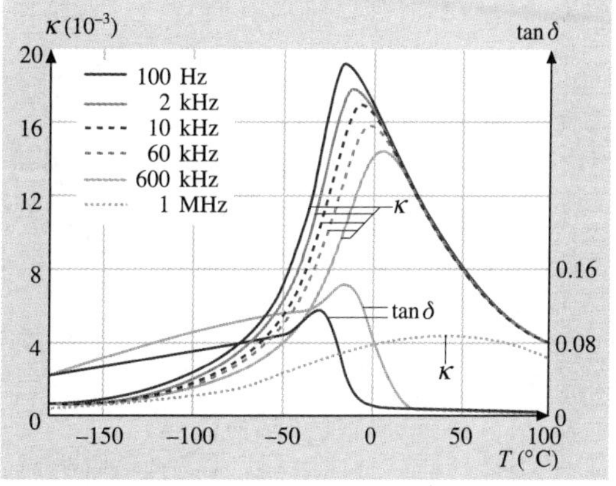

Fig. 4.5-29 $Pb(Mg_{1/3}Nb_{2/3})O_3$ (ceramic). κ and tan δ versus T. Parameter: f

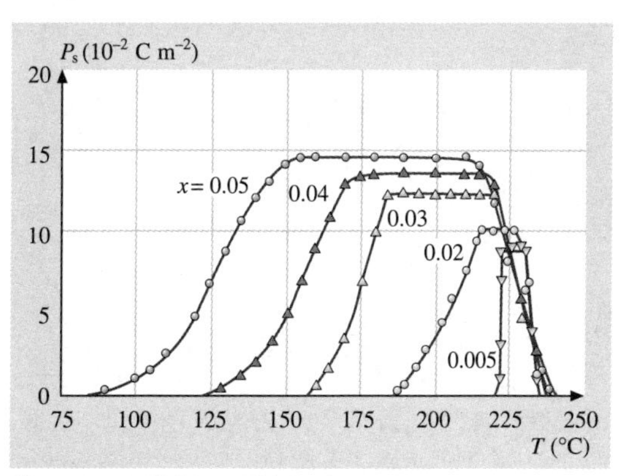

Fig. 4.5-30 $Pb(Zr_{1-x}Ti_x)O_3$ (ceramic). P_s versus T. Parameter: x

dielectric constant occurs around the transition temperature. These is a typical behavior of a relaxor (Fig. 4.5-29).

$Pb(Zr,Ti)O_3$ (LB Number 1C-a62). $PbZrO_3$ is antiferroelectric, as described above, while a small addition of $PbTiO_3$ induces a ferroelectric phase (Fig. 4.5-30).

PZT (LB Number 1C-a63). Solid solutions $Pb(Ti_{1-x}Zr_x)O_3$ with $x = 0.5–0.6$ are commonly called PZT. They have very large electromechanical coupling constants and are widely utilized as piezoelectric elements. Thin ceramic films are useful in memory devices (Fig. 4.5-31).

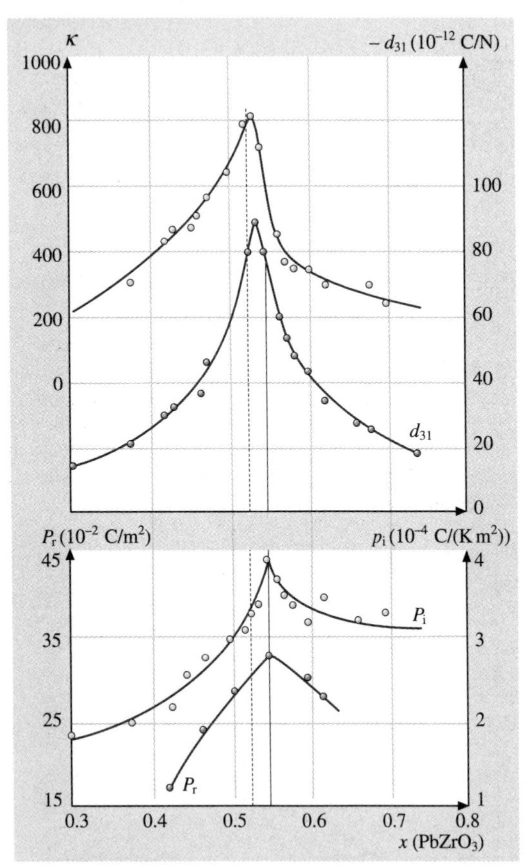

Fig. 4.5-31 $Pb(Ti_{1-x}Zr_x)O_3$ 2% $Zr(Mn_{1/3}Bi_{2/3})O_3$ as an additive (ceramic). κ, p_i, d_{31}, and P_r versus x. p_i is the pyroelectric coefficient, d_{31} is the piezoelectric strain constant, and P_r is the remanent polarization (see Fig. 4.5-1)

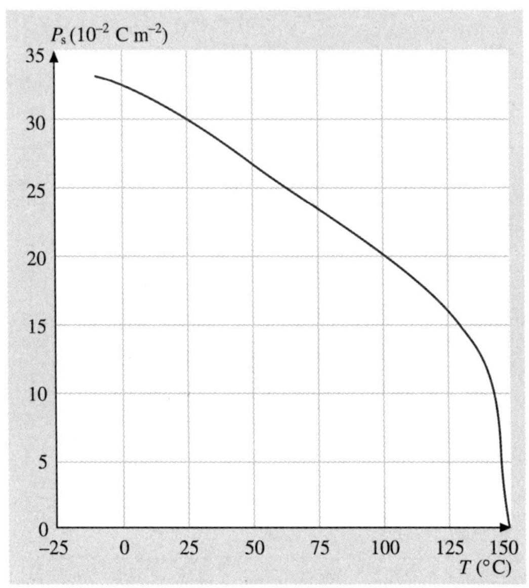

Fig. 4.5-32 $(Pb_{0.92}La_{0.08})(Zr_{0.40}Ti_{0.60})_{0.98}O_3$ (PLZT). P_s versus T

PLZT (LB Number 1C-c66). The acronym PLZT means La-modified PZT. Hot-pressed ceramic PLZT is transparent and useful for optical switches and similar devices (Fig. 4.5-32).

$(PbTiO_3)_x(Pb(Sc_{1/2}Nb_{1/2})O_3)_{1-x}$ (LB Number 1C-b34). These solid solutions have very large electromechan-

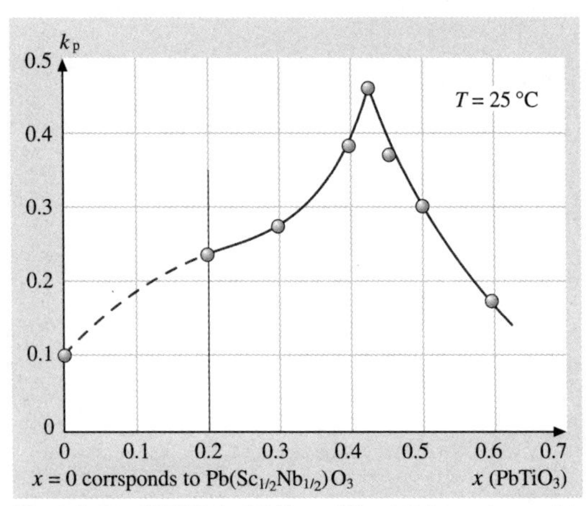

Fig. 4.5-33 $(PbTiO_3)_x(Pb(Sc_{1/2}Nb_{1/2})O_3)_{1-x}$ (ceramic). k_p versus x. k_p is the planar electromechanical coupling factor

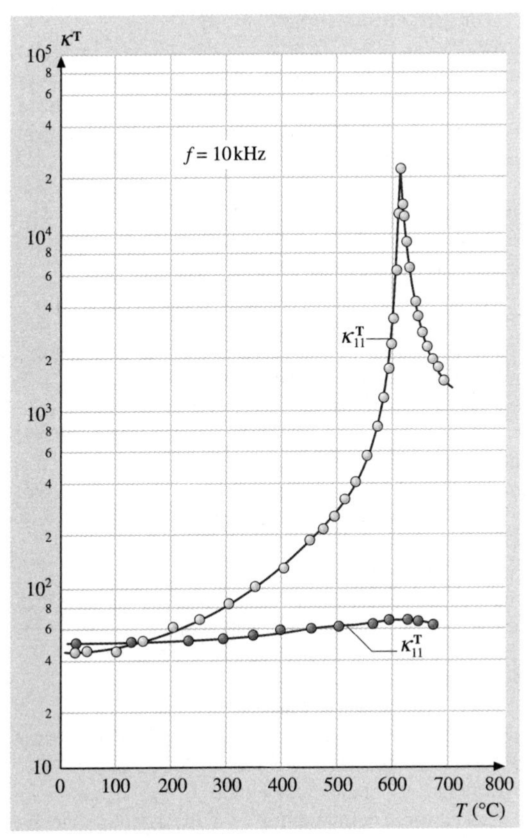

Fig. 4.5-34 $LiTaO_3$. κ_{11}^T and κ_{33}^T versus T. $f = 10$ kHz

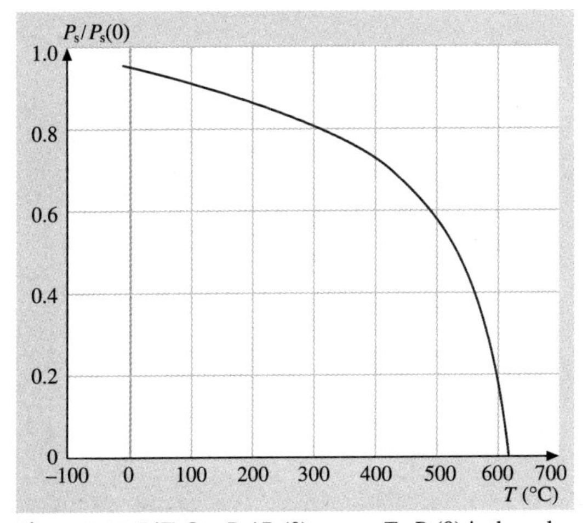

Fig. 4.5-35 $LiTaO_3$. $P_s/P_s(0)$ versus T. $P_s(0)$ is the value of P_s at 0 K (about 0.5 C m^{-2})

ical coupling factors, and are utilized for piezoelectric actuators and similar devices (Fig. 4.5-33).

$LiNbO_3$ Family

$LiTaO_3$ (LB Number 2A-2). This crystal is ferroelectric below 620 °C. The coercive field is large. The crystal is useful for piezoelectric elements, for linear and nonlinear optical elements, and for SAW filters in communications technology (Figs. 4.5-34 and 4.5-35).

$SrTeO_3$ Family

$SrTeO_3$ (LB Number 4A-1). This crystal is ferroelectric between 312 and 485 °C (Figs. 4.5-36 and 4.5-37).

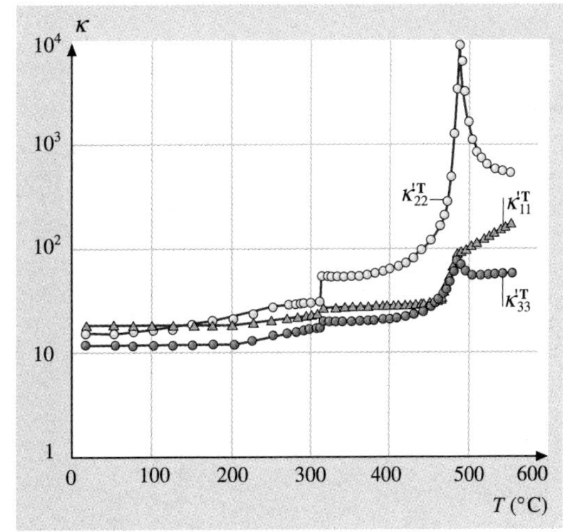

Fig. 4.5-36 $SrTeO_3$. κ versus T. $f = 10$ kHz

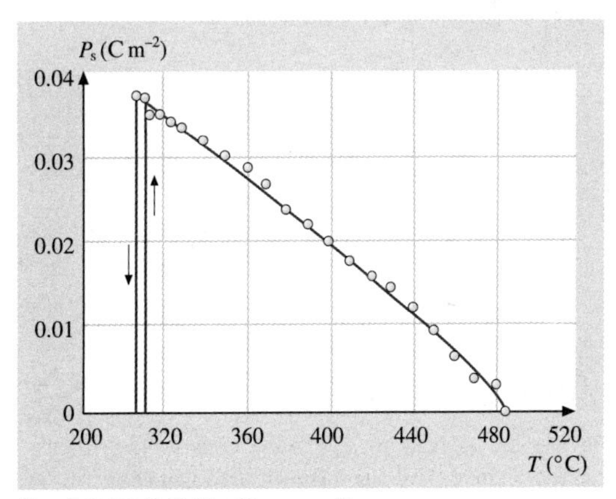

Fig. 4.5-37 $SrTeO_3$. P_s versus T

Part 4 | 5.4

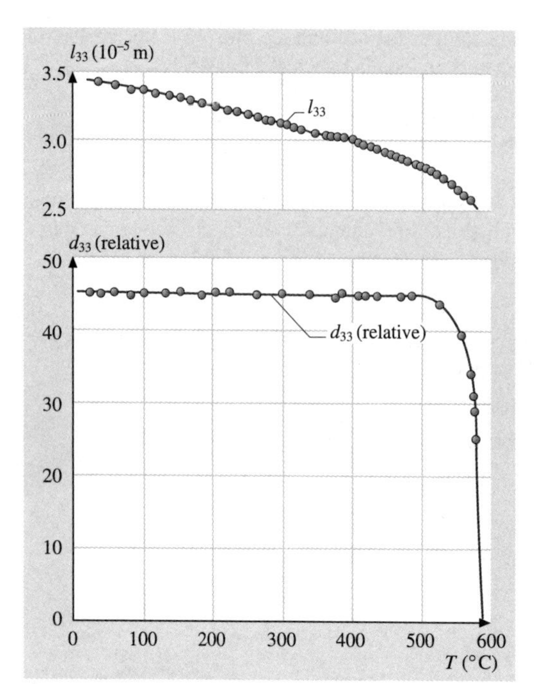

Fig. 4.5-38 $Ba_2NaNb_5O_{15}$. d_{33} and l_{33} versus T. d_{33} is the nonlinear optical susceptibility (relative value), and l_{33} is the coherence length

Tungsten Bronze-Type Family

$Ba_2NaNb_5O_{15}$ (BNN) (LB Number 6B-a7). This crystal is ferroelectric below about 580 °C. The crystal structure is modulated below 300 °C. This material is utilized for optical second-harmonic generation and in optical parametric oscillators (Fig. 4.5-38).

$Ba_2Na(Nb_{1-x}Ta_x)_5O_{15}$ (BNNT) (LB Number 6C-b30). The transition temperature varies over a wide temperature range as x varies from 0 to 1.0 (Fig. 4.5-39).

$Sr_2Nb_2O_7$ Family

$Sr_2(Nb_{1-x}Ta_x)_2O_7$ (LB Number 8B-6). These solid solutions can be made over the whole range of $x = 0$–1.0, and the Curie point varies over a wide temperature range from 1342 °C to −107 °C (Fig. 4.5-40). The solid solutions are very useful as high-temperature dielectric materials, especially because they do not contain Pb, which is volatile at high temperatures.

$Li_2Ge_7O_{15}$ Family

$Li_2Ge_7O_{15}$ (LB Number 13A-1). This crystal is ferroelectric below 283.5 K. The sign of the spontaneous polarization is reversed around 130 K (Figs. 4.5-41 and 4.5-42).

GMO ($Gd_2(MoO_4)_3$) Family

$Gd_2(MoO_4)_3$ (GMO) (LB Number 17A-3). Three crystal structures of GMO are known, α, β, and γ. The β structure is stable above 850 °C but can be obtained at room temperature as a metastable state by

Part 4 | 5.4

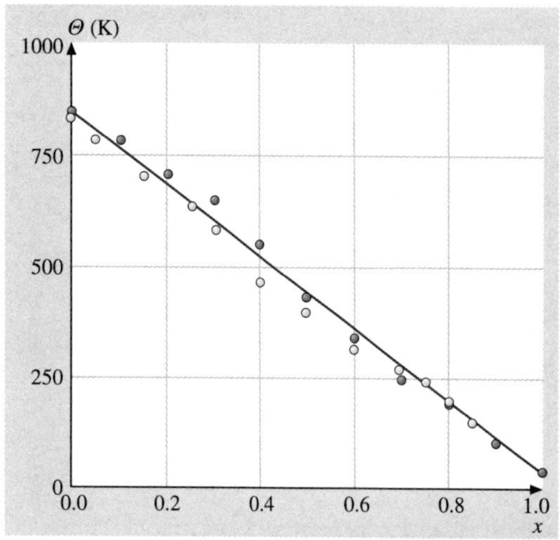

Fig. 4.5-39 $Ba_2Na(Nb_{1-x}Ta_x)_5O_{15}$. Ferroelectric transition temperature Θ versus x. *Brown circles* and *gray circles* represent data measured by different authors

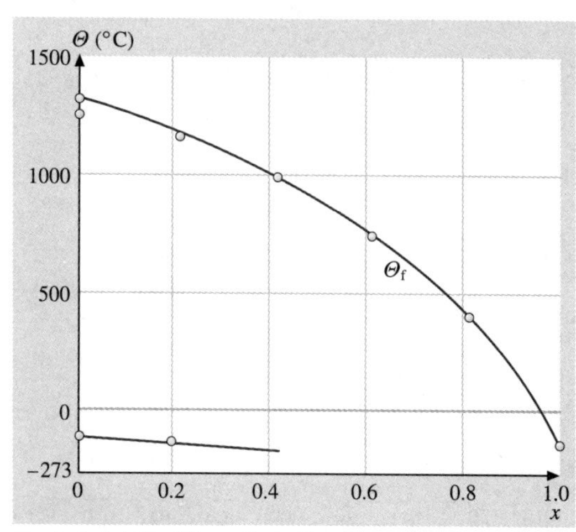

Fig. 4.5-40 $Sr_2(Nb_{1-x}Ta_x)_2O_7$. Ferroelectric transition temperature Θ_f versus x. The *lower curve* shows another phase transition temperature

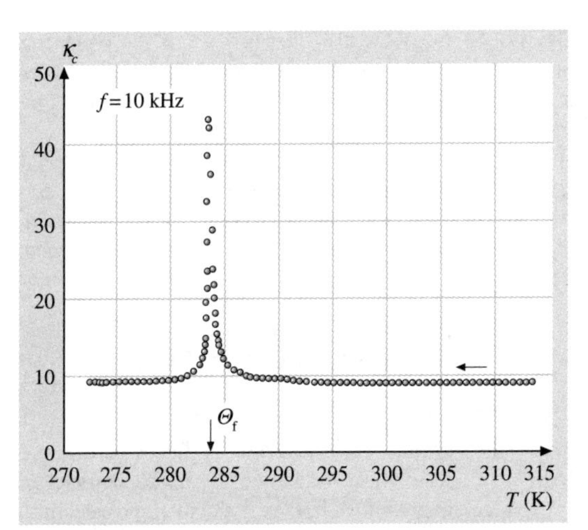

Fig. 4.5-41 $Li_2Ge_7O_{15}$. κ_c versus T. $f = 10\,kHz$

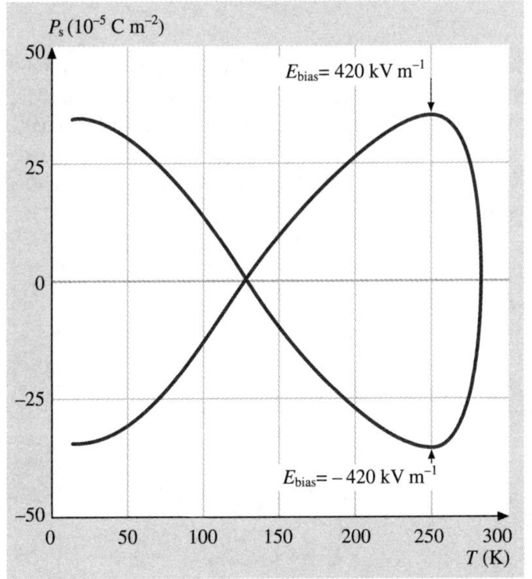

Fig. 4.5-42 $Li_2Ge_7O_{15}$. P_s versus T. P_s was determined by pyroelectric-charge measurement

rapid cooling. Ferroelectric activity takes place in this metastable β-GMO below 159 °C. The phase transition is the indirect type III_{op} discussed in Sect. 4.5.3. The dielectric constant of the clamped crystal (κ_c^S in Fig. 4.5-6) shows no anomaly at the transition point. The ferroelectric phase results from a phonon instability at the (1/2, 1/2, 0) Brillouin zone corner of the parent tetragonal phase. Anharmonic coupling to the

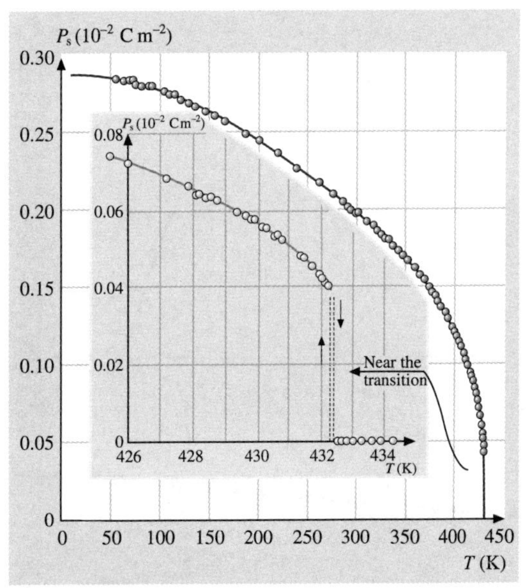

Fig. 4.5-43 $Gd_2(MoO_4)_3$. P_s versus T

resulting antiparallel displacements produces a spontaneous strain, which in turn causes a spontaneous polarization through the normal piezoelectric coupling (Fig. 4.5-43).

Boracite-Type Family

$Ni_3B_7O_{13}I$ (LB Number 18A-23). This crystal is ferroelectric and ferromagnetic below about 60 K (Figs. 4.5-44, 5-45, 5-46): a rare example where both ferroelectric and ferromagnetic spontaneous polarizations

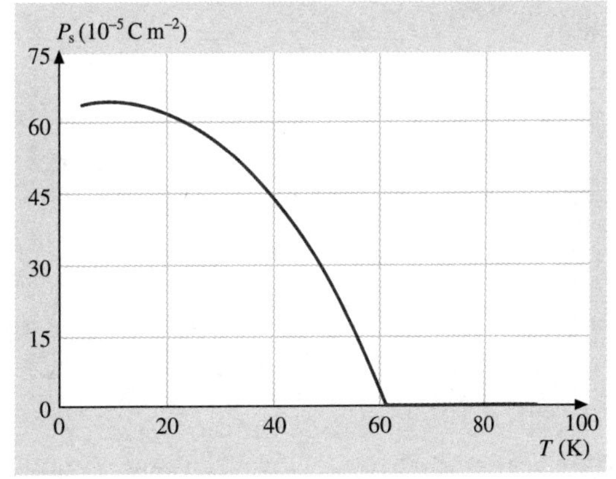

Fig. 4.5-44 $Ni_3B_7O_{13}I$. P_s versus T. P_s was measured with the specimen parallel to the cubic (001) plane

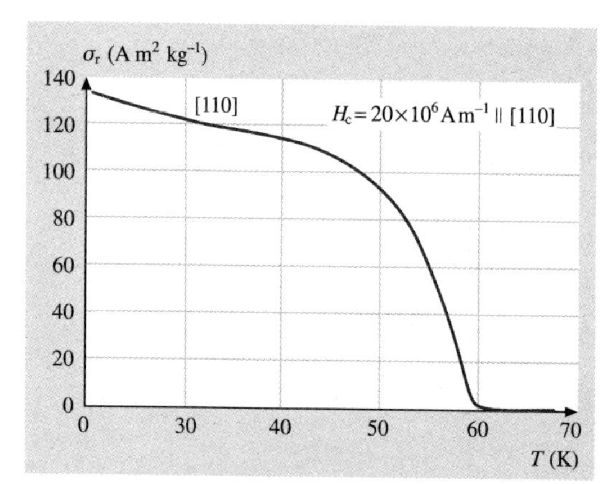

Fig. 4.5-45 $Ni_3B_7O_{13}I$. σ_r versus T. σ_r is the remanent magnetization. The sample was cooled down to 4.2 K in a magnetic field of 1.6×10^6 A/m parallel to [100] prior to the measurement

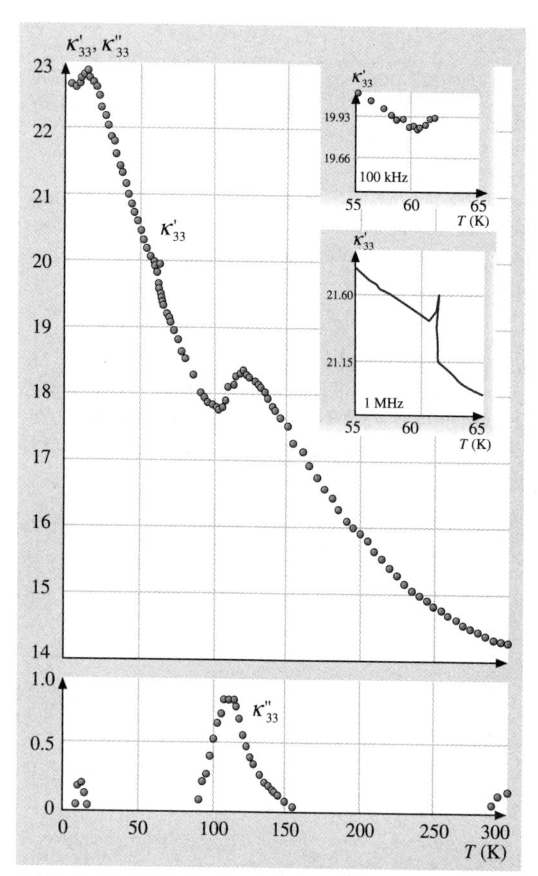

Fig. 4.5-46 $Ni_3B_7O_{13}I$. κ'_{33} and κ''_{33} versus T. $f=$ 100 kHz. The *insets* show details of κ'_{33} versus T at 100 kHz and 1 MHz around Θ_f

take place simultaneously. There is a magnetoelectric effect, where the magnetic polarization is reversed by reversal of the electric polarization and vice versa.

4.5.4.2 Inorganic Crystals Other Than Oxides [5.3]

SbSI Family

SbSI (LB Number 20A-7). SbSI is ferroelectric below 20 °C. The phase transition is of the displacive type, a relatively rare characteristic in nonoxide materials. The crystal is photoconductive (Figs. 4.5-47 and 4.5-48).

$BaMnF_4$ Family

$BaMnF_4$ (LB Number 26A-2). This crystal exhibits a dielectric anomaly at about 242 K (Fig. 4.5-49). The coercive field is very large. The crystal is antiferromagnetic below 25 K (Fig. 4.5-50). The dielectric constant varies depending upon the magnetic field at low temperatures (Fig. 4.5-51).

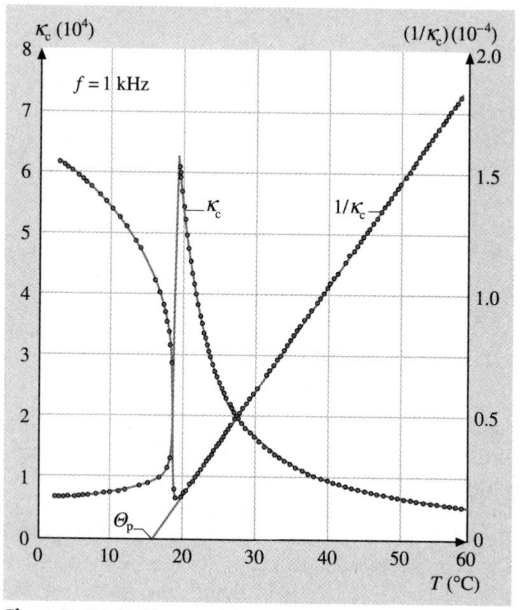

Fig. 4.5-47 SbSI. κ_c and $1/\kappa_c$ versus T

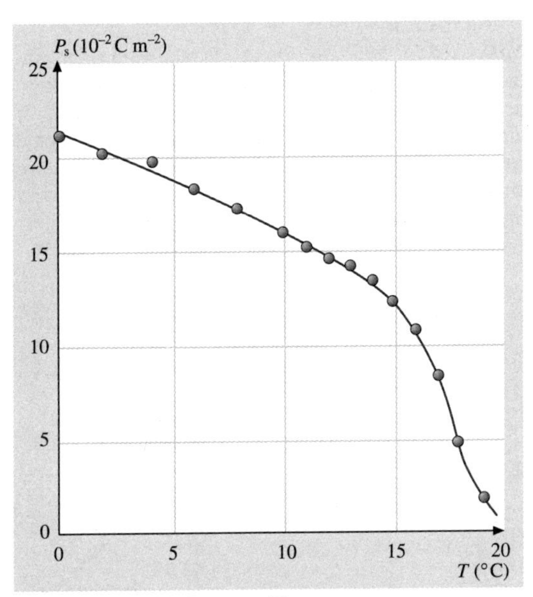

Fig. 4.5-48 SbSI. P_s versus T

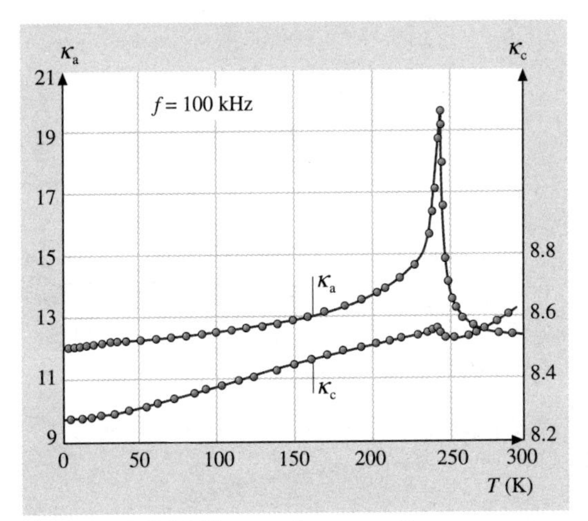

Fig. 4.5-49 $BaMnF_4$. κ_a and κ_c versus T

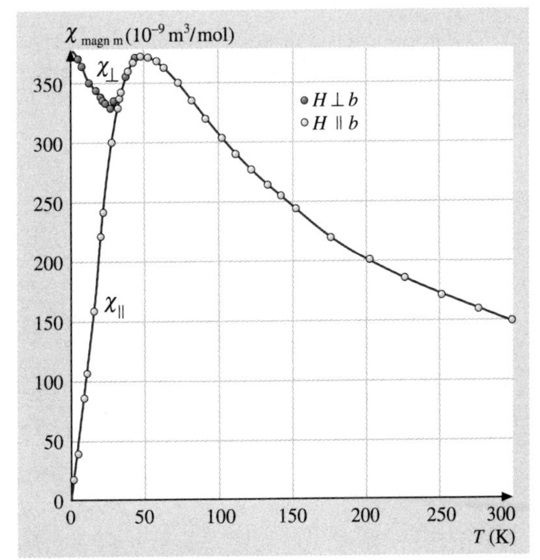

Fig. 4.5-50 $BaMnF_4$. $\chi_{magn\,m}$ versus T. $\chi_{magn\,m}$ is the magnetic susceptibility, and $\chi_\perp$ and $\chi_\parallel$ are the magnetic susceptibilities measured perpendicular and parallel, respectively, to the b axis

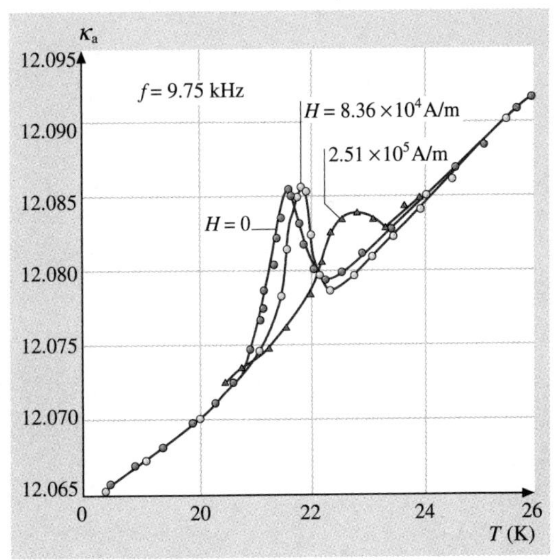

Fig. 4.5-51 $BaMnF_4$. κ_a versus T. Parameter: magnetic field H. $f = 9.75$ kHz

HCl Family

HCl (LB Number 27A-1). This crystal is ferroelectric below 98 K. The chemical formula is the simplest one among all known ferroelectrics. The coercive field is large (Figs. 4.5-52 and 4.5-53).

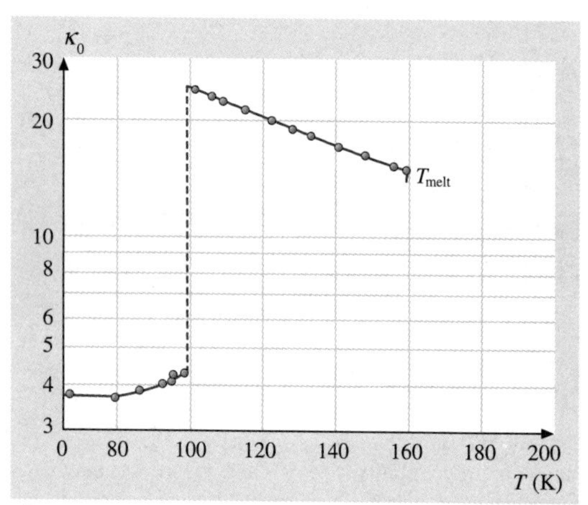

Fig. 4.5-52 HCl (polycrystalline). κ_0 versus T. κ_0 is the static dielectric constant

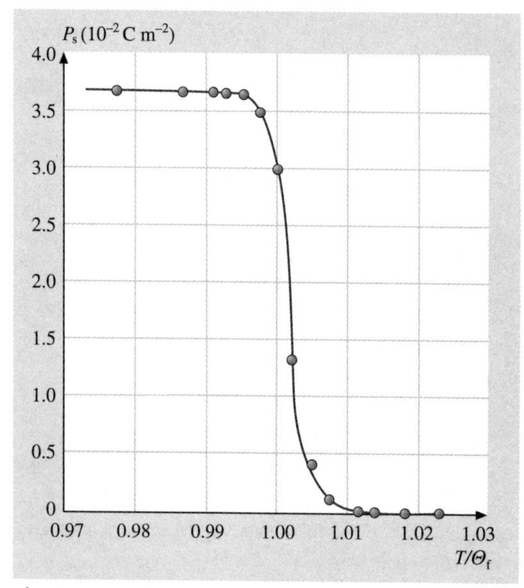

Fig. 4.5-53 HCl. P_s versus T/Θ_f obtained from pyroelectric-current measurement. $\Theta_f = 98$ K

$NaNO_2$ Family

$NaNO_2$ (LB Number 28A-1). This crystal is ferroelectric below 163.9 °C (Figs. 4.5-54 and 4.5-55). The spontaneous polarization results from orientational order of the NO_2^- ions. Between 163.9 and 165.2 °C, the crystal structure is incommensurately modulated with a wave vector $\delta \boldsymbol{a}^*$, where δ varies from 0.097 to 0.120 with increasing temperature.

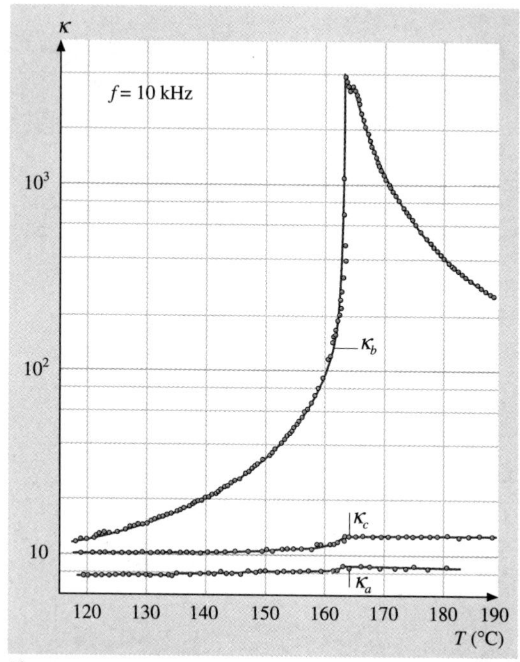

Fig. 4.5-54 $NaNO_2$. κ_a, κ_b, and κ_c versus T

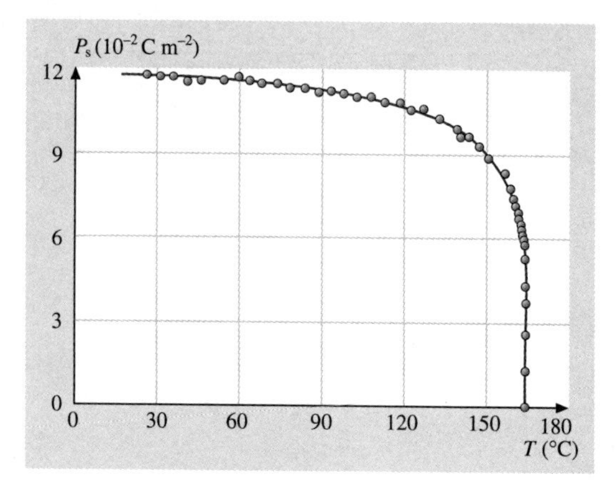

Fig. 4.5-55 $NaNO_2$. P_s versus T, determined by pyroelectric-charge measurement

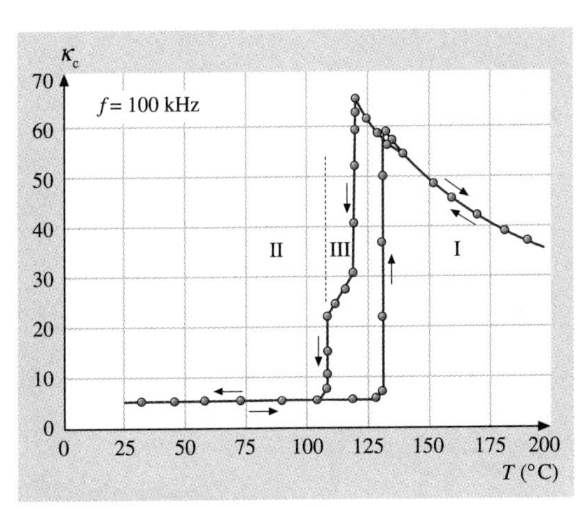

Fig. 4.5-56 KNO_3. κ_c versus T

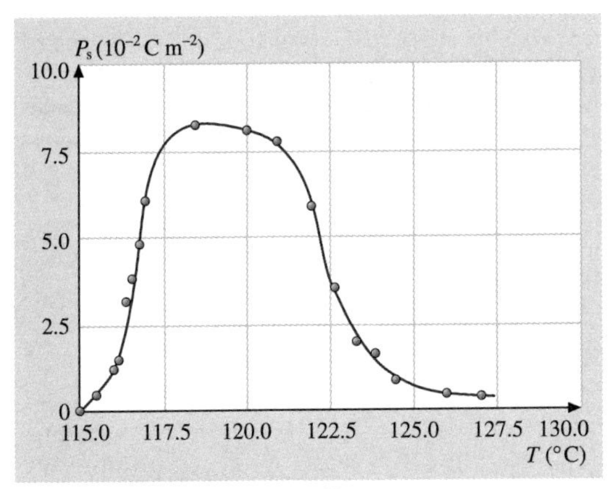

Fig. 4.5-57 KNO_3. P_s versus T

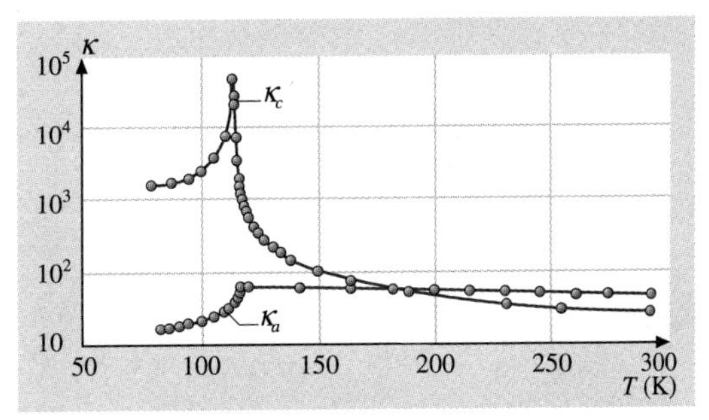

Fig. 4.5-58 KH_2PO_4. κ_a and κ_c versus T. $f = 800$ Hz

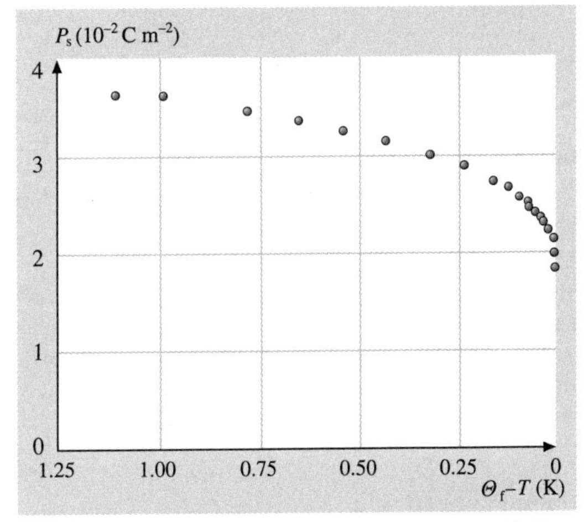

Fig. 4.5-59 KH_2PO_4. P_s versus $(\Theta_f - T)$. $\Theta_f = 123$ K

KNO_3 Family

KNO_3 (LB Number 30A-2). This crystal is ferroelectric between about 115 and 125 °C in a metastable phase III which appears on cooling. Hydrostatic pressure stabilizes this phase (Figs. 4.5-56 and 4.5-57).

KDP (KH_2PO_4) Family

KH_2PO_4 (KDP) (LB Number 33A-1). KH_2PO_4 is a classical and extensively studied ferroelectric crystal. It is ferroelectric below 123 K (Figs. 4.5-58 and 4.5-59). The transition is a typical ferroelectric phase transition, related to a configuration change in a three-dimensional hydrogen-bond network. Figures 4.5-61 and 4.5-60 demonstrate changes in the proton configuration associated with the phase transition. The transitions related to hydrogen atom rearrangement in the hydrogen-bond network are characterized by sensitivity to deuteration and hydrostatic pressure, as demonstrated in Figs. 4.5-62 and 4.5-63, respectively. For theoretical studies of the phase transition, readers should refer to [5.7]. The crystal is useful in nonlinear optical devices.

CsH_2PO_4 (LB Number 33A-3). This crystal is ferroelectric below about 151.5 K. The crystal system of its paraelectric phase (monoclinic) is different from that of KH_2PO_4 (tetragonal). The temperature dependence of the dielectric constant above the Curie point deviates considerably from the Curie–Weiss law, suggesting that the transition is related to one-dimensional ordering of hydrogen atoms in a hydrogen-bond network. Deuteration changes the transition temperature from 151.5 to 264.7 K (Fig. 4.5-64).

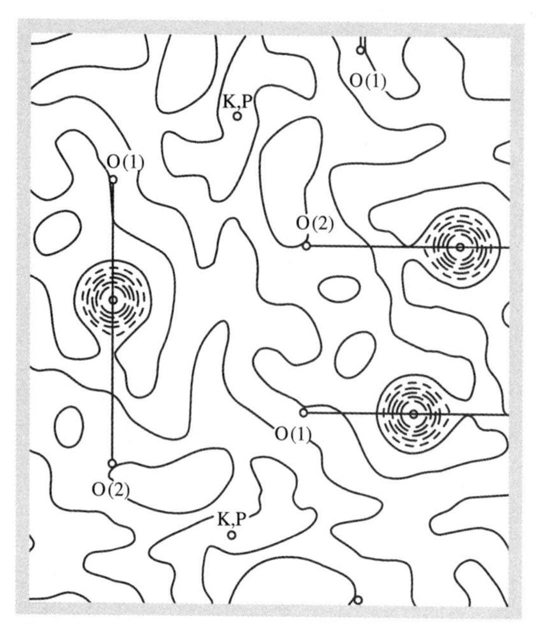

Fig. 4.5-60 KH_2PO_4. Fourier map of the projection of the proton distribution on (001) in the ferroelectric phase (77 K), determined by neutron diffraction. The proton distribution lies approximately on a line joining two oxygen atoms O(1) and O(2) and closer to O(1)

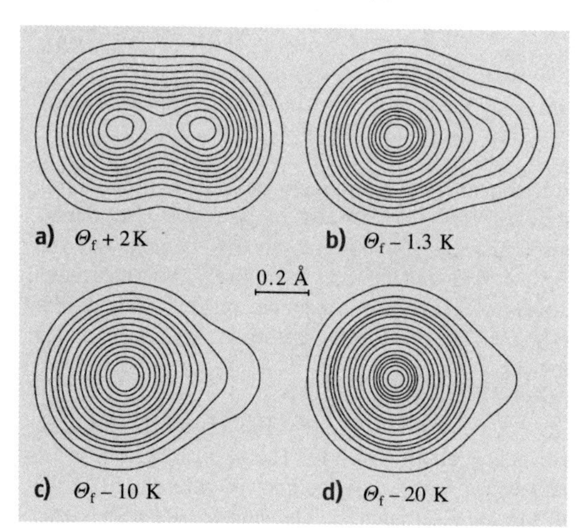

Fig. 4.5-61a–d KH_2PO_4. Change of proton distribution above and below the Curie point Θ_f, determined by neutron diffraction. Contours are all equally spaced. (**a**) $\Theta_f + 2$ K; (**b**) $\Theta_f - 1.3$ K; (**c**) $\Theta_f - 10$ K; (**d**) $\Theta_f - 20$ K

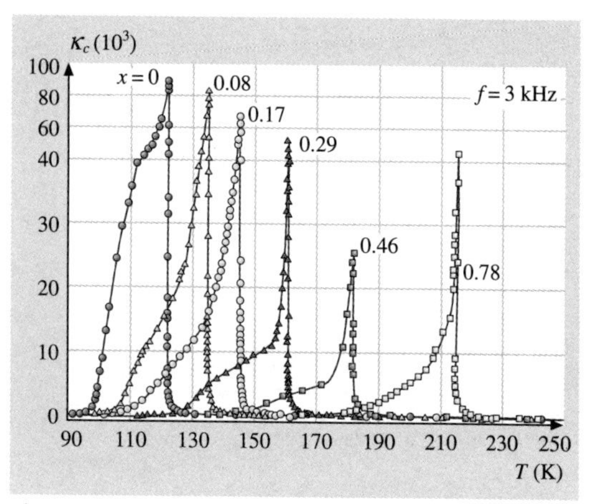

Fig. 4.5-62 $KH_{2(1-x)}D_{2x}PO_4$. κ_c versus T. Parameter: x

RbH_2PO_4–$NH_4H_2PO_4$ (LB Number 33B-5). The mixed crystals $Rb_{1-x}(NH_4)_xH_2PO_4$, where $0.2 < x < 0.8$, show a characteristic temperature dependence of the dielectric constants, suggesting that at low temperatures

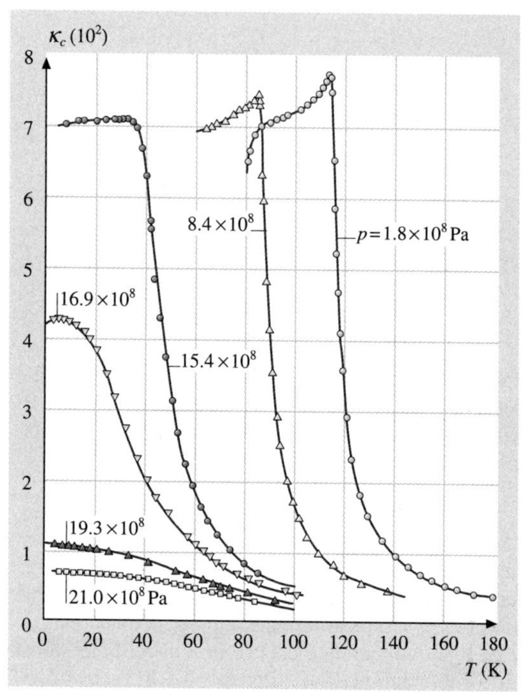

Fig. 4.5-63 KH_2PO_4. κ_c versus T. Parameter: hydrostatic pressure p

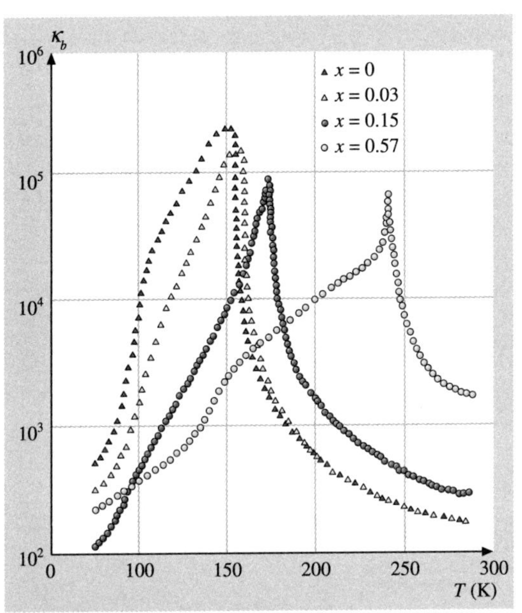

Fig. 4.5-64 CsH_2PO_4 and $CsH_{2(1-x)}D_{2x}PO_4$. κ_b versus T. Parameter: x

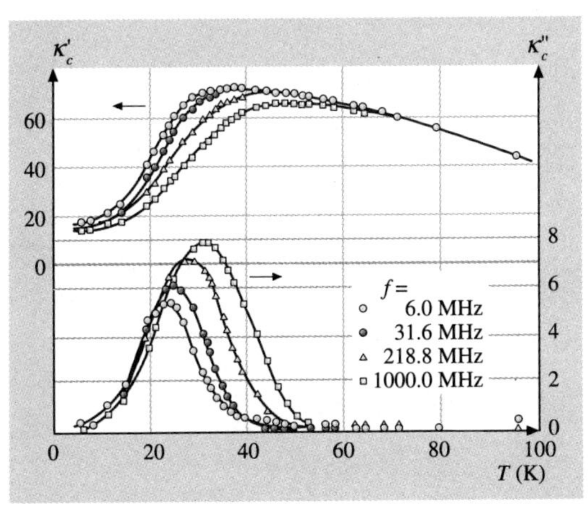

Fig. 4.5-65 $Rb_{0.65}(NH_4)_{0.35}H_2PO_4$. κ'_c and κ''_c versus T. Parameter: f

a local order becomes predominant in a configuration of hydrogen atoms without definite long-range order, i. e. a dipole glass state develops (Fig. 4.5-65).

$PbHPO_4$ Family

$PbHPO_4$ (LB Number 34A-1). This crystal is ferroelectric below 37 °C (Figs. 4.5-66 and 4.5-67). It exhibits

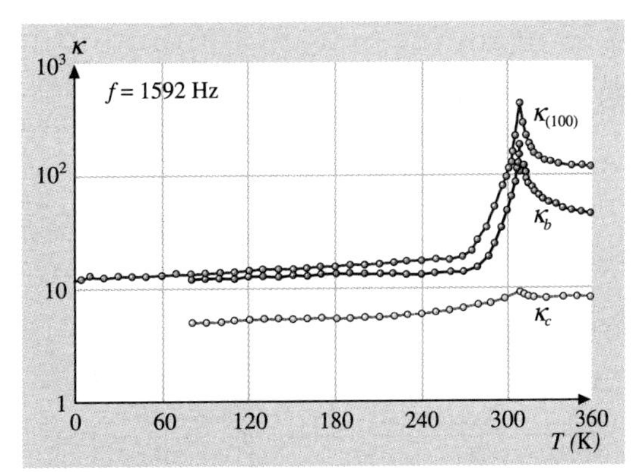

Fig. 4.5-66 $PbHPO_4$. $\kappa_{(100)}$, κ_b, and κ_c versus T. $\kappa_{(100)}$ is the dielectric constant perpendicular to the (100) plane

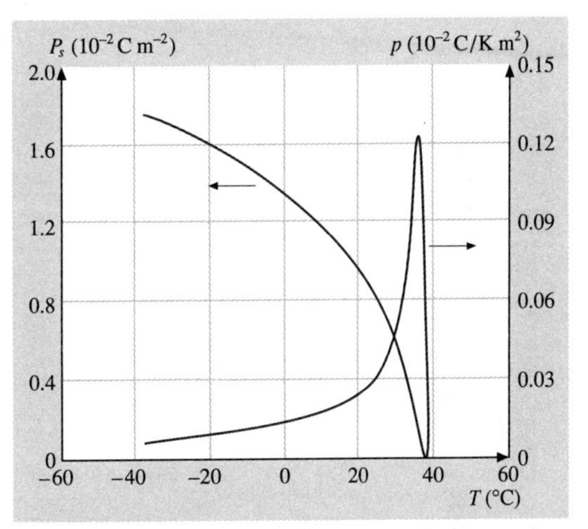

Fig. 4.5-67 $PbHPO_4$. P_s and p versus T, measured on (100) planar specimen. $p = -dP_s/dT$ is the pyroelectric coefficient

characteristic critical phenomena, suggesting that the spontaneous polarization results from an ordered arrangement of hydrogen atoms in a one-dimensional array of hydrogen bonds.

$(NH_4)_2SO_4$ Family

$(NH_4)_2SO_4$ (LB Number 39A-1). This crystal is ferroelectric below −49.5 °C. The dielectric constant is practically independent of temperature above the Curie point (Fig. 4.5-68). The spontaneous polarization changes its sign at about −190 °C (Fig. 4.5-69), suggesting a ferrielectric mechanism for the spontaneous polarization.

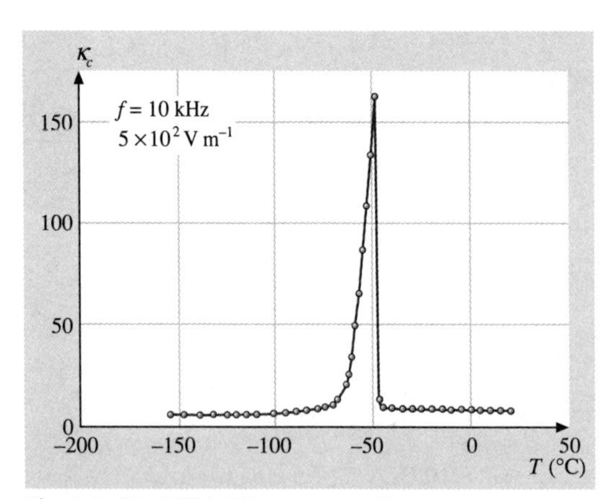

Fig. 4.5-68 $(NH_4)_2SO_4$. κ_c versus T

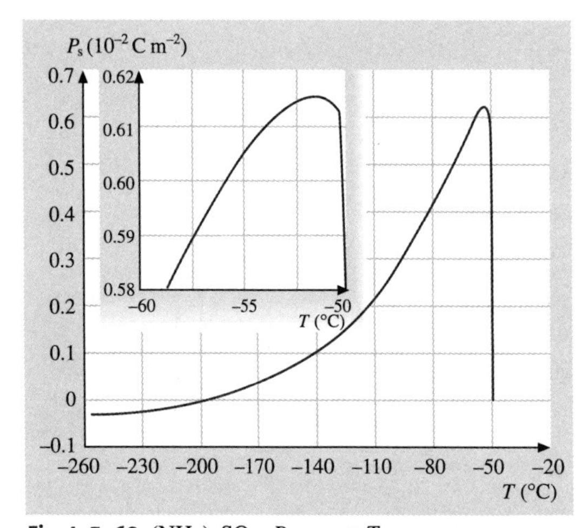

Fig. 4.5-69 $(NH_4)_2SO_4$. P_s versus T

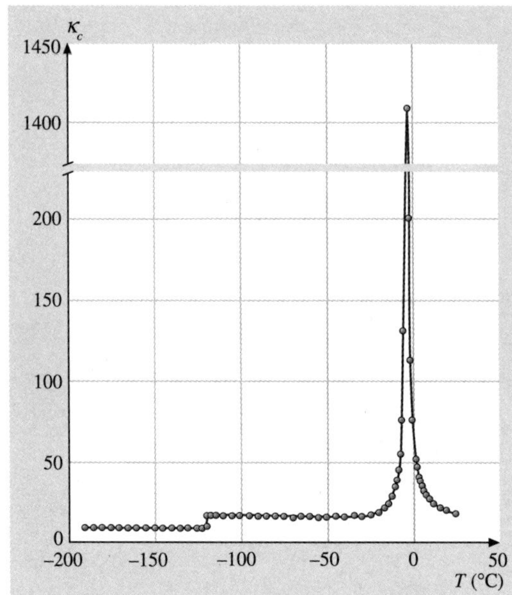

Fig. 4.5-70 NH_4HSO_4. κ_c versus T. $f = 10\,kHz$

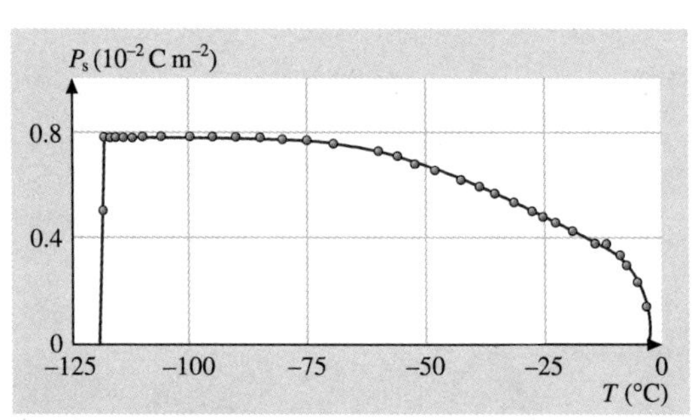

Fig. 4.5-71 NH_4HSO_4. P_s versus T

$(NH_4)HSO_4$ Family

$(NH_4)HSO_4$ (LB Number 40A-5). This crystal is ferroelectric in the temperature range between −119 and −3 °C (Figs. 4.5-70 and 4.5-71).

$(NH_4)LiSO_4$ Family

$(NH_4)LiSO_4$ (LB Number 41A-5). This crystal is ferroelectric in the temperature range between 10 and 186.5 °C (Figs. 4.5-72 and 4.5-73).

$(NH_4)_3H(SO_4)_2$ Family

$(NH_4)_3H(SO_4)_2$ (LB Number 42A-1). This crystal is ferroelectric in its phase VII below −211 °C. Another

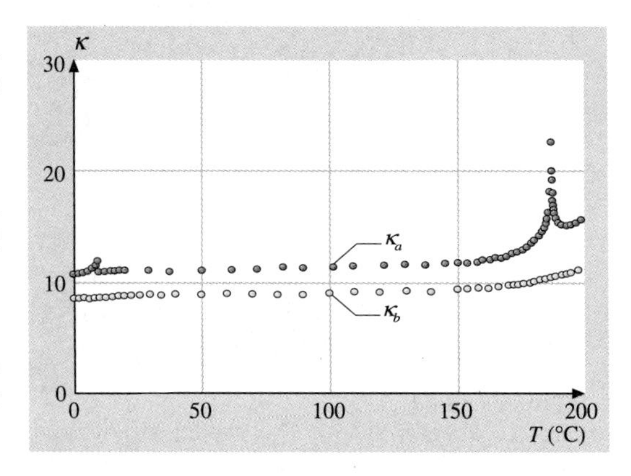

Fig. 4.5-72 NH_4LiSO_4. κ_a and κ_b versus T. $f = 3\,kHz$

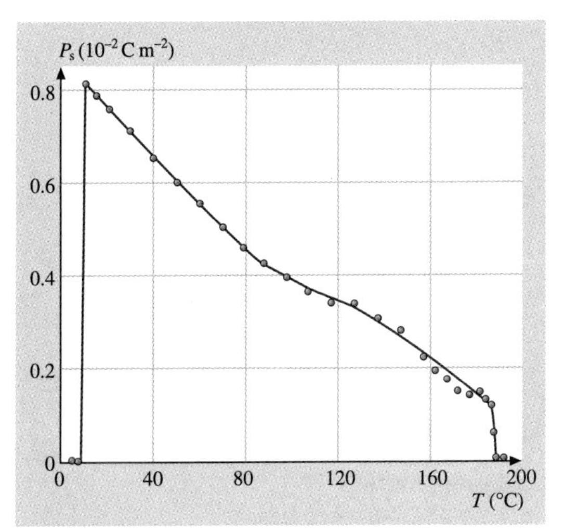

Fig. 4.5-73 NH_4LiSO_4. P_s versus T

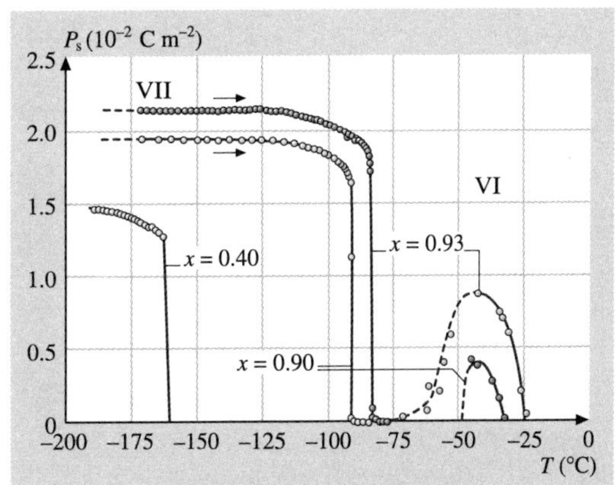

Fig. 4.5-75 $((NH_4)_3H)_{1-x}((ND_4)_3D)_x(SO_4)_2$. P_s versus T. Parameter: x. *Gray circles* (for Phase VI), determined by pyroelectric measurements. *Brown circles* (for phase VII), determined by hysteresis loop measurements

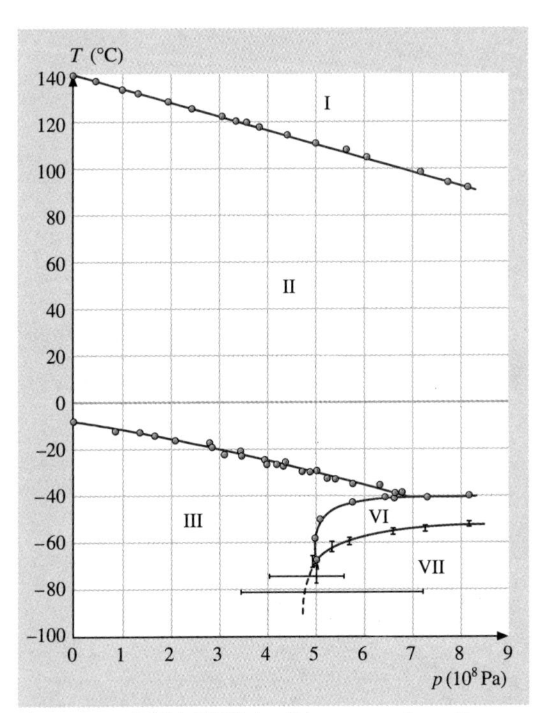

Fig. 4.5-74 $(NH_4)_3H(SO_4)_2$. T versus p phase diagram. p is the hydrostatic pressure. Phases VI and VII are ferroelectric

ferroelectric phase, VI, is induced by hydrostatic pressure (Fig. 4.5-74). When H is substituted by D, phase VI appears at atmospheric pressure and the temperature of the transition to phase VII becomes higher. Figure 4.5-75

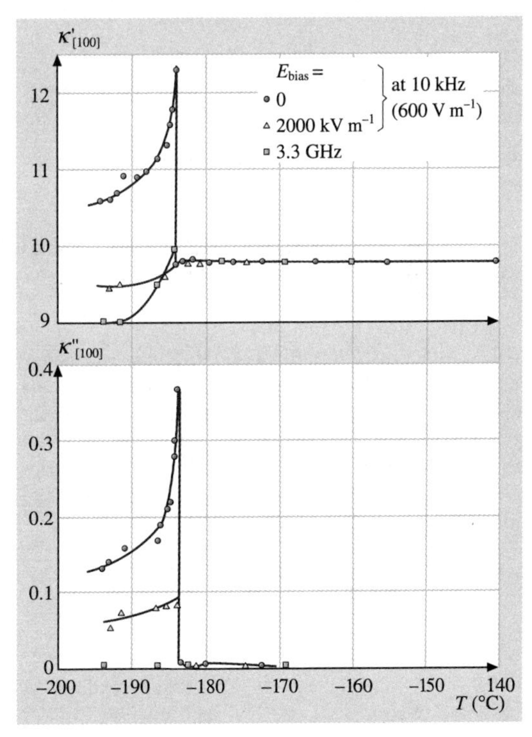

Fig. 4.5-76 $(NH_4)_2Cd_2(SO_4)_3$. $\kappa'_{[100]}$ and $\kappa''_{[100]}$ versus T

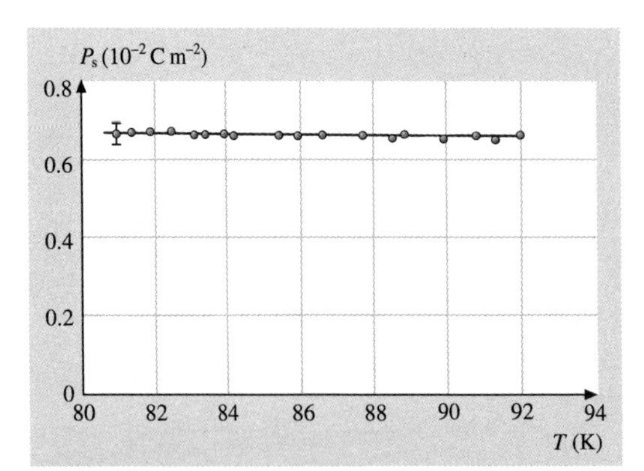

Fig. 4.5-77 $(NH_4)_2Cd_2(SO_4)_3$. P_s versus T

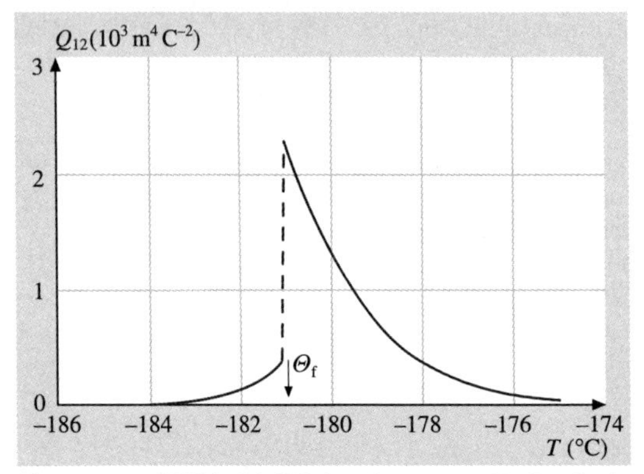

Fig. 4.5-78 $(NH_4)_2Cd_2(SO_4)_3$. Q_{12} versus T. Q_{12} is the electrostrictive constant

shows temperature dependence of P_s for three values of x in $((NH_4)_3H)_{1-x}((ND_4)_3D)_x(SO_4)_2$.

Langbeinite-Type Family

$(NH_4)_2Cd_2(SO_4)_3$ (LB Number 43A-13). This crystal is ferroelectric below about −184 °C. The dielectric constants are insensitive to temperature above the transition point (Fig. 4.5-76), and the spontaneous polarization does not depend upon temperature (Fig. 4.5-77). The electrostrictive constant Q_{12}, however, exhibits an anomaly at the transition point (Fig. 4.5-78).

4.5.4.3 Organic Crystals, Liquid Crystals, and Polymers [5.4]

$SC(NH_2)_2$ Family

$SC(NH_2)_2$ (LB Number 50A-1). This crystal exhibits at least five phases, I, II, III, IV, and V (Figs. 4.5-79

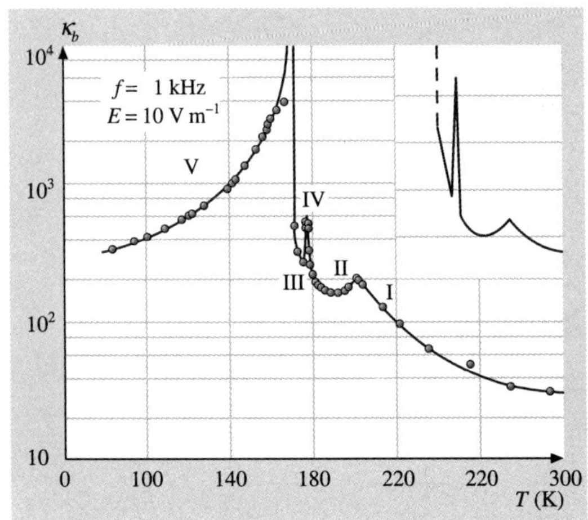

Fig. 4.5-79 $SC(NH_2)_2$. κ_b versus T

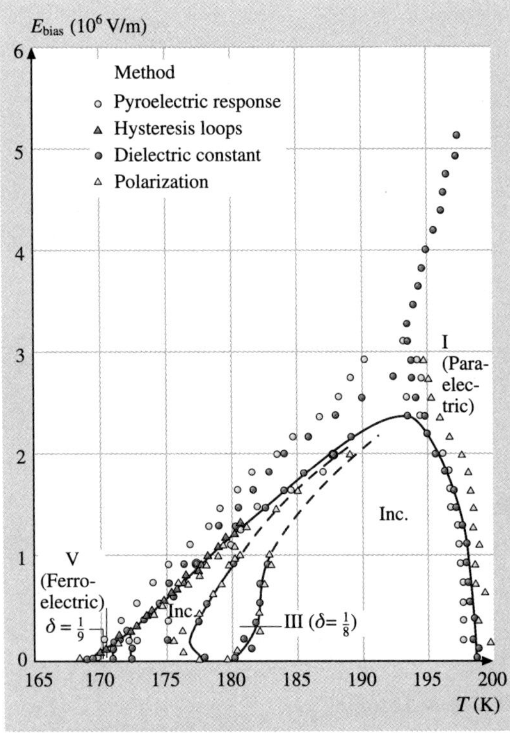

Fig. 4.5-80 $SC(NH_2)_2$. E_{bias} versus T phase diagram. The value of δ means that the phase is commensurately modulated with a vector of wavenumber δc^*. Inc., incommensurate phase

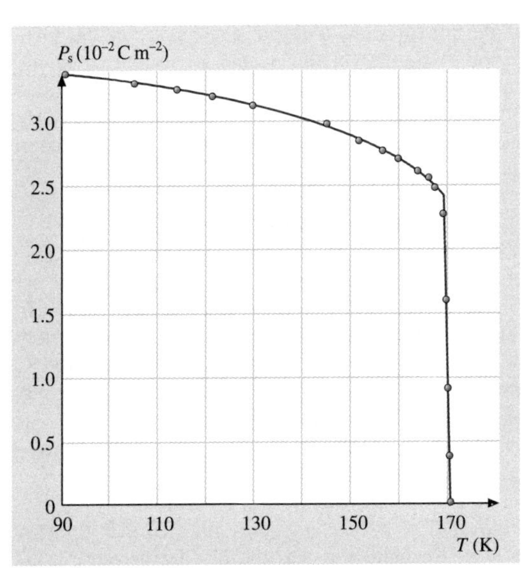

Fig. 4.5-81 $SC(NH_2)_2$. P_s versus T for phase V

and 4.5-80). The crystal is ferroelectric in phase V (Fig. 4.5-81), and slightly ferroelectric with a very small spontaneous polarization in phase III. The crystal structure is modulated commensurately or incommensurately except for phases I and V, as indicated in Fig. 4.5-80.

DSP ($Ca_2Sr(CH_3CH_2COO)_6$) Family

$Ca_2Sr(CH_3CH_2COO)_6$ (DSP) (LB Number 58A-1). This crystal is ferroelectric below about 4 °C (Fig. 4.5-82). The Curie–Weiss constant is small (60 K). Critical slowing-down (see Sect. 4.5.3) takes place (Fig. 4.5-8).

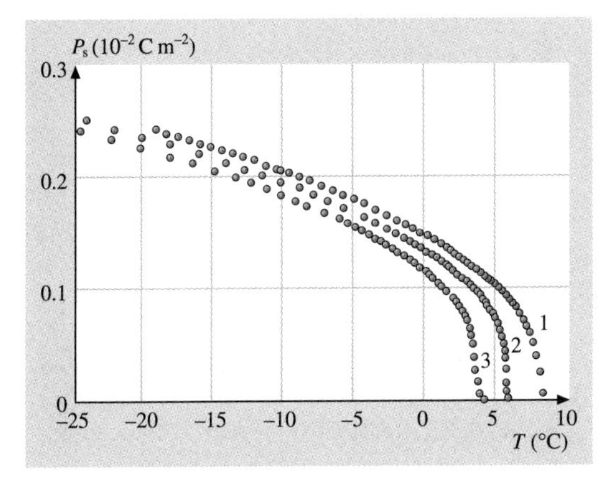

Fig. 4.5-82 $Ca_2Sr(CH_3CH_2COO)_6$. P_s versus T. The three curves show the effect of annealing: 1, unannealed; 2, annealed at 330 °C for 60 h; 3, annealed at 330 °C for 60 h and at 390 °C for 5 h

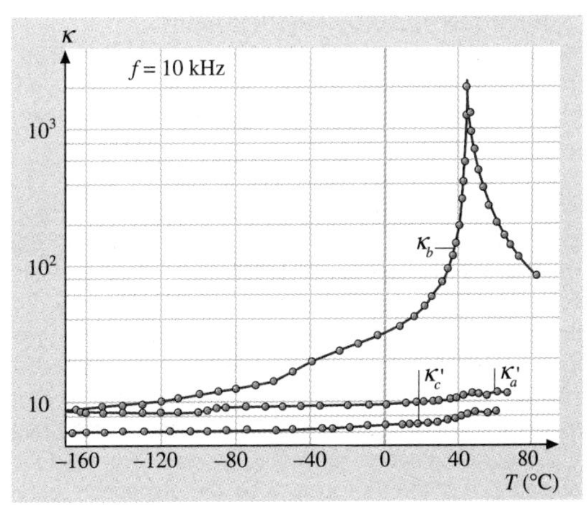

Fig. 4.5-83 $(NH_2CH_2COOH)_3 \cdot H_2SO_4$. $\kappa_{a'}$, $\kappa_{b'}$, and $\kappa_{c'}$ versus T. These quantities are referred to the unit cell vectors: $\boldsymbol{a}' = \boldsymbol{a} + \boldsymbol{c}$, $\boldsymbol{b}' = -\boldsymbol{b}$, and $\boldsymbol{c}' = -\boldsymbol{c}$

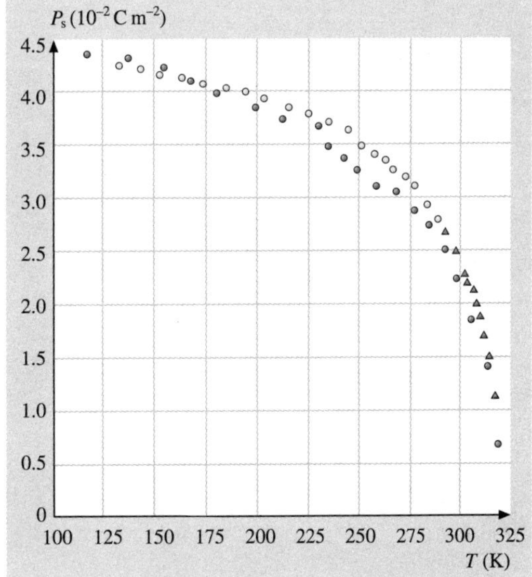

Fig. 4.5-84 $(NH_2CH_2COOH)_3 \cdot H_2SO_4$. P_s versus T. P_s is parallel to the b axis. *Gray circles*: values determined by pyroelectric measurements. *Triangles* and *brown circles*: determined from hysteresis loop by different authors

TGS $((NH_2CH_2COOH)_3 \cdot H_2SO_4)$ Family

$(NH_2CH_2COOH)_3 \cdot H_2SO_4$ (TGS) (LB Number 60A-1). This crystal is ferroelectric below 49.4 °C (Figs. 4.5-83 and 4.5-84). Critical slowing-down takes place (Fig. 4.5-9).

$(CH_3NHCH_2COOH)_3 \cdot CaCl_2$ Family

$(CH_3NHCH_2COOH)_3 \cdot CaCl_2$ (LB Number 64A-1). This crystal is ferroelectric below 127 K (Figs. 4.5-85 and 4.5-86). The Curie–Weiss constant is small (40 K). As discussed in Sect. 4.5.3, two components are expected in the dielectric constants of displacive-type ferroelectrics: one contribution from the soft optical phonon and one contribution from cluster boundary motion. This crystal provides a good example of this situation. Softening of the optical phonon mode B_{2u} occurs (Fig. 4.5-11), indicating that the transition is of the displacive type, while critical slowing-down takes place in the GHz region (Fig. 4.5-10); this seems to take place in the contribution from cluster boundary motion. The value of the dielectric constant at the Curie point at 1.0 GHz seems to contain a contribution from the optical phonon mode.

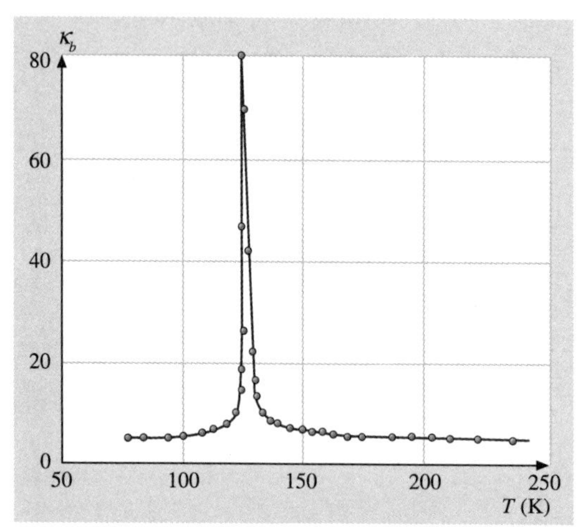

Fig. 4.5-85 $(CH_3NHCH_2COOH)_3 \cdot CaCl_2$. κ_b versus T. $f = 10$ kHz

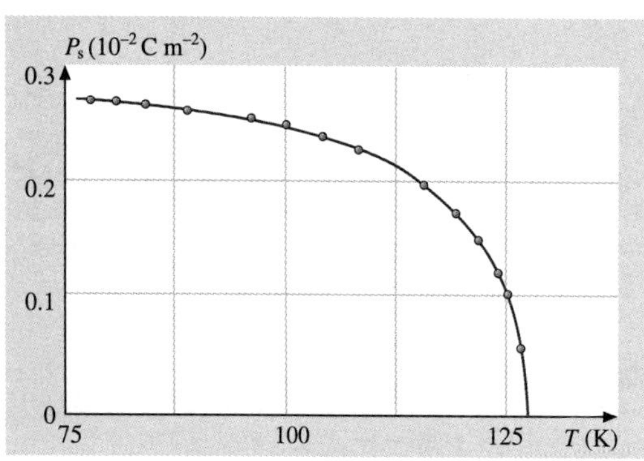

Fig. 4.5-86 $(CH_3NHCH_2COOH)_3 \cdot CaCl_2$. P_s versus T

$(CH_3)_3NCH_2COO \cdot CaCl_2 \cdot 2H_2O$ Family

$(CH_3)_3NCH_2COO \cdot CaCl_2 \cdot 2H_2O$ (LB Number 66A-1). This crystal is ferroelectric below 46 K. It exhibits at least ten phase transitions (eight can be recognized in Fig. 4.5-87), and the crystal structure is commensurately or incommensurately modulated between 46 and 164 K, with a feature known as a devil's staircase. Multiple hysteresis loops are observed in the modulated phases, as shown in Fig. 4.5-88.

Rochelle Salt $(NaKC_4H_4O_6 \cdot 4H_2O)$ Family

$NaKC_4H_4O_6 \cdot 4H_2O$ (Rochelle Salt, RS) (LB Number 67A-1). Rochelle salt was the first ferroelectric crystal to be discovered. It is ferroelectric between −18 and 24 °C (Figs. 4.5-89 and 4.5-90). It is rare that an ordered ferroelectric phase of lower symmetry ($P2_1$ in this case) appears in an intermediate temperature range between disordered phases of the same higher symmetry ($P2_12_12$). Figure 4.5-91 shows X-ray evidence for such space group changes. For theoretical studies, readers should refer to [5.7].

LAT $(LiNH_4C_4H_4O_6 \cdot H_2O)$ Family

$LiNH_4C_4H_4O_6 \cdot H_2O$ (LAT) (LB Number 68A-1). Ferroelectric activity appears along the b axis below 106 K (Fig. 4.5-92). The dielectric constant of the clamped

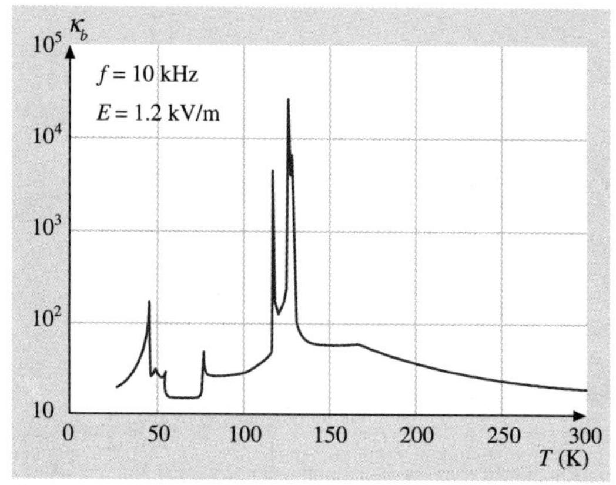

Fig. 4.5-87 $(CH_3)_3NCH_2COO \cdot CaCl_2 \cdot 2H_2O$. κ_b versus T

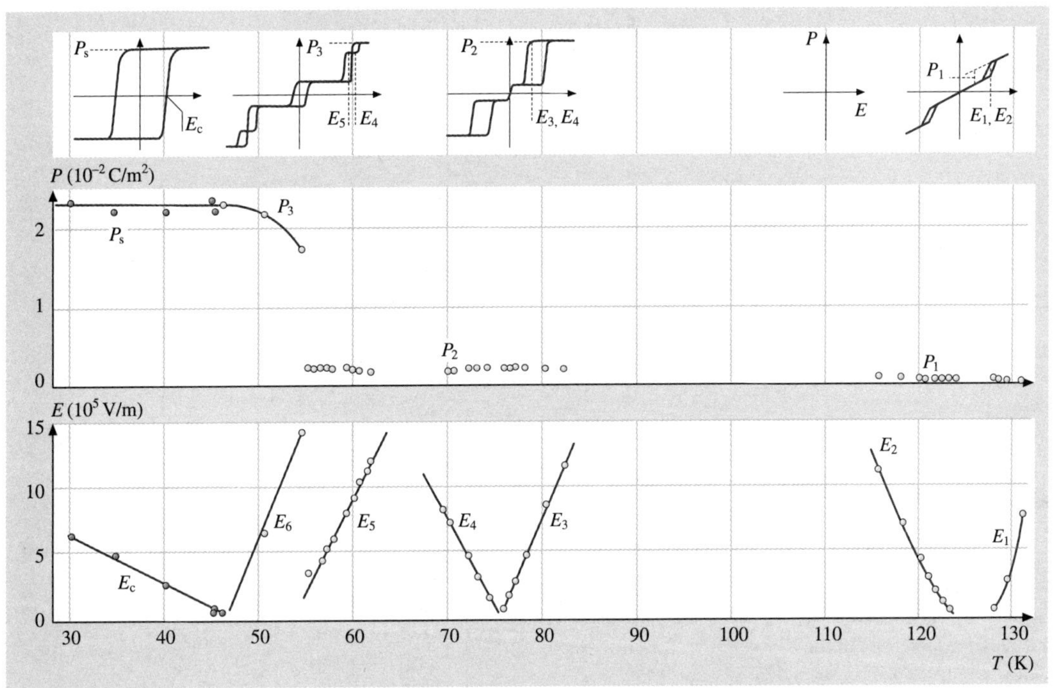

Fig. 4.5-88 $(CH_3)_3NCH_2COO \cdot CaCl_2 \cdot 2H_2O$. P_s, E_c, P_i, and E_i versus T. Definitions of P_i ($i = 1, 2, 3$) and E_i are shown in the *top figure*

crystal (κ_{22}^{S}) shows no anomaly at the transition temperature (Fig. 4.5-7), while the elastic compliances exhibit a pronounced anomaly (Fig. 4.5-93). These data indicate that the ferroelectric phase transition is of the indirect type III_{ac} triggered by an elastic anomaly (see Sect. 4.5.3). The crystal is piezoelectric in the paraelec-

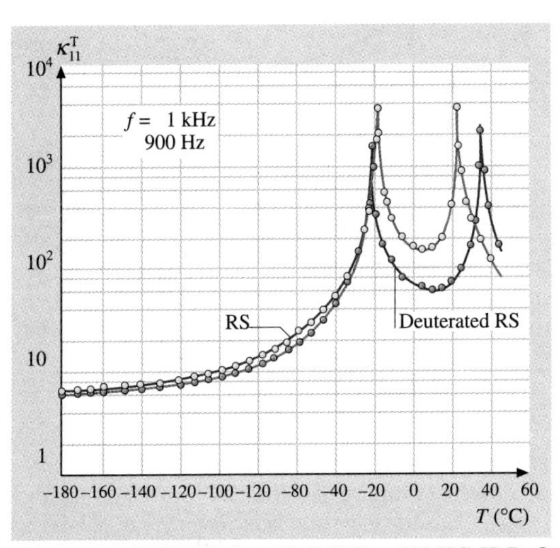

Fig. 4.5-89 $NaKC_4H_4O_6 \cdot 4H_2O$ (RS) and $NaKC_4H_2D_2O_6 \cdot 4D_2O$ (deuterated RS). κ_{11}^{T} versus T

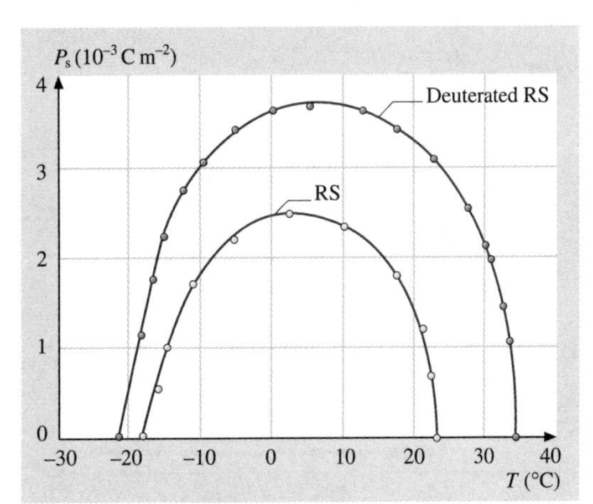

Fig. 4.5-90 $NaKC_4H_4O_6 \cdot 4H_2O$ (RS) and $NaKC_4H_2D_2O_6 \cdot 4D_2O$ (deuterated RS). P_s versus T

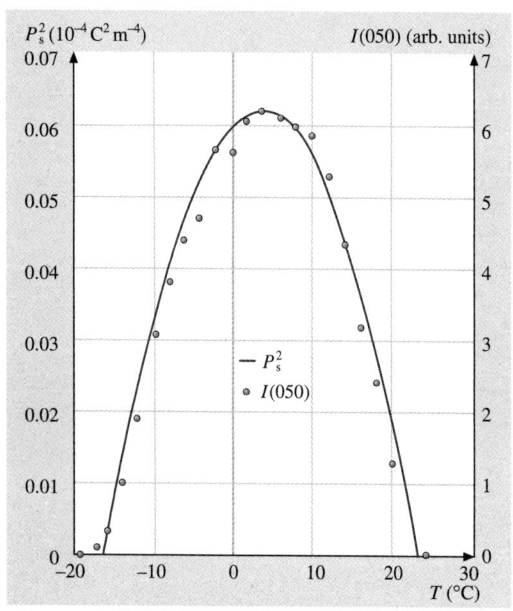

Fig. 4.5-91 $NaKC_4H_4O_6 \cdot 4H_2O$. $I_{(050)}$ versus T. $I_{(050)}$ is the integrated intensity of the X-ray (050) reflection, which should disappear in the space group $P2_12_12$ below −18 °C and above 23 °C. The *solid line* is P_s^2 normalized to fit to $I_{(050)}$

tric phase and the elastic anomaly causes a dielectric anomaly in the free crystal (κ_b in Fig. 4.5-92 and κ_{22}^{T} in Fig. 4.5-7), but no dielectric anomaly in the clamped crystal (κ_{22}^{S} in Fig. 4.5-7).

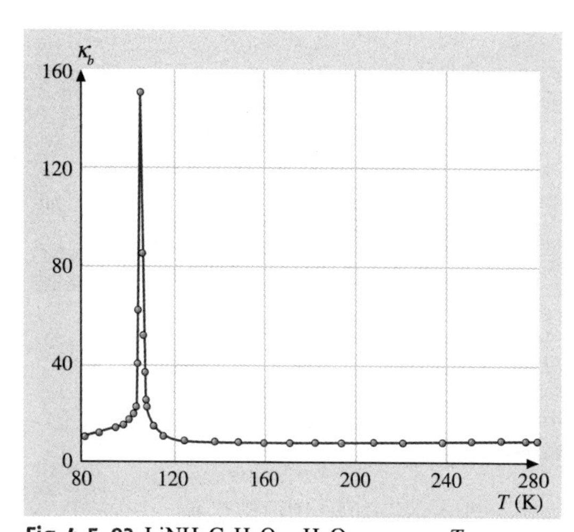

Fig. 4.5-92 $LiNH_4C_4H_4O_6 \cdot H_2O$. κ_b versus T

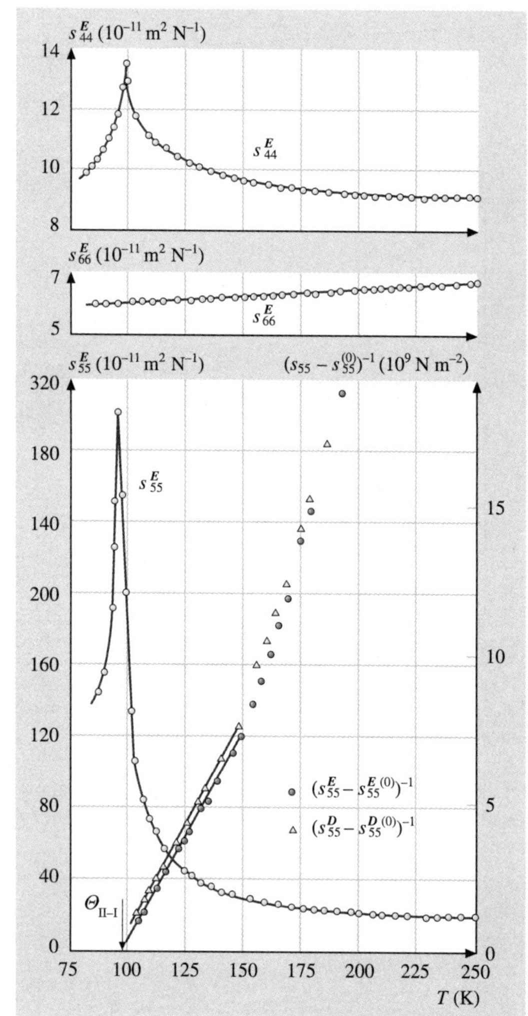

Fig. 4.5-93 $LiNH_4C_4H_4O_6 \cdot H_2O$. s_{44}^E, s_{55}^E, and s_{66}^E versus T. s_{44}^E, s_{55}^E, and s_{66}^E are the shear components of the elastic compliance tensor at constant electric field. The two *straight lines* show $[s_{55}^E - s_{55}^E(0)]^{-1}$ and $[s_{55}^D - s_{55}^D(0)]^{-1}$, where s_{55}^D is s_{55} at constant electric displacement, $s_{55}^E(0) = 18.0\,\mathrm{m^2\,N^{-1}}$, and $s_{55}^D(0) = 17.0\,\mathrm{m^2\,N^{-1}}$

Liquid Crystal Family

DOBAMBC, p-decyloxybenzylidene p′-amino 2-methylbutyl cinnamate (LB Number 71A-1 (A)) (Liquid Crystal). This liquid crystal is ferroelectric in the smectic C* phase between about 76 and 92 °C on heating. On cooling, the C* phase is transformed into a smectic H*

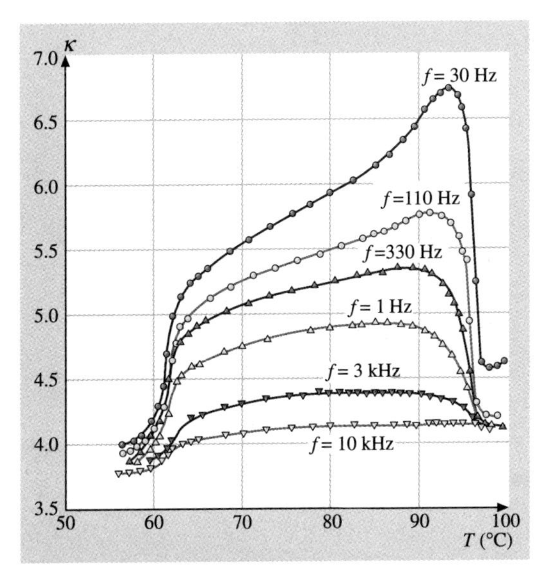

Fig. 4.5-94 DOBAMBC. κ versus T. Parameter: f. Sample cell thickness = 2.3 μm

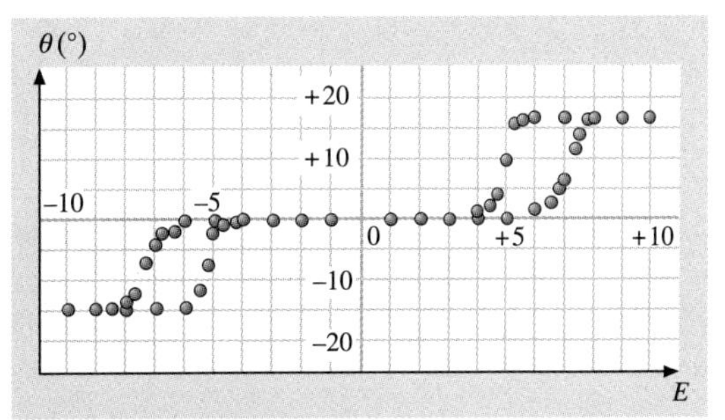

Fig. 4.5-97 MHPOBC. θ versus E. θ is the apparent tilt angle. Sample cell thickness = 3 μm

phase at about 63 °C, retaining the ferroelectric activity (Figs. 4.5-94 and 4.5-95).

MHPOBC, 4-(1-methylheptyl oxycarbonyl)phenyl 4′-octyloxybiphenyl-4-carboxylate (LB Number 71B-1 (A)) (Liquid Crystal). Seven phases are known for this liquid crystal. It exhibits an antiferroelectric hysteresis loop at low frequency, as shown in Fig. 4.5-96a, between

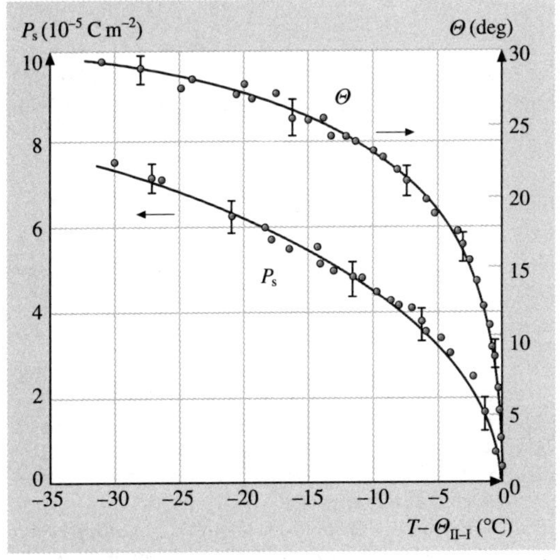

Fig. 4.5-95 DOBAMBC. P_s and θ versus $T - \Theta_{II-I}$. θ is the apparent tilt angle, and Θ_{II-I} is the ferroelectric transition temperature

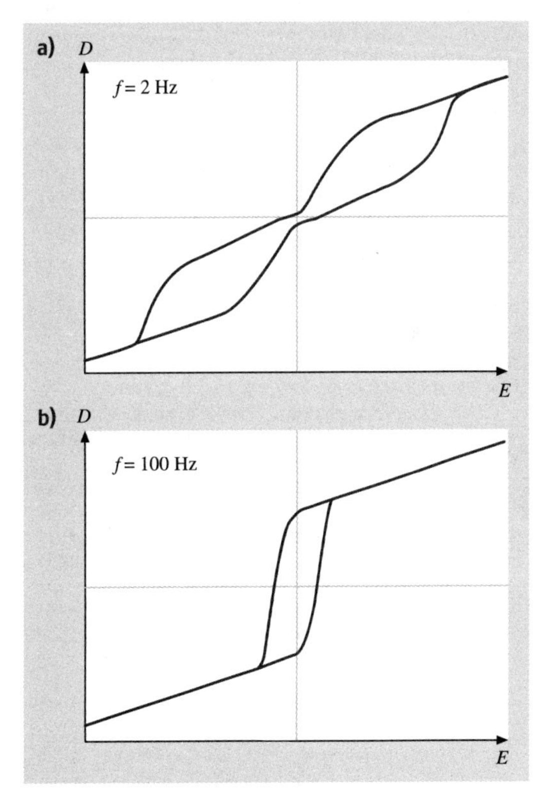

Fig. 4.5-96a,b MHPOBC. $D-E$ hysteresis loops obtained from switching-current measurements. D is the electric displacement. Sample cell thickness = 3 μm. (**a**) Double hysteresis loop at 2 Hz. (**b**) Single hysteresis loop at 100 Hz

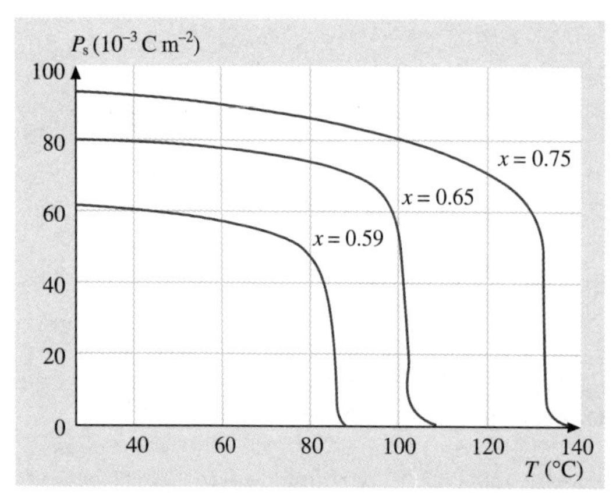

Fig. 4.5-98 $((CH_2CF_2)_x(CF_2CHF)_{1-x})_n$. P_s versus T. Parameter: x. Thickness of specimen film $= 50\,\mu m$

65 and 119.7 °C. The antiferroelectric behavior can be best demonstrated by measuring the apparent tilt angle, as shown in Fig. 4.5-97. The hysteresis loop turns into a triple one between 119.7 and 120.6 °C, and into a normal ferroelectric hysteresis loop between 120.6 and 122.1 °C. The phase exhibiting the triple hysteresis loop is called "ferrielectric" by several authors.

Polymer Family

$((CH_2CF_2)_x(CF_2CHF)_{1-x})_n$, Vinylidene fluoride–trifluoroethylene Copolymer, $((VDF)_x(TrFE)_{1-x})_n$ (LB Number 72-2) (Polymer). Random polymers $(CH_2CF_2)_x$ $(CF_2CHF)_{1-x}$ exhibit a ferroelectric hysteresis loop for $x > 0.5$ (Figs. 4.5-98 and 4.5-99). Ferroelectric activity is preserved up to the melting point for $x > 0.8$.

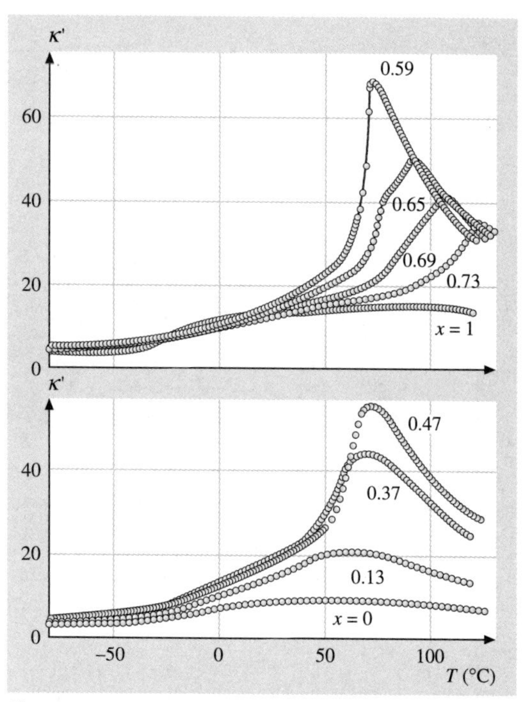

Fig. 4.5-99 $((CH_2CF_2)_x(CF_2CHF)_{1-x})_n$. κ' versus T. Parameter: x. $f = 1$ kHz. Thickness of specimen film $= 50\,\mu m$

References

5.1 Y. Shiozaki, E. Nakamura, T. Mitsui (Eds.): *Ferroelectrics and Related Substances: Oxides: Perovskite-Type Oxides and* LiNbO$_3$ *Family*, Landolt–Börnstein, New Series III/36 (Springer, Berlin, Heidelberg 2001)

5.2 Y. Shiozaki, E. Nakamura, T. Mitsui (Eds.): *Ferroelectrics and Related Substances: Oxides: Oxides Other Than Perovskite-Type Oxides and* LiNbO$_3$ *Family*, Landolt–Börnstein, New Series III/36 (Springer, Berlin, Heidelberg 2002)

5.3 Y. Shiozaki, E. Nakamura, T. Mitsui (Eds.): *Ferroelectrics and Related Substances: Inorganic Crystals Other Than Oxides*, Landolt–Börnstein, New Series III/36 (Springer, Berlin, Heidelberg 2004)

5.4 Y. Shiozaki, E. Nakamura, T. Mitsui (Eds.): *Ferroelectrics and Related Substances: Organic Crystals, Liquid Crystals and Polymers*, Landolt–Börnstein, New Series III/36 (Springer, Berlin, Heidelberg 2005) in preparation

5.5 J. Valasek: Piezoelectric and Allied Phenomena in Rochelle Salt, Phys. Rev. **15**, 537 (1920)

5.6 J. Valasek: Piezo-Electric and Allied Phenomena in Rochelle Salt, Phys. Rev. **17**, 475 (1921)

5.7 T. Mitsui, E. Nakamura, I. Tatsuzaki: *An Introduction to the Physics of Ferroelectrics* (Gordon and Breach, New York 1976)

5.8 F. Jona, G. Shirane: *Ferroelectric Crystals* (Dover Publications, New York 1993)

5.9 B. A. Strukov, A. P. Levanyuk: *Ferroelectric Phenomena in Crystals: Physical Foundations* (Springer, Berlin, Heidelberg 1998)

5.10 M. E. Lines, A. M. Glass: *Principles and Applications of Ferroelectrics and Related Materials* (Clarendon, Oxford 1979)

5.11 Y. Xu: *Ferroelectric Materials and Their Applications* (North-Holland, Amsterdam 1991)
5.12 B. Jaffe, W. R. Cook Jr., H. Jaffe: *Piezoelectric Ceramics* (Academic Press, London 1971)
5.13 J. Zelenka: *Piezoelectric Resonators and Their Applications* (Elsevier, Amsterdam 1986)
5.14 T. Ikeda: *Fundamentals of Piezoelectricity* (Oxford Univ. Press, Oxford 1990)
5.15 R. A. Cowley: *Structural Phase Transitions* (Consultant Bureau, New York 1980)
5.16 K. A. Miller, H. Thomas (Eds.): *Structural Phase Transitions I*, Topics in Current Physics, Vol. 23 (Springer, Berlin, Heidelberg 1981)
5.17 M. Fujimoto: *The Physics of Structural Phase Transitions* (Springer, Berlin, Heidelberg 1997)
5.18 R. Blinc, A. P. Levanyuk (Eds.): *Incommensurate Phases in Dielectrics*, Modern Problems in Condensed Matter Science 14 (Elsevier, Amsterdam 1986)
5.19 F. J. Scott: Soft-mode spectroscopy: Experimental studies of structural phase transitions, Rev. Mod. Phys. **46**, 83 (1974)
5.20 G. Shirane: Neutron scattering studies of structural phase transitions, Rev. Mod. Phys. **46**, 437 (1974)
5.21 C. H. Wang: Raman and Brillouin scattering spectroscopy of phase transitions in solids. In: *Vibrational Spectroscopy of Phase Transitions*, ed. by Z. Iqbal, F. J. Owens (Academic Press, Orlando 1984) pp. 153–207
5.22 J. M. Herbert: *Ceramic Dielectrics and Capacitors*, Electrocomp. Sci. Mon., Vol. 6 (Gordon and Breach, New York 1985)
5.23 B. E. Vugmeister, M. D. Glinchuk: Dipole glass and ferroelectricity in random-site electric dipole systems, Mod. Phys. **62**, 993 (1990)
5.24 L. E. Cross: Relaxor ferroelectrics, Ferroelectrics **76**, 241 (1987)
5.25 S. Singh: Nonlinear optical materials. In: *Handbook of Lasers with Selected Data on Optical Technology*, ed. by R. J. Pressley (The Chemical Rubber Co., Cleveland 1971) pp. 489–525
5.26 V. G. Dmitriev, G. G. Gurzadyan, D. N. Nikogosyan: *Handbook of Nonlinear Optical Crystals*, Springer Series in Optical Science, Vol. 64, ed. by A. E. Siegman (Springer, Berlin, Heidelberg 1991)
5.27 P. Yeh: *Introduction to Photorefractive Nonlinear Optics* (Wiley, New York 1993)
5.28 G. Rosenman, A. Skliar, A. Arie: Ferroelectric domain engineering for quasi-phase-matched nonlinear optical devices, Ferroelectr. Rev. **1**, 263 (1999)
5.29 N. Setter, E. L. Colla: *Ferroelectric Ceramics* (Birkhäuser, Basel 1993)
5.30 B. A. Tuttle, S. B. Desu, R. Ramesh, T. Shiosaki: *Ferroelectric Thin Films IV* (Materials Research Society, Pittsburgh 1995)
5.31 C. P. Araujo, J. F. Scott, G. W. Taylor (Eds.): *Ferroelectric Thin Films: Synthesis and Basic Properties* (Gordon and Breach, New York 1996)
5.32 R. Ramesh (Ed.): *Thin Film Ferroelectric Materials and Devices*, Electronic Materials Science and Technology, Vol. 3 (Kluwer, Dordrecht 1997)
5.33 J. F. Scott: The physics of ferroelectric ceramic thin films for memory applications, Ferroelectr. Rev. **1**, 1 (1998)
5.34 D. Damjanovic: Ferroelectric, dielectric and piezoelectric properties of ferroelectric thin films and ceramics, Rep. Prog. Phys. **62**, 1267 (1998)
5.35 J. F. Scott: *Ferroelectric Memories*, Springer Series in Advanced Microelectronics, Vol. 3 (Springer, Berlin, Heidelberg 2000)
5.36 R. M. White: Surface elastic waves, Proc. IEEE **58**, 1238 (1970)
5.37 G. W. Farnell, E. L. Adler: Acoustic Wave Propagation in Thin Layers. In: *Physical Acoustics*, Vol. 9, ed. by W. P. Mason, R. N. Thurston (Academic Press, New York 1972) pp. 35–127
5.38 A. A. Oliner (Ed.): *Acoustic Surface Waves* (Springer, Berlin, Heidelberg 1978)
5.39 K.-Y. Hashimoto: *Surface Acoustic Wave Devices in Telecommunications, Modeling and Simulation* (Springer, Berlin, Heidelberg 2000)
5.40 J. M. Herbert: *Ferroelectric Transducers and Sensors*, Electrocomp. Sci. Mon., Vol. 3 (Gordon and Breach, New York 1982)
5.41 W. H. de Jeu: *Physical Properties of Liquid Crystalline Materials* (Gordon and Breach, New York 1980)
5.42 J. W. Goodby, R. Blinc, N. A. Clark, S. T. Lagerwall, M. A. Osipov, S. A. Pikin, T. Sakurai, K. Yoshino, B. Zeks: *Ferroelectric Liquid Crystals: Principles, Properties and Applications*, Ferroelectrics and Related Phenomena, Vol. 7 (Gordon and Breach, New York 1991)
5.43 G. W. Taylor (Ed.): *Ferroelectric Liquid Crystals: Principles, Preparations and Applications* (Gordon and Breach, New York 1991)
5.44 L. M. Blinov, V. G. Chigrinov: *Electric Effects in Liquid Crystal Materials* (Springer, Berlin, Heidelberg 1994)
5.45 A. Fukada, Y. Takanishi, T. Isozaki, K. Ishikawa, H. Takezoe: Antiferroelectric Chiral Smectic Liquid Crystals, J. Mater. Chem. **4**, 997 (1994)
5.46 P. J. Collings, J. S. Patel (Eds.): *Handbook of Liquid Crystal Research* (Oxford Univ. Press, Oxford 1997)
5.47 S. T. Lagerwall: *Ferroelectric and Antiferroelectric Liquid Crystals* (Wiley VCH, Weinheim 1999)
5.48 T. Furukawa: Ferroelectric Properties of Vinylidene Fluoride Copolymers, Phase Transitions **18**, 143 (1989)
5.49 D. K. Das-Gupta (Ed.): *Ferroelectric Polymer and Ceramic-Polymer Composites* (Trans Tech Publications, Aedermannsdorf 1994)

5.50 H. S. Nalva (Ed.): *Ferroelectric Polymers: Chemistry, Physics and Applications* (Marcel Dekker, New York 1995)

5.51 H. Kodama, Y. Takahashi, T. Furukawa: Effects of Annealing on the Structure and Switching Characteristics of VDF/TrFE Copolymers, Ferroelectrics **205**, 433 (1997)

5.52 N. A. Clark, S. T. Lagerwall: Submicrosecond Bistable Electro-Optic Switching in Liquid Crystals, Appl. Phys. Lett. **36**, 899 (1980)

5.53 A. D. L. Chandani, E. Gorecka, Y. Ouchi, H. Takezoe, A. Fukuda: Antiferroelectric Chiral Smectic Phases Responsible for the Tristable Switching in MHPOBC, Jpn. J. Appl. Phys. **28**, L1265 (1989)

Subject Index

π-bonded chain geometry 993
π-bonded chain model
– diamond(111)2×1 1005
$\alpha = 0$ glass-ceramics 558
$(CH_3NHCH_2COOH)_3 \cdot CaCl_2$ family 932
5CB
– liquid crystals 948
8CB
– liquid crystals 948
8OCB
– liquid crystals 949

A

Abbe value
– glasses 543, 548
Abrikosov vortices 717
absorption and fluorescence spectra of CdSe 1039
absorption coefficient
– two-photon 826
absorption spectra of spherical particles 1046
Ac actinium 84
acceptor surface level 1023
acceptor surface state 1022
accumulation layer 1020
accuracy 4
acids
– liquid crystals 946
acoustic band 1012
acoustic surface wave 906
acronyms
– solid surface 1026
actinium Ac
– elements 84
adatom 995
adopted numerical values for selected quantities 23
Ag silver 65
Ag-based materials 344
age hardening 198
AISI (American Iron and Steel Institute) 221
Al aluminium 78
Al bronzes 298
alkali aluminium silicates
– electrical properties 434
– mechanical properties 434
– thermal properties 434
alkali halides
– surface phonon energy 1017
alkali–alkaline-earth silicate glasses 530
alkali–lead silicate glasses 530
alkaline-earth aluminium silicates
– electrical properties 436
– glasses 530
– mechanical properties 435
– thermal properties 436
allotropic and high-pressure modifications
– elements 46
alloy
– cast irons 270
– cobalt 272
– elinvar 780
– invar 780
– lead, battery grid 413
– lead–antimony 412
– lead–tin 415
– magnesium 163
– Ti_3Al-based 210
– TiAl-based 213
– titanium 206
– wear resistant 274
alloy systems 296
Allred 46
Alnico 798
Al–O–N ceramics
– dielectric properties 447
– optical properties 447
alum family 911
– ferroelectrics 911
alumina
– electrical properties 446
– mechanical properties 445
– properties 445
– thermal properties 446
aluminium Al
– aluminium alloys 171
– aluminium production 171
– chemical properties 172
– cold working 195
– corrosion behavior 204
– elements 78
– hot working 195
– mechanical properties 172
– mechanical treatment 195
– surface layers 204
– work hardening 195
aluminium alloy
– abrasion resistance 190
– aging 198
– behavior in magnetic fields 194
– binary Al-based systems 174
– classification of aluminium alloys 179
– coefficient of thermal expansion 192
– creep behavior 187
– elastic properties 194
– electrical conductivity 194
– hardness 186
– homogenization 198
– machinability 192
– mechanical properties 180, 182
– nuclear properties 194
– optical properties 194
– physical properties 192, 193
– sheet formability 190
– soft annealing 197
– specific heat 194
– stabilization 197
– stress-relieving 198
– structure 182
– technical property 186
– technological properties 190
– tensile strength 186
– thermal softening 195
– work-hardenable 180
aluminium antimonide
– crystal structure, mechanical and thermal properties 610
– electromagnetic and optical properties 619
– electronic properties 616
– transport properties 618
aluminium arsenide
– crystal structure, mechanical and thermal properties 610
– electromagnetic and optical properties 619
– electronic properties 616
– transport properties 618
aluminium casting alloys
– mechanical properties 184
– structure 184
aluminium compounds
– crystal structure, mechanical and thermal properties 610

– electromagnetic and optical properties 619
– electronic properties 614
– mechanical properties 610
– phonon dispersion curves 612
– thermal conductivity 618
– thermal properties 610
– transport properties 617
aluminium nitride
– crystal structure, mechanical and thermal properties 610
– electromagnetic and optical properties 619
– electronic properties 616
– transport properties 618
aluminium phase diagram
– aluminium alloy phase diagram 174
aluminium phosphide
– crystal structure, mechanical and thermal properties 610
– electromagnetic and optical properties 619
– electronic properties 616
– transport properties 618
aluminothermy 174
Al–Ni phase diagram 285
Am americium 151
americium Am
– elements 151
amorphous alloys
– cobalt–based 774
– iron–based 773
– nickel–based 774
amorphous materials 27, 39
amorphous metallic alloys 772
amount of substance
– definition 14
ampere
– SI base unit 14
amphiphilic compound 941
amphiphilic liquid crystal 942
anisotropic magnetoresistance 1058
annealing coefficient
– glasses 548
annealing of steel 223
antiferroelectric crystal 903
antiferroelectric hysteresis loop 904
antiferroelectric liquid crystal 911
antiferroelectrics
– definition 903
– dielectric properties 903
– elastic properties 903
– pyroelectric properties 903
antimony Sb
– elements 98
aperiodic crystals 27
aperiodic materials 33
apparent tilt angle 935
Ar argon 128
area of surface primitive cell
– crystallographic formulas 986
argon Ar
– elements 128
arsenic As
– elements 98
ARUPS 997
As arsenic 98
astatine At
– elements 118
ASTM 241, 242
ASTM (American Society for Testing and Materials) 330
ASW 906
At astatine 118
atomic moment 755
atomic number *Z*
– elements 45
atomic radius
– elements 46
atomic scattering 1019
atomic, ionic, and molecular properties
– elements 46
atomically clean crystalline surface 979
atom–surface potential
– surface phonons 1019–1020
Au gold 65
austenitizing 224

B

B boron 78
Ba barium 68
back-bond state 1006
bainite 223
$BaMnF_4$ family 922
band bending
– solid surfaces 1023
band gap *see* energy gap 592
band pass filters
– glasses 566
band structure
– aluminium compounds 614
– beryllium compounds 653
– boron compounds 606
– cadmium compounds 679
– group IV semiconductors and IV–IV compounds 589–592
– indium compounds 643
– magnesium compounds 657
– mercury compounds 688
– oxides of Ca, Sr, and Ba 662
– zinc compounds 668
barium Ba
– elements 68
barium oxide
– crystal structure, mechanical and thermal properties 660
– electromagnetic and optical properties 664
– electronic properties 661
– transport properties 663
barium titanate 915
base quantities 12
– ISO 13
base unit
– SI 13
basis
– crystal structure 28
$BaTiO_3$ 915
bcc positions
– surface diagrams 982
Be beryllium 68
becquerel
– SI unit of activity 19
benzene 946
berkelium Bk
– elements 151
beryllium Be
– elements 68
beryllium compounds 652
– crystal structure, mechanical and thermal properties 652
– electromagnetic and optical properties 655
– electronic properties 653
– mechanical and thermal properties 652
– optical properties 655
– thermal properties 652
– transport properties 655
beryllium oxide 447
– crystal structure, mechanical and thermal properties 652
– electrical properties 448
– electronic properties 653
– mechanical properties 447
– thermal properties 448
beryllium selenide
– crystal structure, mechanical and thermal properties 652
– electronic properties 653
beryllium sulfide
– crystal structure, mechanical and thermal properties 652

Subject Index

– electronic properties 653
beryllium telluride
– crystal structure, mechanical and thermal properties 652
– electronic properties 653
Bethe–Slater–Pauling relation 755, 756
Bh Bohrium 124
Bi bismuth 98
biaxial crystals 826
binding energy 998
– metal 999
Bioverit
– glasses 559
BIPM (Bureau International des Poids et Mesures) 3, 11, 12
bismuth Bi
– elements 98
Bi – Sr – Ca – Cu – O (BSCCO) 736
Bi–Sr–Ca–Cu–O
– coherence lengths 736
– London penetration depths 736
– maximum T_c 736
– structural data 736
– superconducting properties 741
– upper critical fields 736
BK 7
– glasses 537
Bk berkelium 151
blue phase 942
BNN
– ferroelectric material 920
Bohrium Bh
– elements 124
boiling temperature
– elements 47
Bondi 46
boracite-type family 911, 921
borides
– physical properties 452
Borofloat
– glasses 528, 529
boron antimonide
– crystal structure, mechanical, and thermal properties 604
– electromagnetic and optical properties 610
– electronic properties 607
– transport properties 608
boron arsenide
– crystal structure, mechanical, and thermal properties 604
– electromagnetic and optical properties 610
– electronic properties 606
– transport properties 608
boron B
– elements 78
boron compounds
– crystal structure, mechanical and thermal properties 604
– electromagnetic and properties 610
– electronic properties 606
– mechanical properties 604
– thermal properties 604
– transport properties 608
boron nitride
– crystal structure, mechanical, and thermal properties 604
– electromagnetic and optical properties 610
– electronic properties 606
– transport properties 608
boron phosphide
– crystal structure, mechanical, and thermal properties 604
– electromagnetic and optical properties 610
– electronic properties 606
– transport properties 608
borosilicate glasses 529, 530
Br bromine 118
Bragg equation 40
brasses 298
Bravais cell 979
– 2D lattice 980
Bravais lattice 32
– elements 47
breathing-mode acoustic oscillations 1040
Brillouin scattering 906
Brillouin zone 913
– aluminium compounds 614
– beryllium compounds 653
– boron compounds 606
– cadmium compounds 678
– gallium compounds 626
– group IV semiconductors and IV–IV compounds 589
– indium compounds 643
– magnesium compounds 657
– mercury compounds 688
– oxides of Ca, Sr, and Ba 661
– zinc compounds 668
Brillouin zone corner 921
bromine Br
– elements 118
bronzes 298
BSCCO
– films 738
– single crystal 739
– tapes 739
– wires 739
buckled dimer 991
bulk electron density 998
bulk glassy alloys 217, 218
bulk mobility 1026
bulk modulus
– elements 47

C

C carbon 88
Ca calcium 68
cadmium Cd
– elements 73
cadmium compounds
– crystal structure, mechanical and thermal properties 676
– electromagnetic and optical properties 683
– electronic properties 678
– mechanical and thermal properties 676
– optical properties 683
– thermal properties 676
– transport properties 682
cadmium oxide
– crystal structure, mechanical and thermal properties 676
– electromagnetic and optical properties 683
– electronic properties 678
– transport properties 682
cadmium selenide
– crystal structure, mechanical and thermal properties 676
– electromagnetic and optical properties 683
– electronic properties 678
– transport properties 682
cadmium sulfide
– crystal structure, mechanical and thermal properties 676
– electromagnetic and optical properties 683
– electronic properties 678
– transport properties 682
cadmium telluride
– crystal structure, mechanical and thermal properties 676
– electromagnetic and optical properties 683
– electronic properties 678
– transport properties 682
calamitic liquid crystal 941, 942

Subject Index

calcium Ca
– elements 68
calcium oxide
– crystal structure, mechanical and thermal properties 660
– electromagnetic and optical properties 664
– electronic properties 661
– transport properties 663
californium Cf
– elements 151
candela
– SI base unit 15
capacitor 903
capillary viscometer 943
carat 23
carbide
– cemented 277
– electrical properties 466
– mechanical properties 466
– physical properties 458
– thermal properties 466
carbon C
– elements 88
carbon equivalent (CE) 268
carbon fibers
– physical properties 477
carbon steels 230
carrier concentration n_i
– gallium compounds 631
– indium compounds 647
cast
– classification 268
cast iron 268
– grades 268
– mechanical properties 270
casting technology 170
catalysis 1020
Cd cadmium 73
Ce cerium 142
cell surface 943
cellulose 481, 509
– cellulose acetate (CA) 509, 510
– cellulose acetobutyrate (CAB) 509, 510
– cellulose propionate (CP) 509, 510
– ethylcellulose (EC) 509, 510
– polymers 509
– vulcanized fiber (VF) 509, 510
cement 432
cemented carbides 277
centering types
– crystal structure 28
centrosymmetric media 1007
Cerabone
– glasses 559
ceramic capacitor 906
ceramic thin film 914
– ferroelectrics 903
ceramics 345, 431
– Al–O–N 447
– applications 432
– non-oxide 451
– oxide 437
– properties 432
– refractory 437
– silicon 433
– technical 437
– traditional 432
Ceran®
– glasses 559
– linear thermal expansion 558
Ceravital
– glasses 559
cerium Ce
– elements 142
cesium Cs
– elements 59
Cf californium 151
CGPM (Conférence Générale des Poids et Mesures) 11, 12
CGS
– electromagnetic system 21
– electrostatic system 21
– Gaussian system 21
cgs definitions of magnetic susceptibility 48
chalcogenide glasses 568, 571
channel conductivity 1020
characterization of optical glasses 547
Charpy impact strength 478
– polymers 478
chemical disorder 38
chemical stability
– glasses 530, 531
– optical glasses 550
chemical symbols
– element 45
chemisorption 1020
chiral smectic C
– liquid crystals 944
chlorine Cl
– elements 118
cholesteric phase 942
cholesteryl (cholest-5-ene) substituted mesogens
– liquid crystals 968
cholesteryl compound 944
chromium Cr
– elements 114
CIP (current in the plane of layer) 1051
CIPM (Comité International des Poids et Mesures) 3, 11
Cl chlorine 118
clamped crystal 907, 921, 934
clamped dielectric constant 907
classifications of liquid crystals 942
climatic influences
– glasses 550
cluster boundaries 907
cluster formation 907
CM
– liquid crystals 970
Cm curium 151
CMOS (complementary metal-oxide-semiconductor) 1059
Co cobalt 135
$Co_{17}RE_2$ 805
Co_5RE 805
coating 943
cobalt
– alloys 272
– applications 273
– hard-facing alloy 274
– mechanical properties 277
– superalloys 274
– surgical implant alloys 277
cobalt Co
– elements 135
cobalt corrosion-resistant alloys 276
cobalt-based corrosion-resistant alloys 276
CODATA (Committee on Data for Science and Technology) 4
coefficient of expansion 478
– polymers 478
coefficient of thermal expansion
– glasses 526
coercive field 904
coherence length 920
coherent phonon and Raman spectra 1041
colloidal synthesis
– nanostructured materials 1065
colored glasses
– colorants 567
– glasses 565
columnar phase 941
commensurate reconstruction 986
commercially pure titanium (cp-Ti) 206
communication technology 912

compacted (vermicular) graphite (CG) 268
complex perovskite-type oxide 909
complex refractive index
– zinc compounds 674
composite medium 1045
– dielectric constant 1045
composite solder glasses
– glasses 563
composite structures
– crystallography 34
compound semiconductor 1003
compressibility 478
– polymers 478
compression modulus
– elements 47
condensed matter 27
– classification 28
conductivity
– frequency-dependent 823
conductivity tensor
– elements 47
conductor
– nanoparticle-based 1043
confined electronic systems
– nanostructured materials 1035
confinement effect
– nanostructured materials 1031
constants
– fundamental 3
container glasses 529
continuous distribution of states 1022
continuous-cooling-transformation (CCT) diagram 238
controlled rolling 240
Convention du Mètre 12
conventional system
– ISO 13
conversion factor 945
– density 945
– diamagnetic anisotropy 945
– dipole moment 945
– dynamic viscosity 945
– kinematic viscosity 945
– molar mass 945
– temperatures of phase transitions 945
– thermal conductivity 945
cooperative interactions 903
coordination number
– elements 46, 47
copolymers
– physical properties 477, 483
copper
– unalloyed 297
copper alloys 296, 297
copper Cu
– elements 65
copper–nickel 300
copper–nickel–zinc 300
Co–RE
– phase equilibria 805
corrosion 1020
– resistance 218
Coulomb blockade
– nanostructured materials 1031, 1044
coupled plasmon modes 1048
Co–Sm 803
CPP (current flows perpendicular to the plane of the layer) 1051, 1054
Cr chromium 114
creep modulus 478
– polymers 478
critical field 904
critical slowing-down 907, 908
critical temperature
– elements 47, 48
– Pb alloys 699
CrNi steels 252
Cronstedt, swedish mineralogist 1032
crystal axes 980
crystal morphology 30
crystal optics 824
crystal structure
– beryllium compounds 652
– cadmium compounds 676
– group IV semiconductors 578
– III–V compounds 610
– indium compounds 638
– IV–IV compound semiconductors 578
– magnesium compounds 655
– mercury compounds 686
– oxides of Ca, Sr, and Ba 660
– zinc compounds 665
crystal structure, mechanical and thermal properties
– III–V compounds 621, 638
– III–V semiconductors 604
– II–VI compounds 652, 655, 660, 665, 676, 686
crystal symmetry
– elements 47
– ferroelectrics 915
crystalline ferroelectric 911
crystalline materials
– definition 27
crystalline surface
– atomically clean 979
crystallization
– glasses 524
crystallographic formulas 986
crystallographic properties
– elements 46, 47
crystallographic space group 946–952, 955–961, 964, 966, 968–971
crystallographic structure
– methods to investigate 39
crystallography
– concepts and terms 27
– rudiments of 27
crystals
– biaxial 826
– cubic 826
– isotropic 826
– uniaxial 826
Cryston
– glasses 559
Cs cesium 59
CS2004
– liquid crystals 975
Cu copper 65
cubic
– dielectrics 828, 838
cubic $BaTiO_3$ 916
cubic boron nitride 451
cubic crystal 47
cubic system 986
Curie point 906
Curie temperature 906
– surface 1009
Curie–Weiss constant 931, 932
curium Cm
– elements 151
current/voltage characteristics of films 1044
Cu–Ni
– electrical conductivity 302
– thermal conductivity 303
Cu–Zn
– electrical conductivity 302
– thermal conductivity 303
Cu–Zn phase diagram 299
cyclohexane 946
cycloolefine copolymer (COC) 486, 487
cyclosilicate 433

D

D deuterium 54
damping constant
– optical mode frequency 916
dangling bonds (DBs) 991

DAS 991
data storage media
– nanostructured materials 1031
Db dubnium 105
de Broglie wavelength 1035
Debye length
– solid surfaces 1020
Debye temperature Θ_D
– boron compounds 605
– cadmium compounds 678
– gallium compounds 623
– group IV semiconductors 584
– indium compounds 640
– IV–IV compound semiconductors 584
– mercury compounds 687
– metal surfaces 1013
– oxides of Ca, Sr, and Ba 661
– solid surfaces 1014
– surface phonons 1012
– zinc compounds 667
decimal multiples of SI units 19
degree Celsius
– unit of temperature 14
density ϱ 478, 945–954, 956–971
– aluminium compounds 611
– beryllium compounds 653
– boron compounds 604
– cadmium compounds 676
– elements 47
– gallium compounds 621
– group IV semiconductors and IV–IV compounds 579
– indium compounds 638
– magnesium compounds 656
– mercury compounds 686
– oxides of Ca, Sr, and Ba 660
– polymers 478
– temperature dependence 943
– zinc compounds 665
density of electronic states
– magnesium compounds 658
density of electronic states *see also* density of states 658
density of phonon states *see also* density of states 658
density of states (DOS)
– nanostructure 1034
dentistry 330
depletion layer 1020
– solid surfaces 1024
derived quantities 12
– ISO 13
derived units
– SI 16
– special names and symbols 16
deuteration 925
deuterium D
– elements 54
devices
– optical 903
– piezoelectric 903
– pyroelectric 903
devil's staircase 932
devitrifying solder glasses 562
DFB (distributed feedback) 1038
DFG (difference frequency generation) 825
diamagnetic anisotropy 943, 945
diamond
– crystal structure, mechanical and thermal properties 578–588
– electromagnetic and optical properties 601
– electronic properties 589–594
– transport properties 595
diamond positions
– surface diagrams 983
diamond-like structure 979
Dicor
– glasses 559
dielectric
– constant (low-frequency) 823
– dissipation factor 823
– elasticity 823
– general properties 822
– lossy 823
– low-frequency materials 822
– stiffness constant 823
dielectric anisotropy 943, 946, 947
dielectric anomaly 922, 934
dielectric constant ε 907, 947–958, 960–966, 968–971, 973, 1045
– aluminium compounds 619
– beryllium compounds 655
– boron compounds 610
– cadmium compounds 683
– elements 46
– gallium compounds 635, 637
– group IV semiconductors and IV–IV compounds 601–603
– indium compounds 650
– magnesium compounds 660
– mercury compounds 691
– oxides of Ca, Sr, and Ba 664
– zinc compounds 672
dielectric dispersion 907
dielectric dissipation factor
– glasses 538
dielectric function 1001
– surface layer 1007
dielectric loss tan δ
– gallium compounds 635
dielectric material
– properties 826
dielectric polarization 825
dielectric properties
– glasses 538
dielectric strength
– glasses 539
dielectric tensor 827
dielectrics
– α-iodic acid, α-HIO_3 868
– α-mercuric sulfide, α-HgS 846
– α-silicon carbide, SiC 840
– α-silicon dioxide, α-SiO_2 846
– α-zinc sulfide, α-ZnS 842
– β-barium borate, β-BaB_2O_4 848
– 2-cyclooctylamino-5-nitropyridine, $C_{13}H_{19}N_3O_2$ 874
– 3-methyl 4-nitropyridine 1-oxide, $C_6N_2O_3H_6$ 868
– 3-nitrobenzenamine, $C_6H_4(NO_2)NH_2$ 878
– 4-($N1N$-dimethylamino)-3-acetamidonitrobenzene 884
– ADA 856
– ADP 856
– aluminium oxide, α-Al_2O_3 844
– aluminium phosphate $AlPO_4$ 844
– ammonium dideuterium phosphate, $ND_4D_2PO_4$ 856
– ammonium dihydrogen arsenate, $NH_4H_2AsO_4$ 856
– ammonium dihydrogen phosphate, $NH_4H_2PO_4$ 856
– ammonium Rochelle salt 870
– ammonium sulfate, $(NH_4)_2SO_4$ 872
– "banana" 872
– barium fluoride, BaF_2 828
– barium formate, $Ba(COOH)_2$ 866
– barium magnesium fluoride, $BaMgF_4$ 872
– barium nitrite monohydrate, $Ba(NO_2)_2 \cdot H_2O$ 842
– barium sodium niobate, $Ba_2NaNb_5O_{15}$ 872
– Barium titanate, $BaTiO_3$ 864
– BBO 848
– berlinite 844
– beryllium oxide, BeO 840
– BGO 838
– BIBO 884
– bismuth germanium oxide, $Bi_{12}GeO_{20}$ 838

– bismuth silicon oxide, $Bi_{12}SiO_{20}$ 838
– bismuth triborate, BiB_3O_6 884
– BK7 Schott glass 828
– BMF 872
– BSO 838
– cadmium germanium arsenide, $CdGeAs_2$ 856
– cadmium germanium phosphide, $CdGeP_2$ 856
– cadmium selenide, CdSe 840
– cadmium sulfide, CdS 840
– cadmium telluride, CdTe 834
– calcite, $CaCO_3$ 844
– calcium fluoride, CaF_2 828
– calcium tartrate tetrahydrate, $Ca(C_4H_4O_6) \cdot 4H_2O$ 874
– CBO 868
– CDA 858
– cesium dideuterium arsenate, CsD_2AsO_4 856
– cesium dihydrogen arsenate, CsH_2AsO_4 858
– cesium lithium borate, $CsLiB_6O_{10}$ 858
– cesium triborate, CsB_3O_5 868
– cinnabar 846
– CLBO 858
– CNB 874, 875
– COANP 874
– copper bromide, CuBr 834
– copper chloride, CuCl 834
– copper gallium selenide, $CuGaSe_2$ 858
– copper gallium sulfide, $CuGaS_2$ 858
– copper iodide, CuI 834
– cubic $m3m$ (O_h) 828
– cubic $\bar{4}3m$ (T_d) 834
– cubic, 23 (T) 838
– D(+)-saccharose, $C_{12}H_{22}O_{11}$ 888
– DADP 856
– DAN 884
– DCDA 856
– deuterated L-arginine phosphate, $(ND_xH_{2-x})_2^+(CND)(CH_2)_3CH(ND_yH_{3-y})^+COO^- \cdot D_2PO_4^- \cdot D_2O$ 884
– diamond, C 830
– dipotassium tartrate hemihydrate, $K_2C_4H_4O_6 \cdot 0.5H_2O$ 886
– DKDA 858
– DKDP 858
– DKT 886
– DLAP 884, 886
– DRDA 860
– DRDP 860
– fluorite 828
– fluorspar 828
– forsterite 866
– gadolinium molybdate, $Gd_2(MoO_4)_3$ 876
– gallium antimonide, GaSb 834
– gallium arsenide, GaAs 834
– gallium nitride, GaN 840
– gallium phosphide, GaP 834
– gallium selenide, GaSe 838
– gallium sulfide, GaS 838
– germanium, Ge 830
– GMO 876
– greenockite 840
– halite 832
– hexagonal, 6 (C_6) 842
– hexagonal, $6mm$ (C_{6v}) 840
– hexagonal, $\bar{6}m2$ (D_{3h}) 838
– high-frequency (optical) properties 817
– Iceland spar 844
– indium antimonide, InSb 836
– indium arsenide, InAs 836
– indium phosphide, InP 836
– Irtran-3 828
– Irtran-6 834
– isotropic 828
– KB5 880
– KBBF 846
– KDA 860
– KDP 860
– KLINBO 864
– KTA 880
– KTP 880
– LBO 878
– L-CTT 874
– lead molybdate, $PbMoO_4$ 854
– lead titanate, $PbTiO_3$ 864
– LFM 876
– list of described substances 818
– lithium fluoride, LiF 830
– lithium formate monohydrate, $LiCOOH \cdot H_2O$ 876
– lithium gallium oxide, $LiGaO_2$ 876
– lithium iodate, α-$LiIO_3$ 842
– lithium metagallate 876
– lithium niobate (5% MgO-doped), MgO:$LiNbO_3$ 848
– lithium niobate, $LiNbO_3$ 848
– lithium sulfate monohydrate, $Li_2SO_4 \cdot H_2O$ 886
– lithium tantalate, $LiTaO_3$ 848
– lithium tetraborate, $Li_2B_4O_7$ 864
– lithium triborate, LiB_3O_5 878
– low-frequency properties 817
– magnesium fluoride, MgF_2 852
– magnesium oxide, MgO 830
– magnesium silicate, Mg_2SiO_4 866
– Maxwell's equations 824
– m-chloronitrobenzene, $ClC_6H_4NO_2$ 874, 875
– mNA 878
– m-nitroaniline 878
– MNMA 874
– monoclinic, 2 (C_2) 884
– N,2-dimethyl-4-nitrobenzenamine, $C_8H_{10}N_2O_2$ 874
– N-[2-(dimethylamino)-5-nitrophenyl]-acetamide 884
– nantokite 834
– numerical data 818, 828
– orthorhombic, 222 (D_2) 866
– orthorhombic, $mm2$ (C_{2v}) 872
– orthorhombic, mmm (D_{2h}) 866
– paratellurite 854
– physical properties 817
– PMMA (Plexiglas) 828
– POM 868
– potassium acid phthalate, $KH(C_8H_4O_4)$ 878
– potassium bromide, KBr 830
– potassium chloride, KCl 830
– potassium dideuterium arsenate, KD_2AsO_4 858
– potassium dideuterium phosphate, KD_2PO_4 858
– potassium dihydrogen arsenate, KH_2AsO_4 860
– potassium dihydrogen phosphate, KH_2PO_4 860
– potassium fluoroboratoberyllate, $KBe_2BO_3F_2$ 846
– potassium iodide, KI 830
– potassium lithium niobate, $K_3Li_2Nb_5O_{15}$ 864
– potassium niobate, $KNbO_3$ 880
– potassium pentaborate tetrahydrate, $KB_5O_8 \cdot 4H_2O$ 880
– potassium sodium tartrate tetrahydrate, $KNa(C_4H_4O_6) \cdot 4H_2O$ 870
– potassium titanate (titanyl) phosphate, $KTiOPO_4$ 880
– potassium titanyl arsenate, $KTiOAsO_4$ 880
– proustite 850
– pyragyrite 850
– quartz 846

– RDA 860
– RDP 862
– Rochelle salt 870
– rock salt 832
– RTP 882
– rubidium dideuterium arsenate, RbD_2AsO_4 860
– rubidium dideuterium phosphate, RbD_2PO_4 860
– rubidium dihydrogen arsenate, RbH_2AsO_4 860
– rubidium dihydrogen phosphate, RbH_2PO_4 862
– rubidium titanate (titanyl) phosphate, $RbTiOPO_4$ 882
– rutile 852
– sapphire 844
– Silicon dioxide, SiO_2 828
– silicon, Si 830
– silver antimony sulfide, Ag_3SbS_3 850
– silver arsenic sulfide, Ag_3AsS_3 850
– silver gallium selenide, $AgGaSe_2$ 862
– silver thiogallate, $AgGaS_2$ 862
– sodium ammonium tartrate tetrahydrate, $Na(NH_4)C_4H_4O_6 \cdot 4H_2O$ 870
– sodium chlorate, $NaClO_3$ 838
– sodium chloride, NaCl 832
– sodium fluoride, NaF 832
– sodium nitrite, $NaNO_2$ 882
– strontium fluoride, SrF_2 832
– strontium titanate, $SrTiO_3$ 832
– sucrose 888
– sylvine 830
– sylvite 830
– TAS 850
– tellurium dioxide, TeO_2 854
– tellurium, Te 846
– tetragonal, $4/m$ (C_{4h}) 854
– tetragonal, $4/mmm$ (D_{4h}) 852
– tetragonal, 422 (D_4) 854
– tetragonal, $4mm$ (C_{4v}) 864
– tetragonal, $\bar{4}2m$ (D_{2d}) 856
– TGS 888
– thallium arsenic selenide, Tl_3AsSe_3 850
– titanium dioxide, TiO_2 852
– tourmaline, $(Na,Ca)(Mg,Fe)_3B_3Al_6Si_6(O,OH,F)_{31}$ 850
– triglycine sulfate, $(CH_2NH_2COOH)_3 \cdot H_2SO_4$ 888
– trigonal, 32 (D_3) 844
– trigonal, $3m$ (C_{3v}) 848
– trigonal, $\bar{3}m$ (D_{3d}) 844
– urea, $(NH_2)_2CO$ 862
– wurtzite 842
– YAG 832
– YAP 866
– YLF 854
– yttrium aluminate, $YAlO_3$ 866
– yttrium aluminium garnet, $Y_3Al_5O_{12}$ 832
– yttrium lithium fluoride, $YLiF_4$ 854
– yttrium vanadate, YVO_4 852
– YVO 852
– zinc blende 842
– zinc germanium diphosphide, $ZnGeP_2$ 862
– zinc oxide, ZnO 840
– zinc selenide, ZnSe 836
– zinc telluride, ZnTe 836
– zincite 840
difference frequency generation (DFG) 825
differential conductance as function of voltage 1043
diffraction method 39
diffusion coefficient 946–949, 956–960, 970
diffusion-controlled growth
– nanostructured materials 1064
dimensionality
– nanostructured materials 1033
dimensions
– physical quantities 4, 13
dimer bond length 995
dioxide
– zirconium 448
dipole glass 906
dipole moment 945, 947–953, 955–963, 968–971
direct gap
– group IV semiconductors and IV–IV compounds 592
direct piezoelectric effect 824
director 943
discotic liquid crystal 941, 942, 972
– physical properties 972
disk-like molecule 942
disordered materials 38
dispersion
– glasses 543
dispersion curves
– electronic structure of surfaces 999
dispersion hardening 329
displacement of atoms
– surface phonons 1012
displacive disorder 38
displacive ferroelectrics 907
display lifetime 944
dissipation factor 826, 828
dissipative dispersion 907
dissociation energy of molecule
– elements 46
DOBAMBC (liquid crystal) 934
domain pattern
– perpendicular magnetization 1062
domain wall
– ferroelectric 907
– ferromagnetic 907
donor surface state 1022
doping
– chemical 576
DOS (density of states) 330, 1049
– nanostructured materials 1034
drain 1024
Drude-model metal 1045
drug delivery 944
DSP family 931
DTA (differential thermal analysis) 39
dubnium Db
– elements 105
ductile iron 269
Duran
– glasses 527, 531, 537
Dy dysprosium 142
dye dopant 944
dynamic viscosity 945–948, 955–959, 961, 962, 967, 969–971, 973
dysprosium Dy
– elements 142

E

E7
– liquid crystals 975
ECS 1011
EDX (energy-dispersive analysis of X-rays) 39
EELS 1013
effect of solute elements conductivity of Cu 297
effective masses
– aluminium compounds 617
– boron compounds 607
– cadmium compounds 681
– group IV semiconductors and IV–IV compounds 593
– indium compounds 646
– mercury compounds 689

– oxides of Ca, Sr, and Ba 661
– zinc compounds 670
effective masses m_n and m_p
– gallium compounds 629
einsteinium Es
– elements 151
elastic compliance 828
elastic compliance tensor 934
– elements 47
elastic constant c_{ik} 823
– aluminium compounds 611
– beryllium compounds 652
– cadmium compounds 676
– gallium compounds 621
– group IV semiconductors and IV–IV compounds 580
– indium compounds 638
– magnesium compounds 656
– mercury compounds 687
– oxides of Ca, Sr, and Ba 660
– zinc compounds 665
elastic modulus 478, *see* elastic constant 580
– elements 46, 47
– polymers 478
elastic stiffness 828
– elements 47
elastic tensor 827
elastooptic coefficient 825
elastooptic constant 828
elastooptic tensor 827
electric strength 478
– polymers 478
electrical conductivity
– aluminium compounds 618
– boron compounds 608
– elements 46
– group IV semiconductors and IV–IV compounds 595
electrical conductivity *see also* electrical resistivity 659
electrical conductivity σ
– indium compounds 647
– magnesium compounds 659
– mercury compounds 690
– oxides of Ca, Sr, and Ba 663
electrical resistivity
– boron compounds 609
– elements 48
– gallium compounds 630
electrical steel 766
electroforming 288
electromagnetic and optical properties
– group IV semiconductors and IV–IV compounds 601
– III–V compounds 610, 619, 635, 650
– II–VI compounds 655, 660, 664, 672, 683, 691
electromagnetic concentration effect 1046
electromagnetic confinement
– nanostructured materials 1044
electromechanical coupling constant 912, 918
electron affinity
– elements 46
electron and hole mobilities
– aluminium compounds 618
electron density of states 1034
– cadmium compounds 680
electron diffraction 41
electron effective mass m_n 593, *see* effective mass 661
electron g-factor g_c
– gallium compounds 629
– indium compounds 646
electron microscope image
– magnetic tunneling junction 1057
electron microscopy 1049
electron mobility μ_n 663, *see* mobility μ 682
– elements 48
– gallium compounds 631
– indium compounds 648
– mercury compounds 690
– oxides of Ca, Sr, and Ba 663
electron transport phenomena
– nanostructured materials 1042
electron tunneling
– phonon-assisted 1043
electronegativity
– elements 46
electronic band gap
– elements 48
electronic conductivity σ
– zinc compounds 671
electronic configuration
– elements 46
electronic dispersion curves 1000
electronic ground state
– elements 46
electronic properties
– group IV semiconductors and IV–IV compounds 589–594
– III–V compounds 606, 614, 626, 643
– II–VI compounds 653, 657, 661, 668, 678, 688
electronic structure
– solid surfaces 996
electronic transport, general description
– aluminium compounds 617
– beryllium compounds 655
– boron compounds 608
– cadmium compounds 682
– gallium compounds 629
– group IV semiconductors and IV–IV compounds 595
– indium compounds 647
– magnesium compounds 659
– mercury compounds 689
– oxides of Ca, Sr, and Ba 663
– zinc compounds 670
electronic work function
– elements 48
electronic, electromagnetic, and optical properties
– elements 46
electron–phonon coupling 1040, 1041
electrooptic coefficients 826
electrooptic modulators
– nanostructured materials 1040
electrooptic tensor 827
electro-optical constants
– zinc compounds 675
electro-optical effect 944
electrostrictive constant 930
elemental semiconductor 1003
elements 45
– allotropic modifications 48
– atomic properties 46
– electromagnetic properties 48
– electronic properties 48
– high-pressure modifications 48
– ionic properties 46
– macroscopic properties 46
– materials data 46
– molecular properties 46
– optical properties 48
– ordered according to the Periodic table 52
– ordered by their atomic number 51
– ordered by their chemical symbol 50
– ordered by their name 49
elinvar alloys 786
– antiferromagnetic 789
elinvar-type alloys
– nonmagnetic 792
energy bands *see* band structure 590

energy diagram for a MIM tunnel junction 1053
energy dispersion curve 1000
energy equivalents 24
energy equivalents in different units 24
energy exchange time τ_{e-ph} 1041
energy gap
– aluminium compounds 616
– beryllium compounds 654
– cadmium compounds 681
– gallium compounds 628
– group IV semiconductors 592, 593
– indium compounds 644
– IV–IV compound semiconductors 592, 593
– magnesium compounds 659
– mercury compounds 688
– oxides of Ca, Sr, and Ba 661
– zinc compounds 669
energy gaps
– boron compounds 607
energy shifts in the luminescence peaks 1037
energy-storage cell 1043
engineering critical current density 741
enthalpies of phase transitions 946–973
enthalpy change
– elements 47
enthalpy of combustion 477
– polymers 477
enthalpy of fusion 477
– polymers 477
entropy of fusion 477
– polymers 477
Er erbium 142
erbium Er
– elements 142
Es einsteinium 151
Eu europium 142
europium Eu
– elements 142
EXAFS (extended X-ray atomic fine-structure analysis) 39
excess carrier density 1023
excitation energy
– nanostructured materials 1038
exciton binding energy
– gallium compounds 628
– zinc compounds 669
exciton Bohr radii
– semiconductors 1037
exciton energy
– group IV semiconductors and IV–IV compounds 593
exciton peak energy
– indium compounds 645
exciton Rydberg series 1038
excitons
– nanostructured materials 1036
external forces 46
external-field dependence 46
extinction coefficient *k*
– gallium compounds 635
– zinc compounds 674

F

F fluorine 118
F2
– optical glasses 551
F5
– glasses 537
facets
– crystallography 27
Fahrenheit 48
families of ferroelectrics 909
fast displays 903
fcc positions
– surface diagrams 981
Fe iron 131
$Fe_{14}Nd_2B$
– commercial magnets 804
– magnetic materials 803
Fe–C(-X)
– carbide phases 224
Fe–Cr alloy 226
Fe–Mn alloys 226
Fe–Ni alloys 225
Fermi energy 998
Fermi function 1022
Fermi level pinning 1025
Fermi surface
– nanostructured materials 1055
Fermi surface shift in an electric field 1050
Fermi surfaces
– nanostructured materials 1049
Fermi wavelength
– nanostructured materials 1035
fermium Fm
– elements 151
ferrielectric triple hysteresis loop 936
ferrielectricity 927
ferrite
– applications 812
– hard magnetic 813
– MnZn 812
– NiZn 813
– soft magnetic 811
ferroelectric ceramics 906
ferroelectric hysteresis loop 904
ferroelectric liquid crystal 906, 911, 945, 967
– physical properties 967
ferroelectric mixtures
– liquid crystals 975
– physical properties 975
ferroelectric phase transition 906, 933
ferroelectric polymers 906
ferroelectric transducer 906
ferroelectrics
– classification 906
– definition 903
– dielectric properties 903
– displacive type 906
– elastic properties 903
– families 909, 911
– general properties 906
– indirect type 906
– inorganic crystals 903
– inorganic crystals other than oxides 922
– inorganic crystals oxides 912
– liquid crystals 903, 930
– order–disorder type 906
– organic crystals 903, 930
– phase transitions 903
– piezoelectric properties 903
– polymers 903, 930
– pyroelectric properties 903
– symbols and units 912
ferromagnetic surface 1008
Fe–Co–Cr 795
Fe–Co–V 797
Fe–Nd–B 800
– phase relations 800
Fe–Ni–Al–Co 798
Fe–Si alloys
– rapidly solidified 768
Fibonacci sequence 35
field-effect mobility 1024
– solid surfaces 1024
first Brillouin zone *see* Brillouin zone 657
flat glasses 528
flat-band condition 1020
fluorinated three-ring LC 944
fluorine F
– elements 118

fluoropolymers 480, 496
– poly(ethylene-co-chlorotrifluoroethylene) (ECTFE) 496, 497
– poly(ethylene-co-tetrafluoroethylene) (ETFE) 496, 497
– poly(tetrafluoroethylene-co-hexafluoropropylene) (FEP) 496, 497
– polychlorotrifluoroethylene (PCTFE) 496, 497
– polytetrafluoroethylene (PTFE) 496, 497
flux flow (FF) 718
Fm fermium 151
formation curve
– glasses 524
formulas
– crystallographic 986
Fotoceram
– glasses 559
Fotoform
– glasses 559
Foturan
– glasses 559
Fourier map 926
four-ring system 964
Fr francium 59
francium Fr
– elements 59
Frantz–Keldysh effect 1040
free dielectric constant 907
frequency conversion 825
Fresnel reflectivity
– glasses 549
Friedel–Creagh–Kmetz rule 943
fundamental constants 3
– 2002 adjustment 4
– alpha particle 9
– atomic physics and particle physics 7
– CODATA recommended values 4
– electromagnetic constants 6
– electron 7
– meaning 4
– most frequently used 4
– neutron 8
– proton 8
– recommended values 3, 4
– thermodynamic constants 6
– units of measurement 3
– universal constants 5
– what are the fundamental constants? 3
fused silica
– glasses 534, 537

G

Ga gallium 78
GaAs positions
– surface diagrams 983
gadolinium Gd
– elements 142
gallium antimonide
– crystal structure, mechanical and thermal properties 621
– electromagnetic and optical properties 635
– electronic properties 626
– transport properties 631
gallium arsenide
– crystal structure, mechanical and thermal properties 621
– electromagnetic and optical properties 635
– electronic properties 626
– transport properties 631
gallium compounds
– crystal structure, mechanical and thermal properties 621
– electromagnetic and optical properties 635
– electronic properties 626
– mechanical and thermal properties 621
– thermal conductivity 634
– thermal properties 621
– transport properties 629
gallium Ga
– elements 78
gallium nitride
– crystal structure, mechanical and thermal properties 621
– electromagnetic and optical properties 635
– electronic properties 626
– transport properties 631
gallium phosphide
– crystal structure, mechanical and thermal properties 621
– electromagnetic and optical properties 635
– electronic properties 626
– transport properties 631
gamma titanium aluminides 213
gas permeation 478
– polymers 478
Gd gadolinium 142
Ge germanium 88
germanium
– band structure 590
– crystal structure, mechanical and thermal properties 578–588
– electromagnetic and optical properties 601
– electronic properties 589–594
– transport properties 598
germanium Ge
– elements 88
g-factor
– cadmium compounds 681
g-factor, conduction electrons
– group IV semiconductors and IV–IV compounds 594
glass designation 544
glass formers 527
glass matrix 1040
glass number 8nnn 534, 537
glass number nnnn
– sealing glasses 563
glass structure
– sodium silicate glasses 524
glass temperature
– glasses 524
glass transition temperature 477
– polymers 477
glass-ceramics 525, 526, 558
– density 558
– elastic properties 558
– manufacturing process 558
glasses 523
– Abbe value 547
– abbreviating glass code 543
– acid attack 532
– acid classes 533
– alkali attack 532
– alkali classes 533
– alkali–alkaline-earth silicate 530
– alkali–lead silicate 530
– alkaline-earth aluminosilicate 530
– amorphous metals 523
– armor plate glasses 529
– automotive applications 529
– band pass filters 566
– Borofloat 528, 529
– borosilicate 529, 530
– borosilicate glasses 529
– brittleness 536
– chemical constants 553
– chemical properties 549
– chemical resistance 549
– chemical stability 530, 531
– chemical vapor deposition 523
– color code 554
– composition 527

– compound glasses 529
– container glasses 528, 529
– crack effects 536
– density 528
– dielectric properties 538
– Duran® 531
– elasticity 536
– electrical properties 537
– engineering material 523
– fire protecting glasses 529
– flat 528
– fracture toughness 537
– frozen-in melt 533
– halide glasses 568
– hydrolytic classes 533
– infrared transmitting glasses 568
– infrared-transmitting 571
– inhomogeneous 525
– internal transmission 554
– linear thermal expansion 536, 556
– long pass filters 566
– major groups 526
– manufacturers, preferred optical glasses 545
– melting range 533
– mixtures of oxide compounds 524
– neutral density filters 566
– optical 551
– optical characterization 547
– optical glasses 543
– optical properties 539, 543
– oxide glasses 568
– passivation glasses 562
– physical constants 553
– plate glasses 528
– properties 526
– quasi-solid melt 533
– refractive index 539
– Schott filter glasses 569
– sealing glasses 559
– short pass filters 566
– silicate based 526
– soda–lime type 528
– solder glasses 562
– strength 534
– stress behavior 535
– stress rate 535
– stress-induced birefringence 539
– supercooled melt 533
– surface (cleaning and etching) 533
– surface modification 532
– surface resistivity 538
– technical 530
– tensile strength 536
– thermal conductivity 556
– thermal strength 537
– transmittance 539
– viscosity 534
– vitreous silica 556
– volume resistivity 537
– wear-induced surface defects 535
glasses, colored
– nomenclature 566
– optical filter 566
glasses, sealing
– ceramic 562
– principal applications 561
– recommended material combinations 560
– special properties 561
glasses, solder and passivation
– composite 563
– properties 564
glassy state
– crystallography 39
GMO family
– ferroelectrics 920
GMR (giant magnetoresistance) 1049, 1050
– mechanism 1051
– thickness dependence 1051
GMR ratio 1050
gold
– alloys 347
– applications 347
– chemical properties 361
– electrical properties 356
– electrical resistivity 356
– intermetallic compounds 350
– magnetic properties 358
– mechanical properties 352
– optical properties 359
– phase diagrams 347
– production 347
– special alloys 361
– thermal properties 359
– thermochemical data 347
– thermoelectric properties 358
gold Au
– elements 65
golden mean 35
granular materials 1043
gray
– SI unit of absorbed dose 19
gray iron 269
gray tin
– band structure 590
– crystal structure, mechanical and thermal properties 578–588
– electromagnetic and optical properties 601
– electronic properties 589–594
– transport properties 599
Griffith flaw
– glasses 534
Group IV semiconductors 576, 578–603
– electron mobility 597
– hole mobility 597
Group IV semiconductors and IV–IV compounds
– crystal structure 578
– electromagnetic and optical properties 601
– electronic properties 589–594
– mechanical properties 578
– thermal properties 578
– transport properties 595–601
groups of elements (Periodic table) 45

H

H hydrogen 54
hafnium Hf
– elements 94
Haigh Push-Pull test 408
Hall coefficient
– elements 48
– group IV semiconductors and IV–IV compounds 595
Hall mobility
– group IV semiconductors and IV–IV compounds 596, 597
Hall mobility *see also* mobility μ 596
halogen-substituted benzene 946
hard disk drive 1060
– limits 1060
– technology 1060
hard ferrites
– magnetic properties 814
hard magnetic alloys 794
hardenability 237
hardmetals 277
hassium Hs
– elements 131
Hatfield steel 226
HATOF 1013
HCl family 924
He helium 54
heat capacities c_p, c_V
– boron compounds 605
– group IV semiconductors 584

– IV–IV compound semiconductors 584
heat capacity 477, 956–960, 970
– cadmium compounds 678
– gallium compounds 623
– indium compounds 640
– mercury compounds 687
– oxides of Ca, Sr, and Ba 661
– polymers 477
– zinc compounds 667
heat-resistant steels 258
HEIS (high-energy ion scattering) 989, 1013
helical structure 942
helium He
– elements 54
Hermann–Mauguin symbols 30
hertz
– SI unit of frequency 19
heterocycles 946
hexagonal $BaTiO_3$ 917
Hf hafnium 94
Hg mercury 73
high copper alloy 297
high-T_c superconductors
– lower critical 716
– upper critical 716
high-frequency dielectric constant ε *see also* dielectric constant ε 601
high-Ni alloys 281
high-pressure die casting (HPDC) 168
high-pressure modifications 48
high-strength low-alloy 240
hip implants 277
HMF (half-metallic ferromagnet) 1055
Ho holmium 142
hole effective mass m_p 594, *see* effective mass 661
hole mobility μ_p *see* mobility μ 682
– elements 48
– gallium compounds 631
– indium compounds 648
hollow-ware
– glasses 528
holmium Ho
– elements 142
holohedry
– crystallography 30
homeotropic alignment 943
Hooke's law 47, 823
hopping mechanism
– nanostructured materials 1043
host–guest effect 944
hot forming
– glasses 524
Hoya code
– glasses 544
HPDC (high-pressure die casting) 168
HRTEM (high-resolution transition electron microscopy) 39
Hs hassium 131
Hume-Rothery phase 333, 350
Hume-Rothery phases 296
hydrogen H
– elements 54
hydrostatic pressure 926
hyper-Raman scattering 915

I

I iodine 118
IBA 991, 1024
ICSU (International Council of the Scientific Unions) 4
ideal surface 979
III–V compound semiconductors 604
II–VI semiconductor compounds 652
image potential 996
image state 996, 997, 999
– effective mass 999
impact strength 478
– polymers 478
impurity elements 206
impurity scattering
– group IV semiconductors and IV–IV compounds 599
In indium 78
incommensurate phase 930
incommensurate phases 906
incommensurate reconstruction 986
index of refraction
– complex 674
indirect gap
– group IV semiconductors and IV–IV compounds 589
indium antimonide
– crystal structure, mechanical and thermal properties 638
– electromagnetic and optical properties 650
– electronic properties 643
– transport properties 647
indium arsenide
– crystal structure, mechanical and thermal properties 638
– electromagnetic and optical properties 650
– electronic properties 643
– transport properties 647
indium compounds
– crystal structure, mechanical and thermal properties 638
– electromagnetic and optical properties 650
– electronic properties 643
– mechanical and thermal properties 638
– optical properties 650
– thermal properties 638
– transport properties 647
indium In
– elements 78
indium nitride
– crystal structure, mechanical and thermal properties 638
– electromagnetic and optical properties 650
– electronic properties 643
– transport properties 647
indium phosphide
– crystal structure, mechanical and thermal properties 638
– electromagnetic and optical properties 650
– electronic properties 643
– transport properties 647
induced phase transition 904
inelastic neutron scattering 906
infrared-transmitting glasses 568, 571
inorganic ferroelectrics 903
inorganic ferroelectrics other than oxides 922
inosilicates 433
insulator
– surface phonon 1017
intercritical annealing 240
interface state density
– solid surface 1026
interlayer distance
– crystallographic formulas 986
International Annealed Copper Standard (IACS) 296
international system of units 11
international tables for crystallography 31
International Union of Pure and Applied Chemistry (IUPAC) 15
International Union of Pure and Applied Physics (IUPAP) 14
internuclear distance
– elements 46
intrinsic carrier concentration

Subject Index

– group IV semiconductors and IV–IV compounds 595, 596
intrinsic charge carrier concentration
– elements 48
intrinsic Debye length 1021
intrinsic Fermi level 1020
invar alloys
– Fe–Ni-based 782
– Fe–Pd base 785
– Fe–Pt-based 784
Invar effect 385
inversion center 30
inversion layer 1020
– solid surfaces 1024
inversion layer channel 1024
iodine I
– elements 118
ionic radius
– elements 46, 49
ionization energy
– elements 46
Ir iridium 135
iridium 393
– alloys 393
– applications 393
– chemical properties 398
– diffusion 398
– electrical properties 397
– lattice parameter 394
– magnetic properties 397
– mechanical properties 395
– optical properties 398
– phase diagram 393
– production 393
– thermal properties 398
– thermoelectrical properties 397
iridium Ir
– elements 135
iron and steels 221
iron Fe
– elements 131
iron miscibility gap 226
iron phase diagram 226
iron-carbon alloys 222
iron-cobalt alloys 772
iron–silicon alloys 763
Ising model 906
ISO (International Organization for Standardization) 12
isothermal transformation (IT) diagram 238
isotropic
– dielectrics 828
isotropic liquid 943
IV–IV compound semiconductors 576, 578–603
– electron mobility 597
– hole mobility 597

J

JDOS 1006
jellium 997
jellium model 998
– work functions 999
jewellery 330
joint density of states 1005
jominy apparatus 238
Josephson vortices 717
joule
– SI unit of energy 19

K

K 50
– glasses 537
K potassium 59
K10
– optical glasses 551
K7
– optical glasses 551
katal
– SI unit of catalytic activity 19
KDP family 925
kelvin 48
– SI base unit 14
Kerr effect 825
– optical 826
Kerr ellipticity 1011
KH_2PO_4 family 925
kilogram
– SI base unit 14
kinematic viscosity 945, 947, 950, 951, 955, 962, 964–967, 969, 975
Kleinman symmetry conditions 825
$KNbO_3$ 913
knee joint replacements 277
KNO_3 family 925
Knoop hardness
– optical glasses 550
Kr krypton 128
KRIPES 997
Kroll process 206
krypton Kr
– elements 128
$KTaO_3$ 913

L

LA
– phonon spectra 915
La lanthanum 84
Lamb theory of elastic vibrations 1040
lamellar (flake) graphite (FG) 268
langbeinite-type family 911, 930
lanthanum La
– elements 84
LASF35
– optical glasses 551
LAT family 932
lattice concept
– crystallography 28
lattice constants *see* lattice parameters 656
lattice dynamics 1012
lattice parameter 980
– aluminium compounds 610
– beryllium compounds 652
– boron compounds 604
– cadmium compounds 676
– gallium compounds 621
– group IV semiconductors and IV–IV compounds 579
– indium compounds 638
– magnesium compounds 656
– mercury compounds 686
– oxides of Ca, Sr, and Ba 660
– zinc compounds 665
lattice scattering
– group IV semiconductors and IV–IV compounds 598
lattice vibration 906
lattices
– planes and directions 28
Laue images 41
lawrencium Lr
– elements 151
layer-structure family 909
LB (Langmuir–Blodgett) film 944
LC display 944
LC materials (LCMs) 944
LC–surface interaction 944
lead 407
– antimony 412
– arsenic alloys 421
– battery grid alloys 413
– bearing alloys 415
– bismuth 419
– cable sheathing alloys 421
– calcium–tin 417
– calcium–tin, battery grid 418
– coper alloys 421
– corrosion 408
– corrosion classification 411
– fusible alloys 420
– gamma-ray mass-absorption 412
– grades 407

Subject Index

– internal friction 408
– low-melting alloys 419
– mechanical properties 408
– quaternary eutectic alloy 420
– recrystallization 409
– silver alloys 420
– solder alloys 415
– solders 416
– tellurium alloys 421
– ternary alloys 413
– tin alloy 415
lead glasses 527
lead Pb
– elements 88
lead–antimony
– phase diagram 413
LEED (low-energy electron diffraction) 987, 1013
LEIS (low-energy ion scattering) 989, 1013
LF5
– optical glasses 551
Li lithium 59
$Li_2Ge_7O_{15}$ family 920
$LiBaO_3$ 919
light transmittance
– glasses 539
light-emitting diode 1043
$LiNbO_3$ family 909, 919
linear thermal expansion
– optical glasses 556
linear thermal expansion coefficient α
– aluminium compounds 611
– beryllium compounds 653
– boron compounds 605
– cadmium compounds 677
– gallium compounds 623
– group IV semiconductors 582, 583
– indium compounds 639
– IV–IV compound semiconductors 582, 583
– magnesium compounds 656
– mercury compounds 687
– oxides of Ca, Sr, and Ba 660
– zinc compounds 666
$LiNH_4C_4H_4O_8$ family 911
liquid crystal
– anisotropy 943
– refractive index 943
– rod-like 943
liquid crystal family 911, 934
liquid crystal ferroelectrics 903
liquid crystal material (LCM)
– degradation 944
liquid crystal salts 973
– physical properties 973
liquid crystal two-ring systems with bridges
– physical properties 955
liquid crystal two-ring systems without bridges
– physical properties 947
liquid crystalline acids
– physical properties 946
liquid crystalline compound 943
– mesogenic group 943
– side group 943
– terminal group 943
liquid crystalline mixtures
– physical properties 975
liquid crystals
– ferroelectric properties 905
liquid crystals (LCs) 941
list of described physical properties 822
lithium Li
– elements 59
lithium niobate 919
lithography
– nanostructured materials 1064
Lithosil™ 551
LLF1
– optical glasses 551
local-field effect 1045
long pass filters
– glasses 566
longitudinal acoustic branch 915
long-range order 39, 927
– glasses 524
long-range order of molecules 943
losses
– dynamic eddy current 763
– hysteresis 763
low dielectric loss glasses 527
low-dimensional system
– nanostructured materials 1034
low-frequency dielectric constant 917
low-temperature annealing 298
Lr lawrencium 151
LSMO ($La_{0.7}Sr_{0.3}MnO_3$) 1056
Lu lutetium 142
lumen
– non-SI unit in photometry 15
luminescence
– nanostructured materials 1036
luminous flux
– photometry 15
luminous intensity I_v
– photometry 15
lutetium Lu
– elements 142
lyotropic liquid crystals 942

M

M. Faraday 1032
Macor
– glasses 559
magnesium alloys 163
– corrosion behavior 169
– heat treatments 169
– joining 169
– mechanical properties 168
– nominal composition 165
– solubility data 163
– tensile properties 167
– tensile property 166
magnesium compounds
– crystal structure, mechanical and thermal properties 655
– electromagnetic and optical properties 660
– electronic properties 657
– mechanical and thermal properties 655
– optical properties 660
– thermal properties 655
– transport properties 659
magnesium Mg
– casting practices 168
– elements 68
– magnesium alloys 162
– melting practices 168
magnesium oxide 444
– applications 444
– crystal structure, mechanical and thermal properties 655
– electrical properties 444
– electromagnetic and optical properties 660
– electronic properties 657
– mechanical properties 444
– thermal properties 444
– transport properties 659
magnesium selenide
– crystal structure, mechanical and thermal properties 655
– electromagnetic and optical properties 660
– electronic properties 657
– transport properties 659
magnesium silicate
– electrical properties 435, 436
– mechanical properties 435, 436
– thermal properties 435, 436
magnesium sulfide

– crystal structure, mechanical and thermal properties 655
– electromagnetic and optical properties 660
– electronic properties 657
– transport properties 659
magnesium telluride
– crystal structure, mechanical and thermal properties 655
– electromagnetic and optical properties 660
– electronic properties 657
– transport properties 659
magnet
– Mn–Al–C 811
magnetic domain 1058
magnetic dots
– nanostructured materials 1049
magnetic dots, arrays of 1061
magnetic field constant
– fundamental constant 14
magnetic layers 1048, 1050
– spin valve 1052
magnetic materials 755
– Co_5Sm based 806
– hard 794
– permanent 794
magnetic nanostructures 1031, 1048, 1050
– information storage 1048
– read heads 1048
– sensors 1048
magnetic oxides 811
magnetic periodic structures 33
magnetic reading head 1053, 1058
magnetic recording
– perpendicular discontinuous media 1060
magnetic sensors 1048, 1058
magnetic surface 1008
magnetic susceptibility 948–950, 956, 957
– elements 46, 48
magnetic tunnel junction 1054
– manganite-based 1057
magnetic tunneling junctions (MTJ) 1056
magnetization
– elements 48
magnetocrystalline anisotropy 756
magnetoelectronic devices 1058
magnetoresistance effect 1050
magnetoresistance of Fe/Cr multilayers 1051
magnetostriction 757
magnets
– 17/2 type 807
– 5/1 type 806
– $TM_{17}Sm_2$ 808
majority carriers 1023
malleable irons 270
manganese Mn
– elements 124
manipulating the dot magnetization 1062
manufacturing
– Fe–Nd–B magnets 801
manufacturing process
– glasses 525
martensite 223
martensitic transformation 225, 226
mass magnetic susceptibility
– elements 48
mass susceptibility
– elements 48
mass-production glasses 526
materials
– semiconductors 576
materials data
– elements 46
material-specific parameters 46
matrix composites 170
Maxwell–Garnett model 1045
Maya blue color 1032
MBBA
– liquid crystals 955
MBE (molecular-beam epitaxy) 991, 1032, 1049, 1063
– 0-D structures 1064
– 1-D structures 1064
– 2-D structures 1063
Md mendelevium 151
mechanical and thermal properties
– III–V compounds 621, 638
– II–VI compounds 652, 655, 660, 665, 676, 686
mechanical properties
– elements 46, 47
– group IV semiconductors 578–588
– III–V compounds 610
– III–V semiconductors 604
– IV–IV compound semiconductors 578–588
– optical glasses 550
– technical glasses 533
MEIS (medium-energy ion scattering) 989, 1013
meitnerium Mt
– elements 135
melt viscosity 478
– polymers 478
melting point T_m
– aluminium compounds 611
– beryllium compounds 653
– boron compounds 605
– cadmium compounds 677
– gallium compounds 622
– group IV semiconductors and IV–IV compounds 580
– indium compounds 639
– magnesium compounds 656
– mercury compounds 686
– oxides of Ca, Sr, and Ba 660
– zinc compounds 666
melting temperature 477
– elements 47, 48
– polymers 477
memory devices 903
mendelevium Md
– elements 151
mercury compounds
– crystal structure, mechanical and thermal properties 686
– electromagnetic and optical properties 691
– electronic properties 688
– mechanical and thermal properties 686
– thermal properties 686
– transport properties 689
mercury Hg
– elements 73
mercury oxide
– crystal structure, mechanical and thermal properties 686
– electromagnetic and optical properties 691
– electronic properties 688
– transport properties 689
mercury selenide
– crystal structure, mechanical and thermal properties 686
– electromagnetic and optical properties 691
– electronic properties 688
– transport properties 689
mercury sulfide
– crystal structure, mechanical and thermal properties 686
– electromagnetic and optical properties 691
– electronic properties 688
– transport properties 689
mercury telluride
– crystal structure, mechanical and thermal properties 686

– electromagnetic and optical properties 691
– electronic properties 688
– transport properties 689
mesogen 943
mesogenic group 943
mesophase 941
mesoscopic material
– conductivity 1044
– nanoparticle doped 1044
– waveguide applications 1045
mesoscopic materials 1031, 1033
– manufacturing 1033
mesoscopic system
– quantum size effect 1035
– thermodynamic stability 1035
metal
– nanoparticle 1040
– resonance state 999
– surface 987
– surface core level shifts (SCLS) 998
– surface Debye temperature 1013
– surface phonon 1012
– surface state 999
– work function 997
metal surface 987
– jellium model 998
metals 997
– vertical relaxation 989
meter
– SI base unit 13
metrologica, international journal 12
MFM image of a written line on an array of dots 1063
MFM image of arrays of dots 1063
MFM image of domain pattern 1062
– sidewalls 1062
Mg magnesium 68
MHPOBC (liquid crystal) 935, 967
microphase separation 941
Mie theory 1045
Miller delta 825
Miller indices 28
M–I–M (metal–insulator–metal) heterostructure 1053
minority carriers 1023
missing-row reconstruction 987
Mn manganese 124
Mn–Al–C 810
– phase relations 810
Mo molybdenum 114
Mo-based alloys 317
mobility μ
– aluminium compounds 618
– cadmium compounds 682
– group IV semiconductors and IV–IV compounds 597–599
– indium compounds 648
MOCVD (metal-organic chemical vapor deposition) 1064
modulated crystal structure 924
modulated structures
– crystallography 34
moduli *see* elastic constant 580
Mohs hardness 822, 826
– elements 47
molar enthalpy of sublimation
– elements 47
molar entropy
– elements 47
molar heat capacity
– elements 47
molar magnetic susceptibility
– elements 48
molar mass 945–972
– glasses 527
molar susceptibility
– elements 48
molar volume
– elements 47
mole
– definition 14
– SI base unit 14
mole fraction
– glasses 527
molecular architecture 477
molybdenum Mo
– elements 114
momentum-conservation rule 1036
monocrystalline material 47
monolithic alloys 170
monophilic liquid crystal 942
MOS devices 979
MOS field-effect transistor 1020
MOSFET
– electron and hole mobility 1025
– equilibrium condition 1024
– schematic drawing 1025
Mott–Wannier exciton 1037
MQW (multiple quantum well) 1038
MRAM (magnetic random access memories) 1058
MRAM cell
– schematic diagram 1059
Mt meitnerium 135
MTJ (magnetic tunnel junction) 1053, 1054
MTJ sensor
– operation principle 1059
multi-component alloys 219
multiphase (MP) alloys 276
multiple hysteresis loops 932
multiple quantum wells 1042
MVA-TFT (multidomain vertical-alignment thin-film transistor) 944

N

N nitrogen 98
N16B
– glasses 537
– sealing glasses 563
N-4
– liquid crystals 959
Na sodium 59
$NaNO_2$ family 924
nanocrystal 1038
nanoimprint lithography 1061
nanoimprinted single domain dots
– images 1061
nanoimprinting 1061
nanolithography 1031
nanomaterial 1032
nanometric multilayers 1049
nanoparticle 1031
– doped material 1045
– local-field 1045
nanopatterning 1061
nanophase materials 1032
nanoporous materials 1032, 1043
nanoscience 1032
nanostructured material 1031, 1032
– classification scheme 1034
– conductance 1043
– definition 1032
– electrical conductivity 1043
– manufacturing 1063
– preparation 1031, 1035
– zeolites 1065
nanostructures
– magnetic 1048
nanotechnology 1032
Nb niobium 105
N-BAF10
– optical glasses 551
N-BAF52
– optical glasses 551
N-BAK4
– optical glasses 551
N-BALF4
– optical glasses 551
N-BASF64
– optical glasses 551

Nb-based alloys 318
N-BK7
– glasses 548
– optical glasses 551
Nd neodynium 142
Nd–Fe–B
– physical properties 802
Ne neon 128
near field microscopy 1049
nematic mixtures
– liquid crystals 975
– physical properties 975
nematic phase
– liquid crystals 941
nematic–isotropic transition 945
nematic-phase director 943
Neoceram
– glasses 559
neodymium Nd
– elements 142
neon Ne
– elements 128
Neoparies
– glasses 559
neptunium Np
– elements 151
nesosilicate 433
neutral density filters
– glasses 566
neutron diffraction 41, 926
neutron scattering 915
neutron spectrometer scan 916
new rheocast process (NRC) 170
newton meter
– SI unit of moment of force 19
N-FK51
– glasses 548
– optical glasses 551
N-FK56
– optical glasses 551
$(NH_4)_2SO_4$ family 927
$(NH_4)_3H(SO_4)_2$ family 928
$(NH_4)HSO_4$ family 928
$(NH_4)LiSO_4$ family 928
Ni nickel 139
Ni superalloys 294
nickel
– alloys 279
– application 279
– carbides 285
– low-alloy 279
– mechanical properties 280
– plating 288
nickel Ni
– elements 139
nickel-based superalloys 284
nickel–iron alloys 769
nickel-silvers 300
NIMs (National Institutes for Metrology 4
niobium Nb
– elements 105
nitride
– electrical properties 467
– mechanical properties 467
– thermal properties 467
nitrides
– physical properties 468
nitrogen N
– elements 98
N-KF9
– optical glasses 551
N-KZFS2
– optical glasses 551
N-LAF2
– optical glasses 551
N-LAK33
– optical glasses 551
N-LASF31
– optical glasses 551
No nobelium 151
nobelium No
– elements 151
noble metals 329
– Ag 329
– alloys 329
– applications 330
– Au 329
– catalysts 330
– corrosion resistance 329
– hardness 329
– Ir 329
– optical reflectivity 330
– Os 329
– Pd 329
– Pt 329
– Rh 329
– Ru 329
– vapour pressure 330
NOL (nano-oxide layer) 1053
noncrystallographic diffraction symmetries 36
nondestructive testing 941
nonlinear field-dependent properties 46
nonlinear optical coefficients
– nanostructured materials 1039
nonlinear optical device 903, 925
nonlinear optical susceptibility 920
nonlinear susceptibility tensors 827
non-oxide ferroelectrics 905
non-SI units 11, 20
normalizing 224
Np neptunium 151
N-PK51
– optical glasses 551
N-PSK57
– optical glasses 551
N-SF1
– optical glasses 551
N-SF56
– optical glasses 551
N-SF6
– glasses 548
– optical glasses 551
N-SK16
– optical glasses 551
N-SSK2
– optical glasses 551
nuclear incoherent scattering 916
nuclear reactor 218

O

O oxygen 108
occupied electron shells
– elements 46
Ohara code
– glasses 544
one-dimensional liquid 943
one-electron potential 999
opal
– nanostructured materials 1032
opals 1048
OPO (optical parametric oscillation) 825
optical constants
– gallium compounds 635
– group IV semiconductors and IV–IV compounds 601–603
– indium compounds 651
optical constants n and k
– cadmium compounds 684
optical glasses 526, 543
– thermal properties 556
optical materials
– high-frequency properties 824
optical mode frequency 916
optical parametric oscillation (OPO) 825
optical parametric oscillator 920
optical phonon scattering
– group IV semiconductors and IV–IV compounds 598
optical phonon scattering *see also* phonon scattering 598
optical phonon softening 913
optical properties

- group IV semiconductors and IV–IV compounds 601
- III–V compounds 610, 619, 635, 650
- II–VI compounds 655, 660, 664, 672, 683, 691
optical second-harmonic generator 903
optical transparency range 826
optical visualization 941
optoelectronic devices
- nanostructured materials 1038
order parameter 946–952, 956–960, 962, 963
order parameter, S 943
order principle for mesogenic groups 946
organic ferroelectrics 903
orientational order 941
Os osmium 131
osmium Os
- alloys 402
- applications 402
- cathodes 404
- chemical properties 406
- electrical properties 404
- elements 131
- lattice parameter 402
- magnetic properties 405
- mechanical properties 404
- phase diagrams 402
- production 402
- thermal properties 405
- thermoelectric properties 404
outer-shell orbital radius
- elements 46
over-aging 201
oxidation states
- elements 46
oxide 437
- beryllium 447
- magnesium 444
- physical properties 438
oxide ceramics
- production of 444
oxide ferroelectrics 903, 905
oxide superconductors
- low-T_c oxide 711
oxides of Ca, Sr, and Ba
- crystal structure, mechanical and thermal properties 660
- electromagnetic and optical properties 664
- electronic properties 661
- mechanical and thermal properties 660
- thermal properties 660
- transport properties 663
oxygen O
- elements 108

P

P phosphorus 98
Pa protactinium 151
PAA
- liquid crystals 958
pair distribution function
- crystallography 40
palladium Pd
- alloys 364
- applications 364
- electrical properties 370
- elements 139
- lattice parameters 366
- magnetic properties 372
- mechanical properties 368
- phase diagrams 364
- production 364
- thermoelectric properties 370
parabolic band 1000
Parkes process 330
partially stabilized zirconia (PSZ)
- electrical properties 449
- mechanical properties 449
- thermal properties 449
particle in a box model 1035
particle intensity I_p
- radiometry 15
passivation glasses 562, 564
- glasses 565
- properties 564
pattern transfer by imprinting 1061
Pauling 46
Pb lead 88
$PbHPO_4$ family 927
$PbTiO_3$ 916
$PbZrO_3$ 918
Pb–Ca–Sn
- battery grid alloys 418
PCH-7
- liquid crystals 950
Pd palladium 139
pearlite 222
Pearson symbol
- elements 47
percolation density
- nanostructured materials 1033
periodes of elements (Periodic table) 45
Periodic table
- elements 45
Periodic table of the elements 53
permanent magnets
- Co–Sm 806
perovskite-type family 909, 911, 912
perovskite-type oxide 909
phase diagram
- Fe–C 222
- Fe–Cr 226
- Fe–Mn 225
- Fe–Ni 225
- Fe–Si 227
- Ti–Al 210
- Zr–Nb 218
- Zr–O 218
phase separation
- glasses 525
phase transition temperature 943
phase-matching angle 826
phase-matching condition 825
phonon confinement
- nanostructured materials 1042
phonon density of states
- gallium compounds 623, 625
- indium compounds 642
phonon dispersion
- surface phonons 1012
phonon dispersion curve
- aluminium compounds 612
- gallium compounds 623–625
- indium compounds 641
- surface phonons 1015–1026
phonon dispersion relation
- group IV semiconductors 585–587
- IV–IV compound semiconductors 585, 588
phonon energies
- nanostructured materials 1040
- solid surfaces 1013
phonon frequencies ν
- cadmium compounds 678
- gallium compounds 624
- group IV semiconductors 585
- indium compounds 640
- IV–IV compound semiconductors 585
- magnesium compounds 657
- mercury compounds 687
- oxides of Ca, Sr, and Ba 661
- zinc compounds 667
phonon frequencies ν *see also* phonon wavenumbers $\tilde{\nu}$ 605
phonon instability 921
phonon mode frequency 908
phonon scattering

Subject Index

– group IV semiconductors and IV–IV compounds 598
phonon wavenumbers $\tilde{\nu}$ *see also* phonon frequencies ν 605
– aluminium compounds 612
– cadmium compounds 678
– gallium compounds 624
– indium compounds 640
– magnesium compounds 657
– mercury compounds 687
– zinc compounds 667
phonon wavenumbers $\tilde{\nu}$/frequencies ν
– boron compounds 605
phonon–phonon coupling 907
phosphorus P
– elements 98
photochromic glasses
– nanostructured materials 1032
photoconductive crystal 922
photoelastic effect 824
photoemission spectra for Ag quantum wells 1036
photoluminescence spectra of CdSe 1039
photometry
– intensity measurements 15
photonic band-gap materials 1048
phyllosilicate 433
physical properties
– liquid crystals 943
physical quantities 12
– base 11, 12
– data 13
– definition 12
– derived 11, 12
– general tables 4
piezoelectric
– element 918
– material 917
– strain constant 918
– strain tensor 824
– tensor 827
piezoelectricity 824, 906
piezooptic coefficient 825
planar alignment 943
planar electromechanical coupling factor 918
Planck radiator 15
plasmon excitations
– nanostructured materials 1031
plasmon oscillation
– nanostructured materials 1032
plasmon peak
– metals 1045
plasmon resonance 1045
plastic crystal 941
platinum group metals (PGM) 363
– alloys 363
platinum Pt
– alloys 376
– applications 376
– catalysis 385
– chemical properties 385
– electrical properties 381
– elements 139
– magnetical properties 384
– mechanical properties 378
– optical properties 385
– phase diagrams 376
– production 376
– thermal properties 385
– thermoelectric properties 382
plutonium Pu
– elements 151
PLZT
– ceramic material 918
Pm promethium 142
Po polonium 108
Pockels effect 825
point groups
– crystallography 30
Poisson equation 1022
Poisson number
– elements 47
Poisson's ratio 478
– optical glasses 550
– polymers 478
polarization microscopy 943
polonium Po
– elements 108
poly(4-methylpentene-1) (PMP) 488, 489
poly(ethylene-co-acrylic acid) (EAA) 486, 487
poly-(ethylene-co-norbornene) 486, 487
poly(ethylene-co-vinyl acetate) (EVA) 486, 487
poly(vinyl chloride) 492, 495
– plastisized (60/40) (PVC-P2) 492, 495
– plastisized (75/25) (PVC-P1) 492, 495
– unplastisized (PVC-U) 492, 494–496
polyacetals 480, 497
– poly(oxymethylene) (POM-H) 497–500
– poly(oxymethylene-co-ethylene) (POM-R) 497, 498, 500
polyacrylics 480, 497
– Poly(methyl methacrylate) (PMMA) 497–499
polyamides 480, 501
– polyamide 11 (PA11) 501
– polyamide 12 (PA12) 501
– polyamide 6 (PA6) 501, 502
– polyamide 610 (PA610) 501, 502
– polyamide 66 (PA66) 501, 502
polybutene-1 (PB) 488, 489
polyesters 481, 503
– poly(butylene terephthalate) (PBT) 503–505
– poly(ethylene terephthalate) (PET) 503–505
– poly(phenylene ether) (PPE) 504–506
– polycarbonate (PC) 503, 504
polyether ketones 481, 508
– poly(ether ether ketone) (PEEK) 508, 509
polyethylene 483–486
– high density (HDPE) 483–486
– linear low density (LLDPE) 484–486
– low density (LDPE) 484–486
– medium density (MDPE) 484–486
– ultra high molecular weight (UHMWPE) 484–486
polyethylene ionomer (EIM) 486, 487
polyimides 481, 508
– poly(amide imide) (PAI) 508
– poly(ether imide) (PEI) 508, 509
– polyimide (PI) 508
polyisobutylene (PIB) 488, 489
polymer
– physical properties 483
polymer blend 515
– physical properties 477, 483
– poly(acrylonitrile-co-butadiene-co-acrylester) + polycarbonate (ASA + PC) 515–517
– poly(acrylonitrile-co-butadiene-co-styrene) + polyamide (ABS + PA) 515–517
– poly(acrylonitrile-co-butadiene-co-styrene) + polycarbonate (ABS + PC) 515–517
– poly(butylene terephthalate) + poly(acrylonitrile-co-butadiene-co-acrylester) (PBT + ASA) 517, 521
– poly(butylene terephthalate) + polystyrene (PBT + PS) 515, 519, 520

– poly(ethylene terephthalate) + polystyrene (PET + PS) 515, 519, 520
– poly(phenylene ether) + polyamide 66 (PPE + PA66) 517, 521
– poly(phenylene ether) + polystyrene (PPE + PS) 520, 521
– poly(styrene-co-butadiene) (PPE + SB) 517, 520, 521
– poly(vinyl chloride) + chlorinated polyethylene (PVC + PE-C) 515, 518
– poly(vinyl chloride) + poly(acrylonitrile-co-butadiene-co-acrylester) (PVC + ASA) 515, 518
– poly(vinyl chloride) + poly(vinyl chloride-co-acrylate) (PVC + VC/A) 515, 518
– polycarbonate + liquid crystal polymer (PC + LCP) 515, 519, 520
– polycarbonate + poly(butylene terephthalate) (PC + PBT) 515, 519, 520
– polycarbonate + poly(ethylene terephthalate) (PC + PET) 515, 519, 520
– polypropylene + ethylene/propylene/diene rubber (PP + EPDM) 515, 516
– polysulfone + poly(acrylonitrile-co-butadiene-co-styrene) (PSU + ABS) 517, 521
polymer family 911, 936
polymer ferroelectrics 903
polymer matrix 1040
polymers 477
– abbreviations 482
– Charpy impact strength 478
– coefficient of expansion 478
– compressibility 478
– creep modulus 478
– crystallinity 477
– density 478
– elastic modulus 478
– electric strength 478
– enthalpy of combustion 477
– enthalpy of fusion 477
– entropy of fusion 477
– ferroelectric properties 905
– gas permeation 478
– glass transition temperature 477
– heat capacity 477
– impact strength 478
– melt viscosity 478
– melting temperature 477
– physical properties 477
– physicochemical properties 477
– Poisson's ratio 478
– refractive index 478
– relative permittivity 478
– shear modulus 478
– shear rate 478
– Shore hardness 478
– sound velocity 478
– steam permeation 478
– stress 478
– stress at 50% strain (elongation) 478
– stress at fracture 478
– stress at yield 478
– structural units 479–481
– surface resistivity 478
– thermal conductivity 478
– Vicat softening temperature 477
– viscosity 478
– volume resistivity 478
polyolefines 480, 483–486
polypropylene (PP) 488, 489
polysulfides 481, 506
– poly(phenylene sulfide) (PPS) 506, 507
polysulfones 481, 506
– poly(ether sulfone) (PES) 506, 507
– polysulfone (PSU) 506, 507
polyurethanes 481, 511
– polyurethane (PUR) 511, 512
– thermoplastic polyurethane elastomer (TPU) 511, 512
polyvinylidene fluoride
– ferroelectrics 911
porous aluminium silicates
– electrical properties 436
– mechanical properties 436
– thermal properties 436
Portland cement
– ASTM types 433
– chemical composition 432
positional order 941
potassium K
– elements 59
potential barrier 1022
powder-composite materials 277
Powder-in-Tube (PIT) 739
power-law dependence of conductivity on film thickness 1042
Pr praseodynium 142
practical superconductors
– characteristic properties 705
praseodynium Pr
– elements 142
prefixes
– decimal multiples of units 19
primitive cell
– crystal structure 28
projected band structure 996
projected bond length 995
promethium Pm
– elements 142
property tensor 46
– independent components 827
protactinium Pa
– elements 151
proton distribution 926
pseudopotential calculation 1006
Pt platinum 139
p-type diamond
– Debye length 1021
Pu plutonium 151
pulsed infrared lasers 1040
pyrochlore-type family 909
pyroelectric coefficient 917
pyroelectric measurement 931
PZT
– piezoelectric material 917

Q

quantum confinement 1036
– nanostructured materials 1042
quantum dots
– nanostructured materials 1031, 1035
quantum size effect
– nanostructured materials 1031, 1035
quantum transport
– nanostructured materials 1053
quantum well
– coupled 1043
– nanostructured materials 1031, 1035
quantum wires
– nanostructured materials 1035
quantum-well superlattices 1033
quasicrystals 34
QWIP (quantum well infrared photodetector) 1042

R

Ra radium 68
radiant intensity I_e
– radiometry 15

radiation sources and exposure techniques in lithography 1065
radiometric and photometric quantities 16
radiometry
– intensity measurements 15
radium Ra
– elements 68
radon Rn
– elements 128
Raman scattering 906
Raman scattering spectroscopy 1040
Raman spectrum 908
RAS 997
Rayleigh mode 1012
Rb rubidium
– elements 59
Re rhenium
– elements 124
real and imaginary parts ε_1 and ε_2 of the dielectric constant *see* dielectric constant ε 602
– gallium compounds 637
reconstruction model 987
– solid surfaces 991
reconstruction of semiconductors 991
reconstruction of surface 986
– metals 987
recording media
– arrays of magnetic dots 1061
reduced surface state energy 1022
reduced wave vector 1012
reduced-dimensional material geometries 1033
references
– solid surfaces 1029
reflectance anisotropy spectroscopy (RAS) 1007
refractive index 478, 829, 946–972
– elements 48
– glasses 539, 543
– polymers 478
– Sellmeier dispersion formula 547
– temperature dependence 548
refractive index n 619
– boron compounds 610
– cadmium compounds 684
– gallium compounds 635
– group IV semiconductors and IV–IV compounds 601–603
– indium compounds 650
– mercury compounds 691
– zinc compounds 672
refractories
– boride-based 452
– carbide-based 458
– nitride-based 468
– oxide-based 438
– silicide-based 472
refractory ceramics 437
refractory metals 303
– alloys 303
compositions 305
dispersion-strengthened 304
– annealing 311
– chemical properties 308
– crack growth behavior 325
– creep elongation 316
– creep properties 327
– dynamic properties 318
– evaporation rate 307
– fatigue data 321
– flow stress 316
– fracture mechanics 322
– grain boundaries 314
– high-cycle fatigue properties 319
– linear thermal expansion 306
– low-cycle fatigue properties 320
– mechanical properties 314
– metal loss 308
– microplasticity 318
– oxidation behavior 308
– physical properties 306
– production routes 304
– recrystallization 311
– resistance against gaseous media 309
– resistance against metal melts 309
– specific electrical resistivity 307
– specific heat 307
– static mechanical properties 315
– stress–strain curves 320
– thermal conductivity 306
– thermomechanical treatment 314
– vapor pressure 307
– Young's modulus 307
refractory metals alloys
– application 306
– products 306
refractory production
– raw materials 444
relative permittivity 478
– polymers 478
relaxation of semiconductors 991
relaxation of surface 986
– metals 987
relaxor 906, 909, 918
remanent magnetization 922
remanent polarization 918
residual resistance ratio (RRR) 397
residual resistivity ratio (RRR) 338
resistivity
– gallium compounds 630
response of material 46
Rf rutherfordium 94
Rh rhodium
– elements 135
RHEED (reflection high-energy electron diffraction) 990
rhenium Re
– elements 124
rhodium
– alloys 386
– applications 386
– chemical properties 392
– electrical properties 390
– magnetical properties 391
– mechanical properties 387
– optical properties 392
– phase diagrams 386
– production 386
– thermal properties 392
– thermoelectrical properties 391
rhodium Rh
– elements 135
ribbon silicates 433
RIE (reactive ion etching) 1061
Rn radon 128
Rochelle salt 904
Rochelle salt family 932
rod-like molecule 942
RT (room temperature) 49
RTP (room temperaure and standard pressure) 49
Ru ruthenium 131
rubidium Rb
– elements 59
ruthenium Ru
– alloys 399
– applications 399
– chemical properties 402
– electrical properties 401
– elements 131
– lattice parameter 400
– magnetic properties 401
– mechanical properties 400
– optical properties 402
– phase diagrams 399
– production 399
– thermal properties 402
– thermoelectric properties 401
rutherfordium Rf
– elements 94
RW (weighted sound reduction) 409

S

S sulfur 108
SAE (Society of Automotive Engineers) 221
SAM (self-assembled monolayer) 944
samarium Sm
– elements 142
SAW (surface acoustic wave) 912
Sb antimony 98
SbSI family 922
Sc scandium 84
$SC(NH_2)_2$ family 930
scandium Sc
– elements 84
scattering
– nanoscale objects 1048
scattering losses of a waveguide 1045
Schoenflies symbol 30
– elements 47
Schott AG 523
Schott code
– glasses 544
Schott filter glasses
– glasses 569
Schott glasses 8nnn 540, 541
SDR 997
Se selenium 108
seaborgium Sg
– elements 114
sealing glasses 527
– glasses 559
second
– SI base unit 14
secondary hardening 263
second-harmonic generation (SHG) 825, 906
second-order elastic constants *see* elastic constant 580
second-order phase transition 907
selenium Se
– elements 108
Sellmeier dispersion formula
– glasses 547
Sellmeier equations 826
SEM (scanning electron microscopy) 39
semiconductor 1003
– covalent 992
– field-effect mobility 1024
– III–V compounds 1004
– intrinsic Debye length 1021
– nanocrystal 1040
– polar 992
– quantum confinement 1036
– reconstruction 1004
– reconstruction model 991
– surface 990
– surface core level shift 1003
– surface Debye temperature 1017
– surface phonon 1017
– surface shift 1004
semiconductor band bending 1020
semiconductor nanostructures 1035
semiconductor surface
– Fermi level pinning 1025
– ionization energy 1003
semiconductors
– aluminium compounds 610
– boron compounds 604
– cadmium compounds 676
– chemical doping 576
– gallium compounds 621
– group IV semiconductors and IV–IV compounds 578–603
– III–V compounds 576, 604
– II–VI compounds 576, 652
– indium compounds 638
– introduction 575
– IV–IV compounds 576
– magnesium compounds 655
– mercury compounds 686
– oxides of Ca, Sr, and Ba 660
– physical properties 577
– table of contents 575
– zinc compounds 665
semi-solid metal processing (SSMP) 170
sensor
– chemiresistor-type 1043
SF1
– optical glasses 551
SF11
– optical glasses 551
SF2
– optical glasses 551
SF6
– glasses 537, 548
– optical glasses 551
SF66
– optical glasses 551
SFG (sum frequency generation) 825
SFM (superfluorinated material) 975
Sg seaborgium 114
shape memory 298
– nickel 279
shape-memory alloys
– TiNi 216
shear modulus 478
– elements 47
– polymers 478
shear rate 478
– polymers 478
SHG (second-harmonic generation) 825, 906
Shore hardness 478
– polymers 478
short pass filters
– glasses 566
short-range order 39
– glasses 524
Shubnikov groups
– crystallography 33
SI (Système International d'Unités) 3, 11
SI (the International System of Units) 12
SI base unit 13
SI definitions of magnetic susceptibility 48
SI derived units 16, 17
– with special names 17, 18
SI prefixes 19
Si silicon 88
SI units
– base quantities 13
– base units 13
Si_3N_4 ceramics 451
Si_3N_4 powders 472
SiC ceramics 451
side group 943
sievert
– SI unit of dose equivalent 19
silica
– glasses 524
silicate 433
silicate based glasses 526
silicide 473
– physical properties 472
silicon
– electromagnetic and optical properties 601
– electronic properties 589–594
– transport properties 598
silicon carbide
– band structure 590
– crystal structure, mechanical and thermal properties 578–588
– electromagnetic and optical properties 601

– electronic properties 589–594
– transport properties 595
silicon nitride 467
silicon Si
– crystal structure, mechanical and thermal properties 578–588
– elements 88
silicon steels
– grain-oriented 765
– non-oriented 763
silicon technology 1036
silicon-based lasers 1036
silicon–germanium alloys
– band structure 590
– transport properties 601
silicon-germanium alloys
– crystal structure, mechanical and thermal properties 578–588
– electromagnetic and optical properties 601
silicon–silicon oxide interface 1025
silver 330
– alloys 330
– application 330
– chemical properties 344
– crystal structures 333
– diffusion 342
– electrical properties 338
– intermetallic phases 333
– magnetic properties 339
– mechanical properties 335
– optical properties 341
– phase diagrams 331
– production 330
– ternary alloys 345
– thermal properties 340
– thermodynamic data 331
– thermoelectric properties 339
silver Ag
– elements 65
simple perovskite-type oxide 909
single hysteresis loop 904
SiO_2
– glasses 524
SK51
– optical glasses 551
Sm samarium
– elements 142
smectic C* phase 934
smectic phase 942
Sn tin 88
SNR (signal-to-noise ratio) 1060
soda lime glasses 528, 529, 534
sodium Na
– elements 59
soft annealing 224
soft magnetic alloys 758
– nanocrystalline 776
soft magnetic materials
– composite 759
– sintered 759
soft-mode spectroscopy 906
solar cell 1043
solder alloy 345
solder glasses 562
sol–gel synthesis
– nanostructured materials 1065
solid material
– structure 27
solid material, structure 27
solid surface energy 944
solid-state polymorphism 943
sorosilicate 433
sound velocity 478, 947–949, 956–962, 968–971
– elements 47
– polymers 478
source 1024
sp^3-bonded crystal 991
space charge function
– solid surfaces 1022
space charge layer
– semiconductor surface 1020
– solid surfaces 1020
space groups
– crystallography 31
SPARPES 1011
speed of light
– fundamental constant 13
spheroidal (nodular) graphite (SG) 268
spheroidite 223
spin accumulation 1049
spin diffusion length 1050
spin electronics
– applications 1057
– nanostructured materials 1031, 1049
spin polarization 1055
– nanostructured materials 1049
spin valve multilayers 1052
spin valve read head
– schematic diagram 1058
spin valve sensor 1053
spin-asymmetric material 1049
spinel structure
– crystallography 33
spin-electronic switch 1055
spin–orbit splitting energy Δ_{so}
– aluminium compounds 617
– cadmium compounds 681
– gallium compounds 628
– group IV semiconductors and IV–IV compounds 593
– indium compounds 644
– mercury compounds 689
– zinc compounds 670
spintronics
– nanostructured materials 1031, 1049
SPLEED 1011
spontaneous electric polarization 903
spontaneous polarization 917, 945
Sr strontium 68
$Sr_2Nb_2O_7$ family 920
SRI (sound reduction index) 409
$SrNb_2O_7$ family 909
$SrTeO_3$ family 919
$SrTiO_3$ 913
stabilized zirconia (PSZ) 448
stacking faults
– crystallography 41
stain resistance
– optical glasses 550
stainless steels 240
– austenitic 252
– duplex 257
– ferritic 246
– martensitic 250
– martensitic-ferritic 250
standard electrode potential
– elements 46
standard entropy
– elements 47
standard temperature and pressure (STP) 46
Stark effect
– nanostructured materials 1040
static dielectric constant 826
– elements 48
static dielectric constant 828–889
STC (sound transmission classification) 409
steam permeation 478
– polymers 478
steel
– austenitic 259
– carbon 227
– ferritic 258
– ferritic austenitic 259
– hardening 237
– heat-resistant 258, 261
– high-strength low-alloy (HSLA) 240

– low-alloy carbon steel 227
– mechanical properties 237
– stainless 240
– tool 262
stibiotantalite family 909
stiffness constant *see* elastic constant 580
STM (scanning tunneling microscopy) 988, 992, 997
STM spectroscopy 1005
STN (supertwisted nematic) effect 944
STO ($SrTiO_3$) 1056
storage capacity of hard disks 1048
storage density evolution of hard disk drives 1048
storage media 1060
– arrays of nanometer-scale dots 1060
– limits 1060
– technology 1060
storing information on the sidewalls of the dots 1062
strain
– polymers 478
strength
– glasses 534
stress 478
– polymers 478
stress at 50% strain (elongation) 478
– polymers 478
stress at fracture 478
stress at yield 478
– polymers 478
stress birefringence
– glasses 539, 549
stress intensity factor
– glasses 534
strong-confinement regime
– nanostructured materials 1037, 1038
strontium oxide
– crystal structure, mechanical and thermal properties 660
– electromagnetic and optical properties 664
– electronic properties 661
– transport properties 663
strontium Sr
– elements 68
strontium titanate 913
structural parameters 990
structural phase transitions 906
structure
– diamond-like 979
structure type
– crystallography 33
Strukturbericht type 47
sublattice 903
sublattice polarization 904
submicrometer magnetic dots 1060
substituted mesogens (liquid crystals)
– physical properties 968
sulfur S
– elements 108
sum frequency generation (SFG) 825
superalloys
– Ni-based cast 288
– nickel 294
superconducting high-T_c
– crystal structure 712
superconducting oxides
– high-T_c chemical composition 712
superconductivity
– elements 48
superconductor 695
– borides 745
– borocarbides 746, 747
– carbides 745
– commercial Nb_3Sn 709
– critical temperature 699
– crystal structure 712
– Debye temperature 696
– device applications 719
– high-T_c cuprates 712, 713, 720
– industrial wire performance 719
– metallic 696
– Nb alloys 702
– non-metallic 712
– Pb alloys 696
– pinning 717
– practical metallic 704
– production Nb_3Sn 708
– Sommerfeld constant 696
– SQUIDs 720
– structural data 720
– thermodynamic properties 696
– Type I 695
– Type II 695
– V alloys 700
– vortex lines 717
– Y−Ba−Cu−O 723
supercooled liquid
– glasses 524
supercooled mesophase 945
superstructures
– crystallography 41
supertwisted nematic (STN) effect 944
Supremax
– glasses 527
surface
– Curie temperature 1009
– diagram 979
– ionization energy 1003
– magnetic 1008
– semiconductor 990
– structure of an ideal 979
surface band structure 996
surface Brillouin zone (SBZ) 996
surface conductivity
– solid surfaces 1024
surface core level shifts (SCLS) 998
– solid surfaces 1003, 1004
surface differential reflectivity (SDR) 1008
surface excess conductivity 1024
surface magnetization 1011
surface mobility 1024
surface of diamond 1004
surface phonon 1012
– dispersion 1019
– metal 1013
– mode 1012
surface plasmon
– absorption of nanoparticles 1046
– dispersion curve 1001
surface resistivity 478
– polymers 478
surface resonance 996
– phonons 1012
surface response
– dielectric theory 1007
surface state
– acceptor 1022
– band 1005
– donor 1022
– transitions 1005
surface state bands
– solid surfaces 1004
surface states 996
surface tension 948–950, 955–962, 968–971, 975
– elements 47
surface tension (γ_{LV}) 943
surfaces 979
surgical implant alloys
– cobalt-based 277
susceptibility

– magnetic 48
– mass 48
– molar 48
– nonlinear dielectric 825, 829
– second-order nonlinear dielectric 826
– third-order nonlinear dielectric 826
SV (spin valve) 1052
symmetry elements of point groups 30
synthesis of clusters
– gas-phase production 1066
– nanostructured materials 1065
synthetic silica
– glasses 557

T

T tritium 54
TA
– phonon spectra 915
Ta tantalum 105
Ta-based alloys 318
tailoring of the electronic wave function 1032
tantalum Ta
– elements 105
Tb terbium 142
TC 12 (Technical Committee 12 of ISO) 12
Tc technetium 124
Te tellurium 108
technetium Tc
– elements 124
technical ceramics 437
technical coppers 297
technical glasses 527, 530
technical specialty glasses 526
tellurium Te
– elements 108
TEM (transmission electron microscopy) 39
TEM image of a superlattice of Au clusters 1066
TEM views of single dots 1062
temper graphite (TG) 268
temperature dependence of carrier concentration
– group IV semiconductors and IV–IV compounds 596
temperature dependence of electrical conductivity
– indium compounds 647
temperature dependence of electronic mobilities
– indium compounds 649
temperature dependence of energy gap
– indium compounds 645
temperature dependence of linear thermal expansion coefficient
– cadmium compounds 677
– magnesium compounds 656
temperature dependence of the lattice parameters
– group IV semiconductors and IV–IV compounds 580–582
temperature dependence of thermal conductivity
– group IV semiconductors and IV–IV compounds 599, 600
– indium compounds 650
temperatures of phase transitions 946–976
tempering of steel 223
template synthesis 944
tensile strength
– elements 47
tensor
– elastooptic 826
– piezoelectric strain 826
terbium Tb
– elements 142
terminal group 943
ternary alloys 298
terne steel coatings 415
TFT (thin-film transistor) 944
TGS family 932
Th thorium 151
thallium Tl
– elements 78
thermal and thermodynamic properties
– elements 46
thermal conductivity κ 478, 945, 947–949, 956, 958
– aluminium compounds 618
– cadmium compounds 682
– elements 46, 47
– gallium compounds 634
– group IV semiconductors and IV–IV compounds 599
– indium compounds 650
– magnesium compounds 659
– mercury compounds 690
– oxides of Ca, Sr, and Ba 663
– polymers 478
– zinc compounds 670
thermal expansion
– glasses 526
thermal expansion coefficient, linear
– elements 47
thermal gap
– indium compounds 645
thermal properties
– group IV semiconductors 578–588
– III–V compounds 610, 621, 638
– III–V semiconductors 604
– II–VI compounds 652, 655, 660, 665, 676, 686
– IV–IV compound semiconductors 578–588
– technical glasses 533, 536
thermal vibrations
– surface phonons 1012
thermal work function
– elements 48
thermally activated flux flow (TAFF) 718
thermochromic material 944
thermodynamic properties
– elements 47
thermoelectric coefficient
– elements 48
thermoelectric power
– oxides of Ca, Sr, and Ba 664
thermography 941, 944
thermomechanical treatment (TMT) 314
thermosets 481, 512
– diallyl phthalate (DAP) 512, 514
– epoxy resin (EP) 514, 515
– melamine formaldehyde (MF) 512, 513
– phenol formaldehyde (PF) 512, 513
– polymers 512
– silicone resin (SI) 514, 515
– unsaturated polyester (UP) 512–514
– urea formaldehyde (UF) 512, 513
thermotropic liquid crystal 942
thin film 906
thin-film transistor (TFT) 944
thixomolding 170
thorium Th
– elements 151
three and four-ring systems
– liquid crystals 964
three-dimensional long-range order 941
three-ring system 964
three-wave interactions
– in crystals 825
thulium Tm
– elements 142

Ti titanium 94
time-temperature-transformation (TTT) diagram 238
tin Sn
– elements 88
titanates 450
titanium 206
– commercially pure grades 207
– creep behavior 208
– creep strength 210
– hardness 207
– high-temperature phase 206
– intermetallic materials 210
– phase transformation 206
– sponge 207
– superalloys 210
– titanium alloys 206
titanium alloys 209
– applications 209
– chemical composition 209
– chemical properties 213
– mechancal properties 213
– mechanical properties 209
– physical properties 213
– polycrystalline 213
– single crystalline 213
– thermal expansion coefficient 214
titanium dioxide
– mechanical properties 450
– thermal properties 450
titanium oxide
– phase diagram 206
titanium Ti
– elements 94
Tl thallium 78
Tm thulium 142
TMR (tunnel magnetoresistance) 1054, 1055
TN (twisted nematic)
– liquid crystals 944
TO 915
tool steels 262
torsional modulus
– optical glasses 550
total losses 763
transformation temperature
– glasses 524
transition range
– glasses 524
transition temperature
– glasses 525
transitions
– surface states 1005
transmission spectra
– colored glasses 566, 567
transmission window
– glasses 524
transmittance
– glasses 548
transmittance of glasses
– color code 549
transport properties
– group IV semiconductors and IV–IV compounds 595–601
– III–V compounds 608, 617, 629, 647
– II–VI compounds 655, 659, 663, 670, 682, 689
transverse acoustic branch 915
transverse optical branch 915
transverse optical mode 906
triple point of water 48
tritium T
– elements 54
truncated crystal 986
tungsten bronze-type family 909, 920
tungsten W
– elements 114
tunnel junction
– magnetic 1053
tunnel magnetoresistance
– function of field and temperature 1057
tunnel magnetoresistance as a function of magnetic field 1055
tunneling
– nanostructured materials 1053
tunneling mechanism
– nanostructured materials 1043
twisted nematic (TN) effect 944
two-dimensional liquid 942
two-photon absorption coefficient 829
two-ring systems with bridges
– liquid crystals 955
two-ring systems without bridges
– liquid crystals 947
Type II superconductors
– anisotropy coefficients 716
– coherence lengths 716
– high-T_c cuprate compounds 716
type metals 414

U

U uranium 151
ultrahigh density storage media 1049, 1060
unalloyed coppers 296
uniaxial crystals 826
Unified Numbering System for Metals and Alloys (UNS) 296
unit cell of Si(111) 7×7 995
units
– amount of substance 14
– atomic 21
– atomic units (a.u.) 22
– candela 15
– CGS units 21
– coherent set of 20
– crystallography 21
– electric current 14
– general tables 4
– length 13
– luminous intensity 15
– mass 14
– natural 21
– natural units (n.u.) 21
– non-SI 22
– non-SI units 20, 21
– other non-SI units 23
– temperature 14
– the international system of 11
– time 14
– used with the SI 20
– X-ray-related units 22
units of physical quantities
– fundamental constants 3
units outside the SI 20
UNS (Unified Numbering System) 221
UPS 998
uranium U
– elements 151
UTS – ultimate tensile strength 219

V

V vanadium 105
van der Waals attraction 1019
vanadium V
– elements 105
vertical nanomagnets 1062
vertical relaxation of metals 989
VFT (Vogel, Fulcher, Tammann) equation
– glasses 533
Vicat softening temperature 477
– polymers 477
Vickers hardness
– elements 47

vinylpolymers 480, 489–492
– poly(acrylonitrile-co-butadiene-co-styrene) (ABS) 492–494
– poly(acrylonitrile-co-styrene-co-acrylester) (ASA) 492–494
– poly(styrene-co-acrylnitrile) (SAN) 489, 490, 492
– poly(styrene-co-butadiene) (SB) 489–491
– poly(vinyl carbazole) (PVK) 492, 493
– polystyrene (PS) 489–491
VIP (viewing-independent panel) 975
viscosity 478, 948, 950–954, 963–965, 967, 968, 975
– dynamic 943
– elements 47
– glasses 524
– kinematic 943
– optical glasses 556
– polymers 478
– technical glasses 533
– temperature dependence 525
viscosity of glasses
– temperature dependence 534
vitreous silica
– electrical properties 557
– gas solubility 557
– glasses 526, 556
– molecular diffusion 557
– optical constants 557
vitreous solder glasses 562
Vitronit
– glasses 559
volume compressibility
– elements 47
volume magnetization
– elements 48
volume of primitive cell
– crystallographic formulas 986
volume resistivity 478
– polymers 478
volume–temperature dependence
– glasses 524
Vycor™
– glasses 527

W

W tungsten 114
wavelength dependence of refractive index n
– indium compounds 651
WDX (wavelength-dispersive analysis of X-rays) 39
weak-confinement regime
– nanostructured materials 1037, 1038
wear-induced surface defects
– glasses 535
Weibull distribution
– glasses 535
weight fraction
– glasses 527
Wood's metal 420
work function Φ
– metal 997
– solid surfaces 997
work hardening wrought copper alloys 300
wrought alloys 298
wrought magnesium alloys 164
wrought superalloys 284
wtppm (weight part per million) 407
Wyckoff position
– crystallography 32

X

Xe xenon 128
xenon Xe
– elements 128
X-ray diffraction 39
X-ray interferences
– crystallography 27

Y

Y yttrium 84
Yb ytterbium 142
Young's modulus 823
– elements 47
– optical glasses 550
YS – yield stress 219
ytterbium Yb
– elements 142
yttrium Y
– elements 84
Y–Ba–Cu–O
– critical current density 733
– crystal defects 728
– crystal structure 723
– electric resistivity 730
– grain boundaries 730
– hole concentration 733
– lattice parameters 726
– lower critical field 734
– oxygen content 724
– pinning 735
– substitutions 725
– superconducting properties 731
– thermal conductivity 730
– transition temperature 733
– upper critical field 734

Z

zeolites
– nanostructured materials 1031, 1065
Zerodur®
– glasses 558
– linear thermal expansion 558
zinc compounds
– crystal structure, mechanical and thermal properties 665
– effective hole mass 670
– electromagnetic and optical properties 672
– electronic properties 668
– mechanical and thermal properties 665
– optical properties 672
– thermal properties 665
– transport properties 670
zinc oxide
– crystal structure, mechanical and thermal properties 665
– electromagnetic and optical properties 672
– electronic properties 667
– transport properties 670
zinc selenide
– crystal structure, mechanical and thermal properties 665
– electromagnetic and optical properties 672
– electronic properties 667
– transport properties 670
zinc sulfide
– crystal structure, mechanical and thermal properties 665
– electromagnetic and optical properties 672
– electronic properties 667
– transport properties 670
zinc telluride
– crystal structure, mechanical and thermal properties 665
– electromagnetic and optical properties 672
– electronic properties 667
– transport properties 670

zinc Zn
– elements 73
zircaloy 219
– irradiation effect 219
zirconium
– alloys 217
– bulk glassy alloys 218
– bulk glassy behavior 220
– low alloy materials 217
– nuclear applications 218
– technically-pure materials 217
zirconium dioxide 448
zirconium Zr
– elements 94
ZLI-1132
– liquid crystals 975
Zn zinc 73
Zr zirconium 94

Periodic Table of the Elements

18 VIIIA	IUPAC Notation / CAS Notation
2 He	Atomic Number / Element Symbol
	Unstable Nuclei

Main Groups

1 IA	2 IIA	13 IIIA	14 IVA	15 VA	16 VIA	17 VIIA	18 VIIIA	Shells
1 H							2 He	K
3 Li	4 Be	5 B	6 C	7 N	8 O	9 F	10 Ne	K–L
11 Na	12 Mg	13 Al	14 Si	15 P	16 S	17 Cl	18 Ar	K–L–M
19 K	20 Ca	31 Ga	32 Ge	33 As	34 Se	35 Br	36 Kr	–L–M–N
37 Rb	38 Sr	49 In	50 Sn	51 Sb	52 Te	53 I	54 Xe	–M–N–O
55 Cs	56 Ba	81 Tl	82 Pb	83 Bi	84 Po	85 At	86 Rn	–N–O–P
87 Fr	88 Ra							–O–P–Q

Subgroups

3 IIIB	4 IVB	5 VB	6 VIB	7 VIIB	8 VIII (1)	9 VIII (2)	10 VIII (3)	11 IB	12 IIB	Shells
21 Sc	22 Ti	23 V	24 Cr	25 Mn	26 Fe	27 Co	28 Ni	29 Cu	30 Zn	–L–M–N
39 Y	40 Zr	41 Nb	42 Mo	43 Tc	44 Ru	45 Rh	46 Pd	47 Ag	48 Cd	–M–N–O
57 La	72 Hf	73 Ta	74 W	75 Re	76 Os	77 Ir	78 Pt	79 Au	80 Hg	–N–O–P
89 Ac	104 Rf	105 Db	106 Sg	107 Bh	108 Hs	109 Mt	110 Ds	111 Rg	112	–O–P–Q

Lanthanides (Shells –N–O–P)

58 Ce	59 Pr	60 Nd	61 Pm	62 Sm	63 Eu	64 Gd	65 Tb	66 Dy	67 Ho	68 Er	69 Tm	70 Yb	71 Lu

Actinides (Shells –O–P–Q)

90 Th	91 Pa	92 U	93 Np	94 Pu	95 Am	96 Cm	97 Bk	98 Cf	99 Es	100 Fm	101 Md	102 No	103 Lr

Most Frequently Used Fundamental Constants

CODATA Recommended Values of Fundamental Constants

Quantity	Symbol and relation	Numerical value	Unit	Relative standard uncertainty
Speed of light in vacuum	c	299 792 458	m/s	Fixed by definition
Magnetic constant	$\mu_0 = 4\pi \times 10^{-7}$	$12.566370614\ldots \times 10^{-7}$	N/A^2	Fixed by definition
Electric constant	$\varepsilon_0 = 1/(\mu_0 c^2)$	$8.854187817\ldots \times 10^{-12}$	F/m	Fixed by definition
Newtonian constant of gravitation	G	$6.6742(10) \times 10^{-11}$	$m^3/(kg\,s^2)$	1.5×10^{-4}
Planck constant	h	$4.13566743(35) \times 10^{-15}$	eV s	8.5×10^{-8}
Reduced Planck constant	$\hbar = h/2\pi$	$6.58211915(56) \times 10^{-16}$	eV s	8.5×10^{-8}
Elementary charge	e	$1.60217653(14) \times 10^{-19}$	C	8.5×10^{-8}
Fine-structure constant	$\alpha = (1/4\pi\varepsilon_0)(e^2/\hbar c)$	$7.297352568(24) \times 10^{-3}$		3.3×10^{-9}
Magnetic flux quantum	$\Phi_0 = h/2e$	$2.06783372(18) \times 10^{-15}$	Wb	8.5×10^{-8}
Conductance quantum	$G_0 = 2e^2/h$	$7.748091733(26) \times 10^{-5}$	S	3.3×10^{-9}
Rydberg constant	$R_\infty = \alpha^2 m_e c/2h$	10 973 731.568525(73)	1/m	6.6×10^{-12}
Electron mass	m_e	$9.1093826(16) \times 10^{-31}$	kg	1.7×10^{-7}
Proton mass	m_p	$1.67262171(29) \times 10^{-27}$	kg	1.7×10^{-7}
Proton–electron mass ratio	m_p/m_e	1836.15267261(85)		4.6×10^{-10}
Avogadro number	N_A, L	$6.0221415(10) \times 10^{23}$		1.7×10^{-7}
Faraday constant	$F = N_A e$	96 485.3383(83)	C	8.6×10^{-8}
Molar gas constant	R	8.314472(15)	J/K	1.7×10^{-6}
Boltzmann constant	$k = R/N_A$	$1.3806505(24) \times 10^{-23}$ $8.617343(15) \times 10^{-5}$	J/K eV/K	1.8×10^{-6} 1.8×10^{-6}
Josephson constant	$K_J = 2e/h$	$483\,597.879(41) \times 10^9$	Hz/V	8.5×10^{-8}
von Klitzing constant	$R_K = h/e^2 = \mu_0 c/2\alpha$	25 812.807449(86)	Ω	3.3×10^{-9}
Bohr magneton	$\mu_B = e\hbar/2m_e$	$927.400949(80) \times 10^{-26}$ $5.788381804(39) \times 10^{-5}$	J/T eV/T	8.6×10^{-8} 6.7×10^{-9}
Atomic mass constant	$u = (1/12)m(^{12}C)$ $= (1/N_A) \times 10^{-3}$ kg	$1.66053886(28) \times 10^{-27}$	kg	1.7×10^{-7}
Bohr radius	$a_0 = \alpha/4\pi R_\infty$ $= 4\pi\varepsilon_0\hbar^2/m_e e^2$	$0.5291772108(18) \times 10^{-10}$	m	3.3×10^{-9}
Quantum of circulation	$h/2m_e$	$3.636947550(24) \times 10^{-4}$	m^2/s	6.7×10^{-9}